AF389417

Design and Post Processing for Metal Additive Manufacturing

Design and Post Processing for Metal Additive Manufacturing

Editors

Bartłomiej Wysocki
Joseph Buhagiar
Tomasz Durejko

Basel • Beijing • Wuhan • Barcelona • Belgrade • Novi Sad • Cluj • Manchester

Editors

Bartłomiej Wysocki
Multidisciplinary Research
Center
Cardinal Stefan Wyszynski
University in Warsaw
Dziekanow Lesny
Poland

Joseph Buhagiar
Department of Metallurgy
and Materials Engineering,
Faculty of Engineering
University of Malta
Msida
Malta

Tomasz Durejko
Institute of Materials Science
and Engineering, Department
of Materials Technology
Military University of
Technology
Warsaw
Poland

Editorial Office
MDPI
St. Alban-Anlage 66
4052 Basel, Switzerland

This is a reprint of articles from the Special Issue published online in the open access journal *Materials* (ISSN 1996-1944) (available at: www.mdpi.com/journal/materials/special_issues/Des_Post_Metal_AM).

For citation purposes, cite each article independently as indicated on the article page online and as indicated below:

Lastname, A.A.; Lastname, B.B. Article Title. *Journal Name* **Year**, *Volume Number*, Page Range.

ISBN 978-3-0365-9886-4 (Hbk)
ISBN 978-3-0365-9885-7 (PDF)
doi.org/10.3390/books978-3-0365-9885-7

Contents

About the Editors

Bartłomiej Wysocki

Dr. Bartlomiej Wysocki is an assistant professor at the Multidisciplinary Research Center of Cardinal Stefan Wyszynski University in Warsaw (MCB UKSW). He manages the 3D Printing/Additive Manufacturing Laboratory and is responsible for transferring its research results to industry. Dr. Wysocki is skilled in various AM techniques (PBF/LB: SLM/DMLS, PBF/EB: EBM, SLA, FDM, CJP) and materials characterization (microscopic observations, X-ray diffraction, and μCT analysis). His current research interests focus on the fabrication and post-processing of metallic alloys (titanium, copper, metallic glasses, tungsten, high entropy alloys, etc.) for specific applications and implementing AI/ML algorithms in process optimization. Dr. Wysocki received his PhD from the Warsaw University of Technology in 2018 for a thesis entitled "Fabrication of the titanium cellular structures by selective laser melting for medical applications". He has written several dozen articles and been granted numerous patents related to additive technologies. His exceptional scientific and implementation activities have been recognized and awarded by various prestigious organizations, including the Polish Minister of Education and Science in 2019 and the Kosciuszko Foundation in 2021. In recognition of his continued excellence, he was also the proud recipient of the Rector's Award of Cardinal Stefan Wyszynski University in 2022. Dr. Wysocki collaborates with universities worldwide, including those in the USA, Japan, South Africa, and Malta. He works closely with companies interested in implementing or improving additive technologies in their operations. From 2015 to 2021, he also served as the President of the Management Board of MaterialsCare Sp. z o. o. The company specializes in providing advanced 3D printing solutions for medicine and helps in the operations of oncology patients.

Joseph Buhagiar

Prof Ing Joseph Buhagiar is the Head of Department of Metallurgy and Materials Engineering and an Associate Professor of the University of Malta. Prof Buhagiar received his B.Eng (Hons) degree in Mechanical Engineering from the University of Malta in 2003. He joined the Department of Metallurgy and Materials Engineering of the University of Malta as an Assistant Lecturer in 2004, and in 2008 he received his PhD in "Plasma Surface Engineering and Characterisation of Biomedical Stainless Steels", from the University of Birmingham (UK). Following his PhD studies, he became a University of Malta Lecturer in 2008. He was promoted to Senior Lecturer in 2013 and Associate Professor in 2016. He was appointed as Head of Department in October 2022. He is on the Editorial Board of the *Journal of Surface Engineering* and represents Malta on the Coal and Steel Committee. His current research interests are in the fields of biomaterials, biodegradation, and surface engineering. He is the project leader of MaltaHIP-II (Development of a Small Diameter Low-Wearing Hip Joint Prosthesis) funded by the project TRAKE. He is currently a collaborator on the projects: GO2MALTI3D (Developing a Go-To-Market Strategy for an Innovative MALTI3D Printing System) and MaltaKnee2 (Inflatable Arthroscopic Device as a Therapy for Early-Stage Knee Osteoarthritis). Both projects were financed by the Malta Council for Science and Technology (MCST) under the Go To Market: Accelerator Programme. He was the project leader of the BioSA project and a collaborator in the MALTAHIP, MaltaKnee, MALTI3D, and SEAM projects. In 2018, Prof Buhagiar was conferred the National Order of Merit of the Republic of Colombia in the grade of Knight. The decoration was bestowed since Prof. Buhagiar's academic profile was fundamental for the permanent support that he provides to Colombian students in Malta and for the links that he has forged between universities in both countries.

Tomasz Durejko

Dr. Tomasz Durejko graduated in 1997 with an M.Sc. in Mechanics and Machine Building from the Faculty of Engineering, Military University of Technology, Warsaw, Poland. In 2006, he received his PhD in Mechanics and Machine Building from the Military University of Technology, Warsaw, Poland. In the years 2006–2019, he was employed as an assistant professor at the Department of Advanced Materials and Technologies, Faculty of New Technologies and Chemistry, Military University of Technology, where he focused his research mainly on the implementation of additive technology in Poland. In 2007, 2013, and 2015, he completed short-term internships at the University of Wollongong, Max Planc Institute and Warwick University, respectively.

Currently, Dr. Tomasz Durejko leads the Department of Materials Technology, Military University of Technology, and his research activities have encompassed fields such as materials for additive technology/powder metallurgy, the characterization of intermetallics, and advanced superconducting materials. He is also the coordinator of the Additive Manufacturing Centre of Military University of Technology, a member of the Polish Society of Materials Science, a multiple member of the National Centre for Research and Development expert team, and he holds a number of other positions also.

Preface

The additive manufacturing (AM) of metals has received significant attention as it enables the fabrication of functional, net-shape parts in various industrial sectors using laser, electron beam, or binder jetting methods. Recent advances in AM techniques offer many opportunities in terms of design freedom and allow for the fabrication of complex geometries like cellular solids, metamaterials, or biomimetic materials which could not be easily manufactured using conventional techniques. Today, these objects can be fabricated from elemental or alloyed metallic powders based on computer-aided design (CAD) models.

This Special Issue of *Materials*, entitled "Design and Post-Processing for Metal Additive Manufacturing", invited submissions on the design of elements with predicted microstructure and mechanical properties, artificial intelligence/machine learning (AI/ML) in AM, numerical algorithms for AM, and μ-CT imagining for quality control.

While AM manufacturing in a powder bed provides the possibility of fabricating objects of any shape in one production step, it carries some disadvantages. One drawback is the need to generate support for the fabricated parts to dissipate the heat generated during the 3D printing process from metallic powders and minimize the geometrical distortions induced by internal stresses. This Special Issue also covers computer simulations and improved fabrication protocols that can decrease these issues.

During AM processes, not all particles are melted, and removing any unmelted particles must be performed with mechanical or chemical post-processing methods. Therefore, this Special Issue is dedicated to the various research areas relevant to metal AM. We are interested in the processing parameters and post-processing methods, including annealing and chemical modifications. Finally, the design of materials dedicated to metal AM and the description of modifications to commercial machines were welcomed in the submissions.

Bartłomiej Wysocki, Joseph Buhagiar, and Tomasz Durejko
Editors

Article

Design Rules for Hybrid Additive Manufacturing Combining Selective Laser Melting and Micromilling

David Sommer [1], Babette Götzendorfer [1,*], Cemal Esen [2] and Ralf Hellmann [1]

[1] Applied Laser and Photonics Group, University of Applied Sciences Aschaffenburg, 63743 Aschaffenburg, Germany; david.sommer@th-ab.de (D.S.); ralf.hellmann@th-ab.de (R.H.)
[2] Applied Laser Technologies, Ruhr-University Bochum, 44801 Bochum, Germany; esen@lat.rub.de
* Correspondence: Babette.Goetzendorfer@th-ab.de

Abstract: We report on a comprehensive study to evaluate fundamental properties of a hybrid manufacturing approach, combining selective laser melting and high speed milling, and to characterize typical geometrical features and conclude on a catalogue of design rules. As for any additive manufacturing approach, the understanding of the machine properties and the process behaviour as well as such a selection guide is of upmost importance to foster the implementation of new machining concepts and support design engineers. Geometrical accuracy between digitally designed and physically realized parts made of maraging steel and dimensional limits are analyzed by stripe line projection. In particular, we identify design rules for numerous basic geometric elements like walls, cylinders, angles, inclinations, overhangs, notches, inner and outer radii of spheres, chamfers in build direction, and holes of different shape, respectively, as being manufactured by the hybrid approach and compare them to sole selective laser melting. While the cutting tool defines the manufacturability of, e.g., edges and corners, the milling itself improves the surface roughness to Ra < 2 μm. Thus, the given advantages of this hybrid process, e.g., space-resolved and custom-designed roughness and the superior geometrical accuracy are evaluated. Finally, we exemplify the potential of this particular promising hybrid approach by demonstrating an injection mold with a conformal cooling for a charge socket for an electro mobile.

Keywords: hybrid additive manufacturing; high-speed milling; selective laser melting; construction rules

Citation: Sommer, D.; Götzendorfer, B.; Esen, C.; Hellmann, R. Design Rules for Hybrid Additive Manufacturing Combining Selective Laser Melting and Micromilling. *Materials* **2021**, *14*, 5753. https://doi.org/10.3390/ma14195753

Academic Editor: Bartlomiej Wysocki

Received: 30 August 2021
Accepted: 21 September 2021
Published: 2 October 2021

Publisher's Note: MDPI stays neutral with regard to jurisdictional claims in published maps and institutional affiliations.

1. Introduction

Additive manufacturing (AM), especially selective laser melting (SLM), is continuing to move towards broader acceptance in industry in manifold applications. However, at the same time fundamental limitations remain a challenge, one of them being the surface roughness of SLM built parts, which may be of unsatisfactory quality, in turn restricting usage of such parts in applications that demand close-fitting [1,2], exigent fatigue strength [3], or distinct sterilizability for bio-medical appliances [4]. Thus, SLM parts are typically post-processed by, e.g., either shot-blasting, milling, turning or chemical treatment in order to meet the desired requirements [5,6]. This necessity has led to the development of different hybrid processes and machine concepts (not only for SLM), combining additive techniques with conventional machining. For example, in several approaches CNC-machines are combined with laser wire deposition or metal arc welding to form a hybrid process [7–11]. Another hybrid approach that can vanquish the afore mentioned challenges, combines SLM with in-situ milling [12], allowing for the freedom of design offered by additive manufacturing [13–15] combined with the geometric accuracy and surface quality of milling [16–18] within a single, automated process.

The advancement of hybrid machines that combine additive and subtractive processes indeed paves the way towards new concepts in both product design and manufacturing [8], thus enabling the construction of innovative components that had previously been beyond reach [5,9]. However, this also demands for a throughout understanding of the fundamental properties of these new technologies, their restrictions and requirements.

While most of the research on hybrid additive manufacturing focusses on the development of processes and parameter optimization [19,20], on materials and product design [9,21,22], studies on appropriate design guidelines are, however, rare. Nonetheless, for the advancement of any AM technology, from a design engineering point of view, specific design rules and an elaborate selection guide for an appropriate additive manufacturing are a prerequisite for successful industrial implementation. Such rules have recently been specified for, e.g., fused deposition modelling [23,24], selective laser sintering [24], SLM [25–28], electron beam melting [29] and binder jetting [30]. For a hybrid AM approach as addressed here, however, such design rules are not yet available.

This particular hybrid SLM process itself has previously been demonstrated in different application oriented case studies, such as, e.g., the manufacturing of dental implants. In Refs. [31,32], the generation of an Akers clasp with improved fitting accuracy, retentive force and surface roughness of the cobalt-chromium alloy clasp as compared to a cast clasp is reported. Ohkubo et al., in turn, manufacture titanium removable partial dentures [33].

However, in these studies details about the fabrication process, fundamental properties as well as general design limitations are not discussed. Without this fundamental knowledge, the described hybrid process may not be used to full capacity. Wüst et al. report on an experimental study to optimize the surface roughness of hybrid additive manufactured parts [34], with particular focus on optimization of both SLM and milling to improve the vertical and horizontal surfaces. As the described hybrid process bears the opportunity to adapt the surface properties in dependence of the local required quality, it is crucial to define precise threshold values.

Against this fundamental background, we report on a comprehensive study to define the geometrical accuracy and design limits of selected fundamental geometrical elements to form a catalogue of design rules and compare these to the sole SLM process. In particular, we design 3D models of walls, cylinders, different inner angels, inclinations, overhangs, notches, inner and outer radii of spheres, chamfers in build direction, and holes of different shape and compare these structures in term of geometrical accuracy and feasible minimum dimensions to sole SLM. This shall ultimately support design engineers to purposefully apply hybrid additive manufacturing. Further the role of oversized grain by powder characterization of the performed maraging steel (1.2709), a low-carbon and high-nickel alloy, is analysed. Such aspects that have not been comprehensively reported before, yet define the fundamental basis for any further process optimization and component development.

2. Materials and Methods

2.1. Machine and Process

For the study of the hybrid manufacturing approach, we employed a Lumex Avance-25 (Matsuura Machinery GmbH, Wiesbaden, Germany), combining SLM and three-axis high speed milling, as schematically illustrated in Figure 1. The SLM process (step 1) follows the conventional procedure of additive manufacturing, embracing the layer-wise building of components by the selective laser additive process, which has been extensively examined by various studies [35–37]. The high speed milling (step 2) is performed with a 3-axis milling system, machining the contour surfaces between a preset number of selectively laser molten layers (typically ten).

For SLM, the machine is equipped with a $P_L = 500\,\text{W}$ yttrium fiber laser YLM-500 (IPG Laser GmbH, Burbach, Germany) with an operating wavelength of $\lambda = 1070\,\text{nm}$ and a spot size of $d_{\text{spot}} = 200\,\mu\text{m}$ at focus position. The machine processes under nitrogen atmosphere with less than 3.0% oxygen level. To maintain the machined part at an elevated temperature, as to avoid deformation by curling due to residual stress, the build platform is kept at $\vartheta_{\text{plate}} = 50\,°\text{C}$. The maximum build volume is 250 mm × 250 mm × 185 mm (width × depth × height). The study is performed processing maraging tool steel 1.2709. As the focus of this study is on design rules, we applied previously optimized, standard process parameters for the SLM (see Table 1) [34].

Table 1. Process parameters SLM.

	Laser Power [W]	Scan Speed [mm/s]	Hatch Distance [µm]
Area	320	700	0.12
Contour	320	1400	—
Support	320	700	0.12

The three-axis milling system is basically not distinct from standard industrial high speed milling machines. The high speed spindle operates with a maximum of 45,000 revolutions per minute, working with a maximum turning moment of 1.31 Nm, accessing a tool magazine with 20 different tools. However, it is worthwhile to note, that compared to a conventional milling system no cooling lubricant can be used while milling within the powder bed. Thus, the high speed milling process constitutes a dry machining concept, facing the challenges like, e.g., exceeding temperature conditions or increased tool wear characteristics [38]. In our study standard process parameters are used (see Table 2). Please note, the used milling cutters are solid carbide cutting tools with a nano alloy composed of aluminium, titan and silicon for the reduction of wear characteristics (Mitsubishi Materials Corporation GmbH, Meerbusch, Germany).

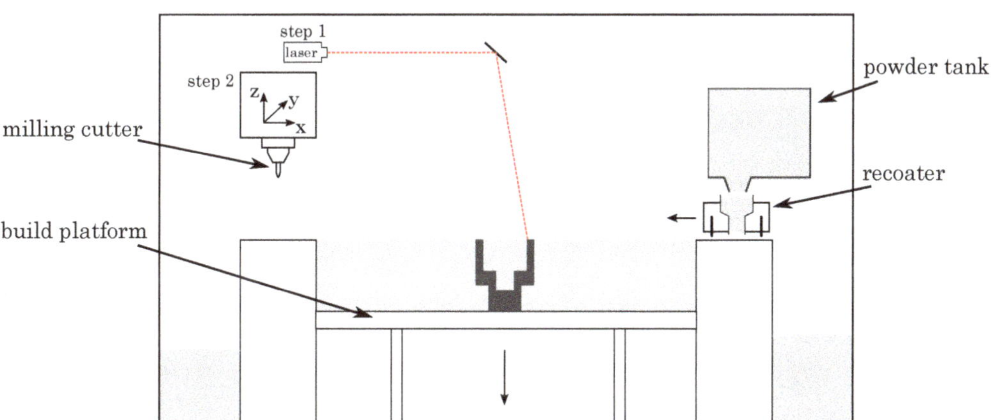

Figure 1. Hybrid manufacturing process.

The milling process is, as mentioned before, directly integrated into the additive process, implying that the SLM process is disrupted after several layers for the milling of the previous built surfaces. Figure 2 depicts the single steps of the hybrid additive manufacturing approach. The SLM builts, in the standard process, ten layers with a thickness of $h_{layer} = 50$ µm each, introducing a material allowance for the milling process (cf. Figure 2). The milling process is partitioned into a roughing and a finishing step using two different milling cutters. The two steps of the milling process gradually remove the material allowance of $d_{x/y} = 150$ µm which is added on the nominal constructed geometry by the SLM. The cutting depth of the roughing cutter represents 120 µm, getting removed working downwards the geometry, starting at the last built layer. Afterwards, the finishing cutter detaches the remaining 30 µm of the allowance (cf. Table 2), providing the final dimensions and a smooth surface, thus warranting the final quality of the build part.

However, it is worthwhile to note that the finishing cutter starts millling from the bottom to the top and also terminates beneath the last built layers, sparing several layers for the next process cycle (cf. Figure 2 (2.2), where the finishing cutter leaves five layers). Due to the selective laser melting, a thermal gradient within the built part developes, as the top layers retain a higher temperature compared to lower built levels, where the heat has dissipated via the subjacent parts or support structures. As it has been shown by [25,39,40], under such thermal conditions it is preferential to start milling from the bottom to the top and to also stop finishing beneath the upper surface, leaving several layers beyond. By virtue of this geometrically shift, the milling process avoids thermal deformations and geometrical deviations, warranting a high surface quality.

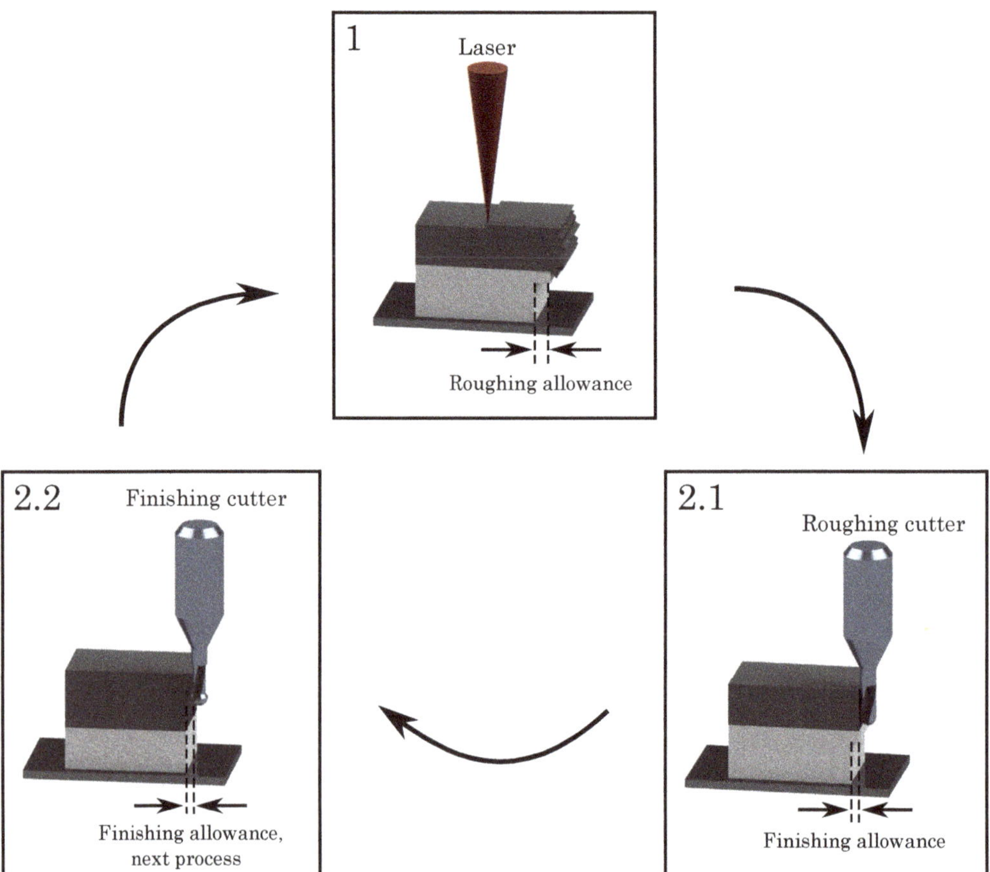

Figure 2. Two-stage milling process of hybrid manufacturing.

Table 2. Milling process parameters.

	Z-Pitch [mm]	**Spindle Speed [rot/s]**	**Feed Rate [mm/min]**
Roughing cutter	0.15	30,000	2000
End mill	0.1	30,000	1600

2.2. Powder Characterization

In this study, we processed 1.2709 maraging steel II (Matsuura Machinery GmbH, Wiesbaden, Germany), which is a low-carbon and high-nickel steel. The term maraging refers to the fact that the material has a martensitic microstructure and that it can be hardened and its strength be increased by aging. Due to its high strength, the alloy is used in tooling, structural, and aerospace applications [41]. The chemical composition of this powder is as listed in Table 3.

Table 3. Chemical composition of Maraging-steel powder.

Element	Fe	Ni	Co	Mo	Ti	Cr	Mn	Si	Al	C	S
wt%	Balance	17–19	8.5–9	4.5–5.2	0.6–0.8	$\leq$0.3	$\leq$0.1	$\leq$0.1	0.05–0.15	$\leq$0.03	$\leq$0.01

The particle size distribution of new and recycled powder was analyzed by a Camsizer X2 based on dynamic digital image analysis methods according to ISO 13322-2 (Retsch Technology GmbH, Haan, Germany).

2.3. Optical Characterization Tools

Different optical characterization tools are used to measure the shape deviation of the test geometries and for closer inspection of small parts. In particular, the geometrical deviation between the digitally designed and physically realized specimens, i.e., verification of the shape accuracy, was determined by stripe line projection using AtosCore 3D-scanner and employing shape variance analysis using ATOS Professional V8-SR1 software (GOM GmbH, Braunschweig, Germany).

For closer inspection of small parts and for validaton of the measured shape deviations, the digital microscope DVM6 (Leica Microsystems GmbH, Wetzlar, Germany) with a PlanAPO FOV 12.55 (maximum zoom of 675:1) is used. The specimens are captured with several images, using the z-stack function to generate a good depth of field.

3. Results and Discussion

3.1. Powder Characterization

During the SLM process, the particle size distribution of the employed powder and the shape of the individual particles typically change and get deteriorated due to adherence and mechanical deformation of particles and grains. This in turn necessitates sieving of used powder to ensure proper size distribution with good flowability. For the hybrid process discussed here, however, chips of milled material and abrasive wear of the milling cutters result in additional debris, in turn influencing the powder bed. Figure 3 compares the particle size distribution of new and recycled, i.e., sieved powder, as well as the oversize particles originating from adhered powder, milled material and wear debris after sieving. For sieving we used a mesh size of $d_x = 63$ µm (cf. vertical line in Figure 3).

Apparently, the pristine and the recycled powder exhibit negligible differences with almost identical size distribution peaking at 31 mm. The sieved oversized particles, however, reveal a broader size distribution extending beyond 500 µm.

For comparison, Scanning Electron Microscope (SEM) images of the pristine, recycled and of the oversized particles are shown in Figure 4, confirming that new and recycled powder exhibit similar shape. The oversize particles, however, clearly reveal larger chips of different size and shape which can be associated to splints of the cutter and flakes of the milled steel.

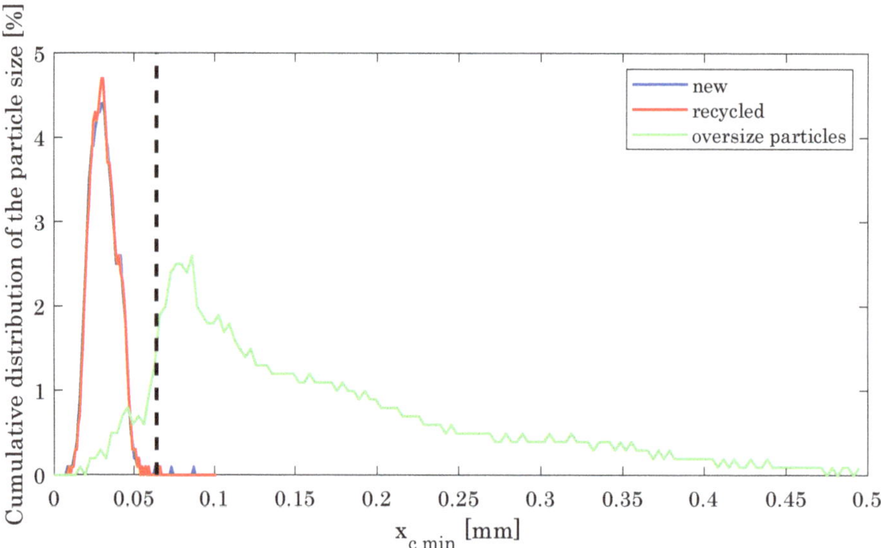

Figure 3. Powder distribution of new powder, recycled powder and the oversize particles.

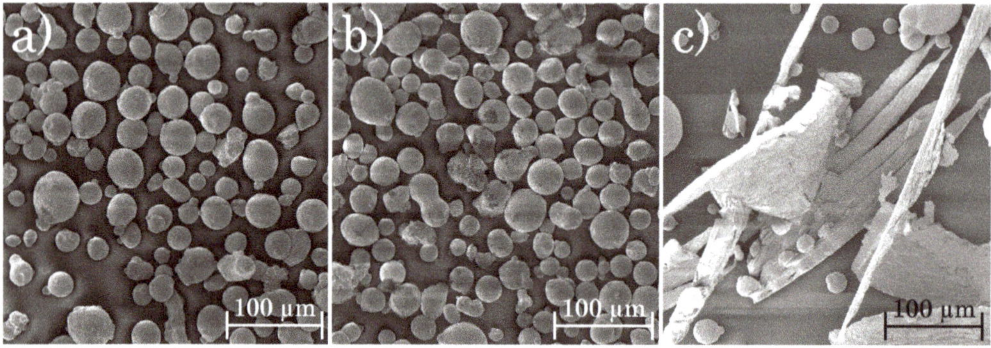

Figure 4. SEM images of (**a**) new, (**b**) recycled and (**c**) oversize particles.

3.2. Geometric Aspects

While the overall potential of AM, in general, as well as design, material and process limitation have been subject of numerous studies, from an engineering point of view, specific design rules are a necessity and prerequisite to fully exploit the opportunities of each additive and hybrid manufacturing technology. Such design rules may outline aspects of geometrical accuracy and accessible surface roughness. For the hybrid approach under study, however, such design rules are yet unpublished.

In general, the methodical approach for the systematic creation of construction rules is realized similarly for all geometrical structures, as schematically illustrated in Figure 5. For every geometry, the construction has to be prepared in a CAM-software and sliced for the building process. After that, for both, the sole SLM and the hybrid process, a batch of specimens is manufactured, typically consisting 35 to 50 specimens. The geometrical accuracy is measured with the 3D-scanner and depicted in a false colour representation, additional measurements are carried out with a microscope and images are taken. Concluding, the analysis of the geometrical deviations and the performance of the milling system is executed and the findings are represented in a construction rule for both parts of the process and summarized in Figure A1 at the end of the publication.

As one of the most challenging basic geometries, inclined structures, generally, pose limits to SLM processes. In order to exemplify the methodical approach of this study (cf. Figure 5), in the following it is thus comprehensively outlined and exemplified for the fabrication of inclined structures. Specifically for this geometry, the most peculiarities occur and the most intense analysis has to be realized.

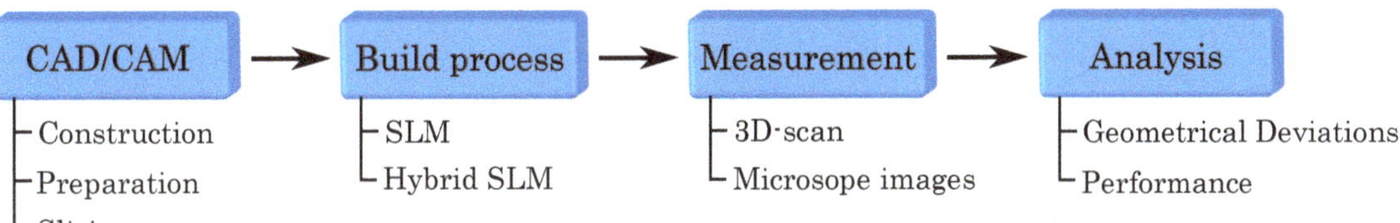

Figure 5. Methodical approach of the systematic creation of construction rules.

3.2.1. Inclined Structures

The manufacturing of inclined structures is one of the limiting criterions in the SLM process. To allow sufficient heat dissipation, inclinations have to be built with additional support structures to avoid thermal deformation[14,42–44]. Depending on material and process parameters, the minimum inclination angle for the manufacturing of unsupported structures is defined for SLM in several studies between 20° [25,45,46] and 45° [15,24].

As a consequence, the primary limitation for the manufacturing of inclined structures in the hybrid process is given by the necessity of support structures in the SLM process. Another limitation is given by the three-axis milling system, enabling the milling of inclined structures with the use of T-slot milling cutters. The milling system permits a post-processing from an inclination angle of $\alpha = 52°$ upwards, determined by the geometry of the T-slot cutter (cf. Figure 6a). With that a non-millable area between the support structure and the finishing area arises.

Apart from the necessity of support structures, the geometrical accuracy of inclined structures is affected by the formation of the stair-case effect [47,48]. Due to the layerwise building process, the layers are geometrically shifted at the manufacturing of small inclination angles, as shown in Figure 6b. With the hybrid manufacturing approach, the staircase effect can be minimised, respectively excluded, and the surface is smoothened by the milling cutter after the building process [49].

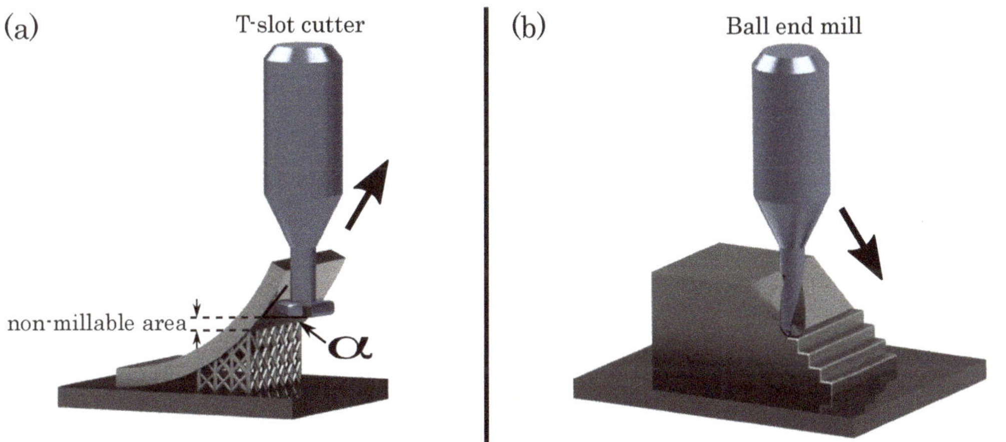

Figure 6. (**a**) Limitation of the hybrid manufacturing process of inclined structures, moving upwards (**b**) Removing of the staircase-effect by the milling cutter, moving downwards.

For the study of the geometrical accuracy and the machinability of inclined structures, walls with a varying inclination angle are manufactured, as depicted in Figure 7. The angles are specified in relation to the base plate and vary between 10° and 90° without the usage of support to examine the machinability of the bottom side of the specimens. The dimensions of the specimens are 10 mm in height, 6 mm in depth and 3 mm in width, remaining constantly.

Figure 7. Test specimen for the study of inclined structures without support structures.

Figure 8 illustrates and compares the SLM and hybrid built parts by a false colour representation provided by stripe line projection, indicating the dimensional deviation of the printed and constructed geometries.

Apparently, the geometrical deviation of inclined structures is, in general, larger for the SLM-built parts (colour code yellow and orange) as compared to the hybrid manufactured parts (colour code green). Please note, the SLM-built parts get manufactured without the additional material allowance for the milling process, the geometric forming is exclusively defined by the SLM process. The results reveal that the geometrical deviations of the SLM-built parts vary for different inclination angles. The top side of specimens show better accuracy in z-direction with nearly vertical shaped inclinations (10°–30°) than in x/y-direction with more horizontal aligned specimens (40°–90°). The hybrid manufactured parts show, as mentioned before, less geometrical deviations, persisting constantly, independent of the inclination angle. The colour code in the false colour representation does not vary significantly within the different specimens. In summary, the geometrical deviation of the SLM process varies between the z- and the x/y-direction, while the hybrid manufacturing approach shows omnidirectional consistent geometrical accuracy.

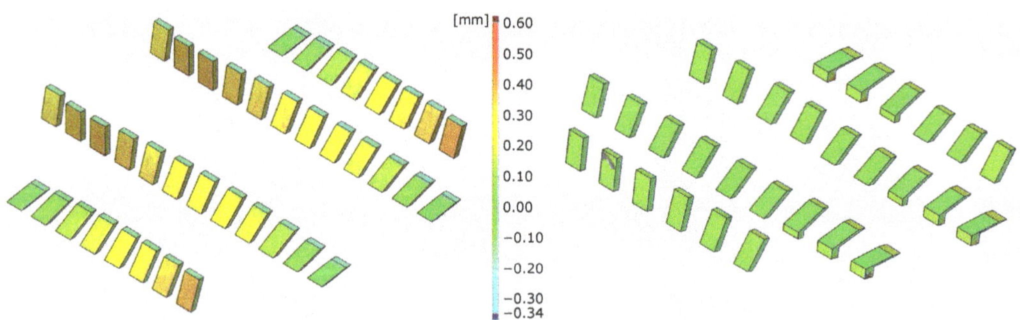

Figure 8. Comparison of SLM –built (**left**) and hybrid manufactured (**right**) inclinations (false colour representation).

In addition to the measurement of the geometric deviation by the stripe line projection, Scanning Electron Microscope (SEM) images of the surfaces are shown in Figure 9 to characterize the staircase-effect of highly inclined parts. The figure depicts the top and the front side of specimens with 10° inclination angle; the unmachined bottom side is presented shortly on the right side of the specimens. The SLM built part on the left side of the figure exhibits misalignments between the single layers caused by the additive process, while the hybrid manufactured part on the right side shows a smooth surface, the staircase-effect itself is not identifiable and is excluded by milling.

Figure 9. Comparison of highly inclined (inclination angle = 10°) SLM-built (**left**) and hybrid manufactured (**right**) parts.

From a design engineering point of view, this concludes a design rule for the hybrid SLM process, which is shown in Figure 10. For both, the sole SLM and the hybrid SLM process regions of inclination angles are specified considering the necessity of support structures and the machinability by the milling cutters.

	SLM	Hybrid SLM
	$0° \leq \alpha \leq 30°$: Support structures necessary	$0° \leq \alpha \leq 30°$: Support structures necessary
	$30° < \alpha \leq 90°$: manufacturable	$30° < \alpha < 52°$: Machinable by the roughing cutter
		$52° < \alpha \leq 90°$: Completely machinable

Figure 10. Contruction rule for the manufacturing of inclined structures.

3.2.2. Wall Thickness

A fundamental geometrical element of any construction that has to be reliably fabricated by SLM is a rectangular shaped wall, for which SLM typically has its limits in terms of wall thickness. The lower limit is defined by the laser spot size [50,51] plus potentially wall broadening due to lateral adherent powder [52].

In the SLM process the minimum printed wall thickness is $d = 200\,\mu m$, which appears to be governed by the laser spot diameter ($d_{spot} = 200\,\mu m$ in our case) rather than the particle size [48]. It is also lower as previous reported construction rule studies for pure SLM, reporting minimum feasible geometrical dimensions between $d = 400\,\mu m$ [15,25] and $d = 600\,\mu m$ [24].

The smallest wall thickness achievable with the hybrid system, in turn, is 400 µm. Thinner SLM built walls break under the milling process conditioned by the impact of the milling cutter. The milling cutter induces vibrations of the thin walls and by turning around the corner, the walls with smaller thickness get destroyed. Figure 11 compares the measured thickness deviation as a function of wall thickness for both the SLM and hybrid built parts. While the thickness deviation for SLM built parts is between 0.06 mm and

0.095 mm, the additional milling process improves the thickness deviation significantly. With a deviation of about 10 µm for thickness $d \geq 600$ µm, the hybrid milling approach can eliminate the geometrical inaccuracy of the SLM process, one of its biggest disadvantages compared to conventional manufacturing methods [5].

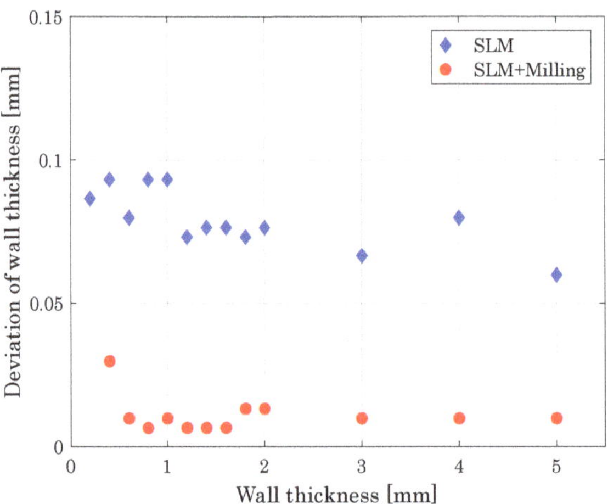

Figure 11. Deviation of wall thickness versus intended wall thickness, giving a measure of the dimensional accuracy of the processes.

3.2.3. Cylinder Diameter

Beyond thin walls, as a further fundamental geometrical element, we evaluated cylinders with respect to the minimum diameter achievable by the hybrid manufacturing approach (Figure 12). The minimum diameter is expected to be higher than the minimal printable wall thickness. Due to the smaller base area, the connection to the base plate is more difficult and the resistance against the impact of the milling cutter can be problematic for smaller diameters.

The minimum printable diameter in the SLM process is $d = 1$ mm, with cylinders having smaller diameters gaining insufficient connection to the building plate (200 µm–600 µm) or being damaged during powder recoating (800 µm). Employing the hybrid system, the smallest achievable cylinder diameter is $d = 1.4$ mm. Due to the low resistance of the specimens, thinner cylinders, reliably built by the SLM down 1.0 mm, break during the milling process.

Figure 12 depicts the deviation of the measured from the designed diameter for both the SLM and hybrid built parts. While for SLM built cylinders this deviation is in the range of 0.11 mm and 0.14 mm, it is significantly reduced to about 15 µm for the hybrid process, confirming that the hybrid manufacturing ensures better geometrical accuracy as provided by the sole SLM process.

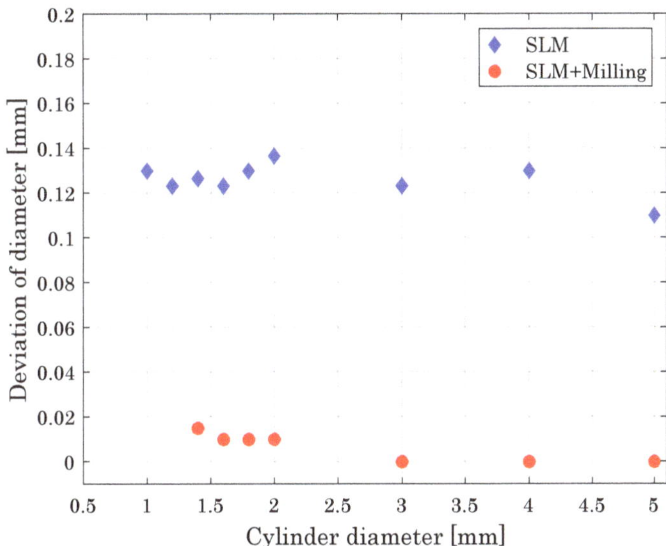

Figure 12. Deviation of cylinder diameter versus intended cylinder diameter, giving a measure of the dimensional accuracy of the processes.

3.2.4. Overhanging Structures

As a consequence of insufficient heat dissipation, in SLM it is recommended to avoid overhanging structures without the use of support structures [53–55]. The hybrid manufacturing does not allow for the machining of horizontal structures underneath, in consequence the fabrication of classic constructed overhangs cannot be improved by the high speed milling process. For both reasons, the construction of overhanging structures should include self-supporting structures [29,56,57]. Apart from that, the length and the thickness of an overhanging structure are essential for the successfull manufacturing of unsupported overhangs.

Our study shows that overhanging structures can be built without support structures with a maximum lenght of $l = 5\,$mm and a minimum thickness of $d = 0.3\,$mm in the SLM process, concluding a design rule for unsupported overhangs as depicted in Figure A1. In addition, it can be specified that an overhang, supported by an inclined structure, ensures the heat dissipation and enables a surface finish with the hybrid manufacturing.

3.2.5. Gap Width

The heat input in SLM is the decisive faktor for the gap width between two elements [58], while in the hybrid manufacturing approach it is given by the radius of the milling cutter adding the allowance for the finishing process for every flank of the gap.

With this, Equation (1) develops as a construction rule for the hybrid manufacturing of gap widths or notches. Due to the diameter of the used milling cutter of $d = 2\,$mm and the general finishing allowance of $a = 30\,\mu$m, the minimum gap width for this study is $b = 2.06\,$mm using the hybrid manufacturing approach.

$$b = d_{\text{milling cutter}} + 2 * a_{\text{finish}} \tag{1}$$

Within SLM, the minimum gap width can be defined by $b = 0.6\,$mm (maximum operated height $h = 3\,$mm). Due to heat dissipation, smaller widths get sealed or adhesion of powder particles deranges the aperture. Furthermore, it is founded that the rectangular transition at the bottom of the notch has not been machined completely, the milling cutter was not able to remove the entire allowance. By virtue of its dimensions, especially in view of the radius r (cf. Figure 13) and the three axis milling system, material transitions in build direction have to be adapted to this geometry. Thus, a continuative study, examining chamfers in build direction, has been carried out to define the limitations of the milling system for material transitions (cf. Section 3.2.10).

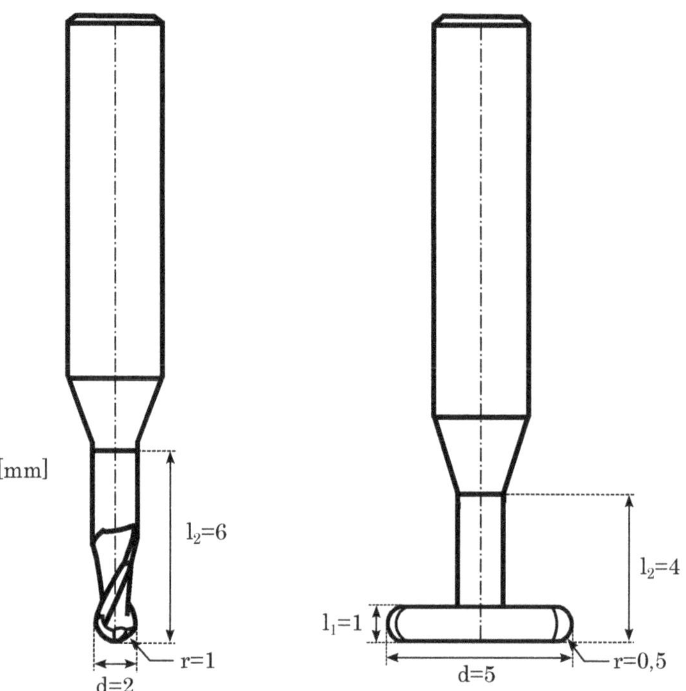

Figure 13. Dimensions of the 1 mm ball end mill and the T-slot cutter.

3.2.6. Outer Diameter of Spheres

While spherical structures are, in general, contingent for SLM, challenges arise for manufacturing of complete spheres. The use of support structures is essential, because, compared to the manufacturing of inclined structures, the heat has to be dissipated across a very small base area with a quickly enlarging geometry [59,60].

For the hybrid manufacturing approach, the milling of sperical structures has to be divided in two sub processes. The overhanging structures are machined with the T-slot cutter, while the upwards facing surfaces get milled by the ball endmill to assure the best possible result in surface roughness.

As mentioned before, the heat dissipation determines the successfull manufacturing of outer spherical structures. Thus, the usage of support structures in the SLM process is elementary to avoid thermal deformation of the spheres in z-direction. Within the hybrid manufacturing, the remaining geometrical deformation is corrected by the milling cutter afterwards and provides a superior geometrical accuracy. As depicted in Figure 14, the deviation of SLM built parts lays between 0.1 mm and 0.18 mm, whereas the deviation of hybrid built parts can be diminished to underneath 0.1 mm in general.

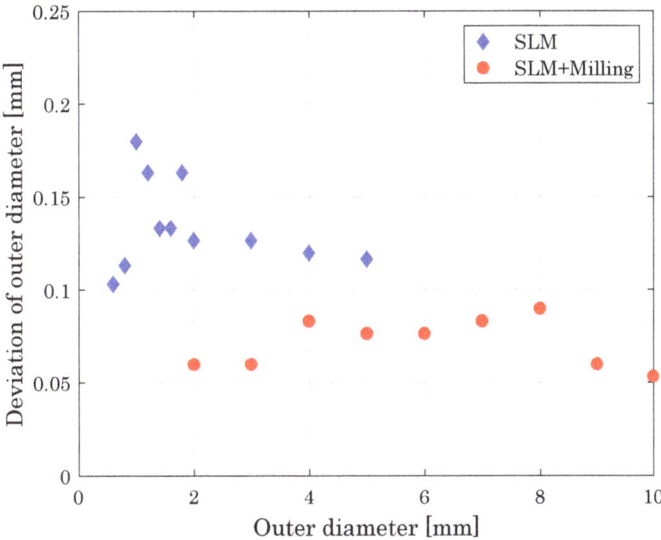

Figure 14. Deviation of outer diameter of spheres versus intended outer diameter, giving a measure of the dimensional accuracy of the processes.

For both processes, the minimum diameter of spheres with reliably built circular geometries can be realized down to $d = 2\,\text{mm}$, below this, the specimens get deformed by the thermal impact of the SLM process.

3.2.7. Inner Diameter of Spheres

Analogue to the manufacturing of outer diameters, the use of support structures is necessary for the manufacuring of inner radii of spheres to guarantee a sufficient heat dissipation [61,62].

The hybrid manufacturing of inner diameters is, again, limited by the diameter of the milling cutter, as Equation (1) defines. Different to the manufacturing of notches, the spheres eliminate the necessity of an additional rounding in the lower part of the component, caused by the radius r of the milling cutter.

With that, the manufacturability of inner diameter of spheres is given at a minimum diameter of 1 mm in the SLM process and 2.06 mm, using the hybrid manufacturing approach and the 1 mm ball end mill (cf. Figure A1).

As shown in Figure 15, the geometrical accuracy of the hybrid manufacturing approach exceeds the generated accuracy in the sole SLM. In addition, Figure 16 depicts the milled spheres in contrast to the SLM-built objects to distinguish the superior smoothness of the hybrid process. Furthermore, the results reveal that the minimum manufacturable inner diameter of 2.06 mm is fabricated reliably without high deviations.

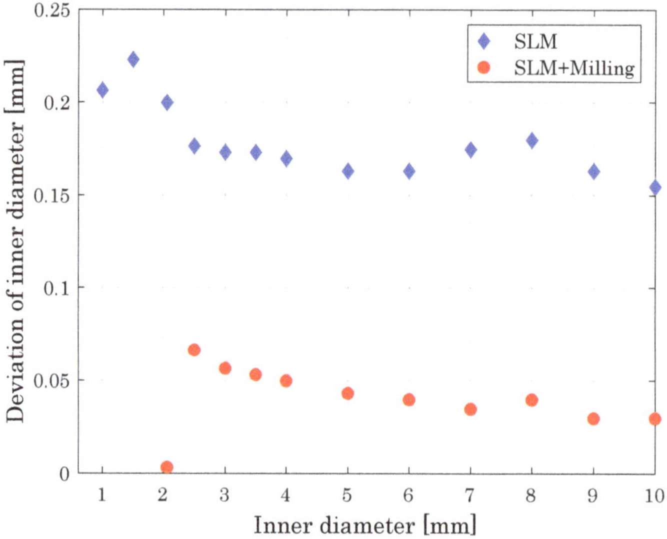

Figure 15. Deviation of inner diameter of spheres versus intended inner diameter, giving a measure of the dimensional accuracy of the processes.

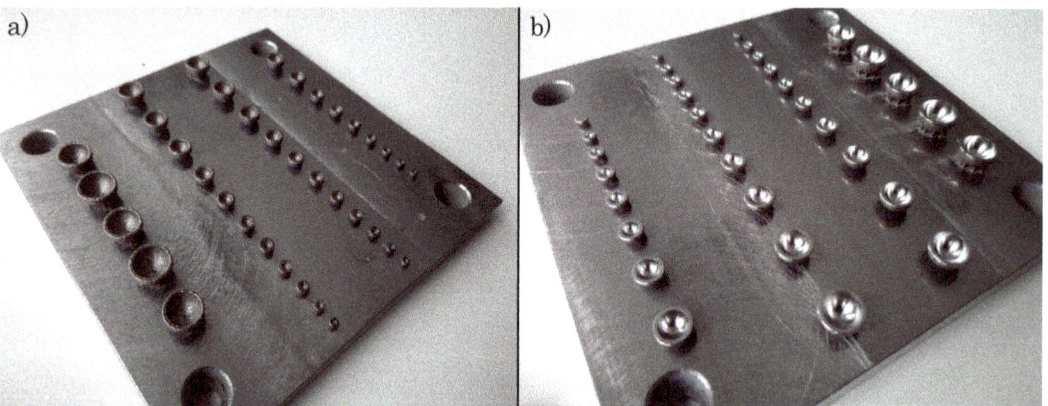

Figure 16. Comparison of (**a**) SLM-built and (**b**) hybrid built inner diameter of spheres.

3.2.8. External Edges

For SLM, outer edges cannot be realized as sharp edges and it is generally recommended to use rounded edges for avoiding geometrical deviations [15]. In the hybrid approach, in addition, the impact of the milling cutter limits the fabrication of sharp edges. As seen before, the milling cutter can induce vibrations and may destroy the structure, turning around a sharp corner (cf. Section 3.2.2).

Nevertheless, as long as the minimum wall thickness is guaranteed (cf. Section 3.2.2), this hampering impact of the milling cutter is negligible and the fabrication of sharp edges can significantly be improved by the hybrid approach.

3.2.9. Inner Angles

Material transitions, in general, should be rounded in the SLM process, conditioned by the laser additive process itself [63,64]. In the hybrid manufacturing the fabrication of inner angles is limited by the geometry of the milling cutter. Due to converging of the legs, inner angles lead to a unmillable geometry at a certain point, undercutting the diameter of the milling cutter.

Figure 17 depicts different inner angles after the milling process. Apparently, the milling cutter is not capable to accurately machine angles below 75°. Only the 90°-angle can be completely milled, exhibiting an unconstructed rounding of the milling cutter.

As a consequence and as a design rule, inner angles have to be defined such that the vertex of the angle has to be rounded according to the diameter of the milling cutter (cf. Figure 13) as to guarantee a complete hybrid processing including an entire removal of the finishing allowance.

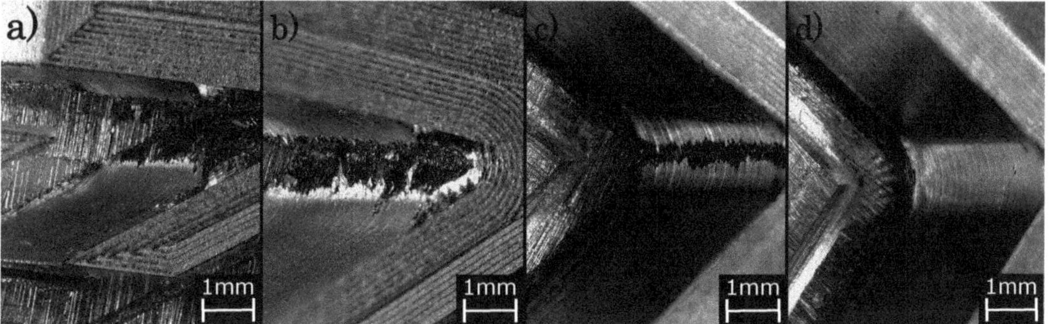

Figure 17. Microscope images of different inner angles with (**a**) 45°, (**b**) 60°, (**c**) 75° and (**d**) 90° after the milling process.

3.2.10. Chamfers in Build Direction

Analogue to the manufacturing of inner angles, it is favourable to prevent sharp edges in build direction to minimize geometrical deviations [63,64]. By means of hybrid manufacturing, the milling cutter limits the creation of sharp chamfers by virtue of his geometry, as shown in Figure 13, illustrating the dimensions of the used ball end mill and the T-slot cutter. In the three-axis milling system, the radius r defines the minimal rounding of material transitions in build direction, especially for directly rectangular transitions.

While for sole SLM the manufacturing of chamfers in build direction is feasible, for the hybrid approach the radius of the milling cutter defines the minimum rounding of the material tranition. In the range of 90°–105° the transition has to be rounded minimal in this radius, from 105° the transition gets manufactured completely, including an entire removal of the finishing allowance.

3.2.11. Bore Holes

The manufacturing of bore holes is, with particular respect to the integration of cooling channels, one of the most interesting application of the SLM [65–67]. Geometrically, bore holes incorporate overhanging parts and to circumvent the necessity of support structures at higher diameters, the use of self-supporting structures is recommended to avoid thermal deformation [14,68].

In hybrid manufacturing, the post-processing is, similar to the manufacturing of outer spheres, divided in two steps, milling the upper surfaces with the ball end mill, while the overhanging part gets machined by the T-slot cutter. Hence, the minimum diameter is defined to $d = 5.06$ mm, following Equation (1) with a diameter of $d = 5$ mm for the T-slot cutter. However, a non millable part of the bore hole remains at the top, depending in its pecularity on the geometry of the bore hole.

For the study of the manufacturing of bore holes, different geometries are examined, embracing the classic circular hole, an elliptic hole and a self-supporting geometry, as depicted in Figure 18. In SLM, the classic circular bore hole has to be supported at a diameter larger than 7.5 mm, excluding the usage as a channel. The elliptic and the self-supporting geometry enable a manufacturing without the necessity of support structures independent on the diameter of the bore hole.

The hybrid manufacturing approach is, as mentioned before, limited to a minimum diameter of $d = 5.06\,\text{mm}$, remaining a non millable part at the top of the bore hole depending on the geometry. The self supporting geometry and the elliptical holes converge rapidly and the non millable part increases in comparison to the the circular bore hole.

Figure 18. Different types of hybrid built bore holes.

3.2.12. Inner Radius of Cylinders (Channels)

Analogue to the manufacturing of notches, the inner radius of cylinders depends, using the hybrid manufacturing approach, on the diameter of the used milling cutter, following Equation (1). Different to the manufacturing of notches, at the inner radius of cylinders, the expulsion of powder can lead to problems and increased signs of wear at the flank of the milling cutter.

In the SLM process, the heat input of the laser is decisive for the manufacturability of channels. They can get sealed by the thermal influence of the surrounding material and adhesion of powder particles can derange the aperture.

For confirming the specification in Equation (1), studying the performance of major cutters and observing the powder expulsion, the manufacturing of inner radii has been examined with the 1 mm ball end mill and the T-slot cutter. As depicted in Figure 13, the T-slot cutter has a diameter of $d = 5\,\text{mm}$ in a completely plane geometry, increasing the difficulty of the powder expulsion.

In the SLM process, the minimum inner diameter is experimentally evaluated with $d = 0.6\,\text{mm}$, smaller channels get sealed or choked by powder adhesion.

In the hybrid process, the Equation (1) is confirmed with both milling cutters, so the minimum diameter is determined with $d = 2.06\,\text{mm}$ for the ball end mill and with $d = 5.06\,\text{mm}$ for the T-slot milling cutter. Furthermore, it is observed that the powder expulsion does not disturb the milling process and the tool wear is not increased significantly by the finishing work in this geometry.

3.3. Industrial Application—Injection Mould

For the exemplification of the potential of this hybrid approach, an injection mold for a charge socket for an electro mobile has been constructed, according the afore discussed design rules and manufactured. Using the advantages of the hybrid manufacturing, injection molds are one of the most interesting industrial application.

With the hybrid addtive manufacturing, injection molds can be built including cooling channels on the one hand and, with a good surface quality and a high geometrical accuracy on the other hand. Due to this, the quality of the die cast components is ensured and the heat dissipation is speeded up, lowering the process duration.

The injection mould is manufactured with a spared part for exhibiting the cooling channels, as shown in Figure 19. As visible, a conformal cooling with a diameter of $d = 2\,\text{mm}$ for the cooling channels was realized, warranting an optimal heat dissipation in all parts of the injection mold. Further, the milling cutter provides a smooth surface, disabling asperities and with that, assuring an improved quality of the die cast components.

Figure 19. Injection mold of a charge socket for an electro mobile with a spared part, confirming the conformal cooling.

The manufacturing of such an unique component can only be permitted by this new manufacturing technique, combining the two fabrication processes. It synergises the geometrical freedom of the additive manufacturing with the surface quality and the geometrical accuracy of the milling process.

In regard to the quality of the die cast components, the surface roughness of an injection mould is one of the decisive factors for the process. Thus, injection molds get manufactured by conventional milling so far, ensuring a high surface quality. Though, the objective of this study is to identify design rules, it can be stated that, in general, the surface roughness Ra of the hybrid approach is better than 1 μm for vertical and horizontal surfaces [34]. In case of curved geometries, like shown in Figure 19, this value raises slightly to Ra = 2 μm.

4. Conclusions

We have evaluated fundamental properties of a hybrid SLM process, which encompasses selective laser melting as an additive process and high speed milling as a subtractive process. Though the milling process takes place within the powder bed, introducing additional splints of the milled steel and the cutter, used powder exhibits comparable properties in particle shape and size as the pristine powder after sieving, i.e., oversized particles with a length of below 0.5 mm can successfully be removed in the recycling chain of the powder. In addition, during the SLM process we observe no adverse impact on accessible geometrical shapes and parts of these oversized particles.

With respect to design engineering, based on the distinctive features of the hybrid additive manufacturing approach, design rules for basic elements like walls, cylinders, angels, inclinations, overhangs, notches, channels, inner and outer radii of spheres, chamfers in build direction, and holes of different structures have been derived and compared to the sole selective laser melting process. The milling process itself, with regard to its specific properties and limitations in design engineering points, originating from the 3-axis milling system, was investigated. Moreover, the geometrical deviations, contingent on the different shapes of the elements, have been determined and with that, the advantages of the hybrid manufacturing approach in comparison to the sole SLM process have been carved out.

Concluding, the specified design rules are consolidated in a comprehensive catalogue for the ultimate support of design engineers to purposefully apply hybrid additive manufacturing. With that, the hybrid approach can be exploited to its full capacity as well as the advantages of this particular system can be capitalised and the existing disadvantages of the sole SLM process can be eliminated. However, by defining the constructional limitations of reliably reproducable parts, common drawbacks of quality assurance issues concerning the acceptance of AM processes as industrial procuction technology are overcome.

Exemplifying the potential of this particular hybrid approach, an injection mold with conformal cooling for a charge socket of an electro mobile has been manufactured. The component summarizes the advantages of both parts of the hybrid manufacturing. Due to the freedom of design in the selective laser melting, an injection mold of every shape can be built, integrating a conformal cooling in a single fabrication step. By virtue of the high speed milling system, the surface finishing can be assured, providing a good quality of the die cast components.

Author Contributions: Conceptualization, D.S. and R.H.; methodology, B.G. and R.H.; validation, D.S.; formal analysis, D.S.; investigation, D.S.; data curation, D.S.; writing—original draft preparation, D.S., B.G. and R.H.; writing—review and editing, D.S., B.G. and R.H.; visualization, D.S.; supervision, C.E. and R.H.; project administration, R.H.; funding acquisition, R.H. All authors have read and agreed to the published version of the manuscript.

Funding: This research received no external funding.

Institutional Review Board Statement: Not applicable.

Informed Consent Statement: Not applicable.

Data Availability Statement: Not applicable.

Conflicts of Interest: The authors declare no conflict of interest.

Appendix A

		SLM	Hybrid SLM	Characteristics
inclinations		$0° \leq \alpha \leq 30°$: Support structures necessary $30° < \alpha \leq 90°$: manufacturable	$0° \leq \alpha \leq 30°$: Support structures necessary $30° \leq \alpha \leq 52°$: Machinable by the roughing cutter $52° \leq \alpha \leq 90°$: completely machinable	• Due to the heat dissipation, highly inclined structures need to be supported • By virtue of the 3-axis milling system, the machinability is limited to an inclination angle of 52°
thin walls		$d \geq 0{,}2$ mm $\Delta d_{max} = 227$ µm $Ra = 12$ µm	$d \geq 0{,}4$ mm $\Delta d_{max} = 35$ µm $Ra = 1$ µm	• Smaller walls break during the milling process • Maximum deviation is reduced significantly by the hybrid approach • Surface Roughness reduced to about 1µm by the hybrid approach
cylinders		$d \geq 1$ mm $\Delta d_{max} = 153$ µm	$d \geq 1{,}4$ mm $\Delta d_{max} = 26$ µm	• The connection to the base plate necessitates higher diameters • Smaller diameter break during the milling process
overhangs	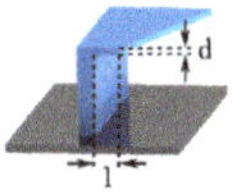	$d \geq 0.3$ mm $l \leq 5$ mm	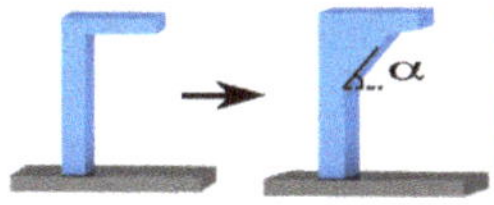	• Self-supporting structures recommended for the SLM process • For the 3-axis milling system, overhanging structures have to be designed with a maximum inclination angle of 52°

		SLM	Hybrid SLM	Characteristics
notches	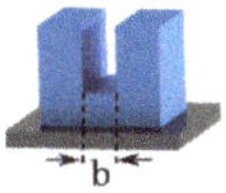	$b \geq 0.6$ mm $\Delta b_{max} = 153$ µm	$b \geq d_{\text{milling cutter}} + 2 \cdot a_{\text{finish}}$ $b \geq 2.06$ mm $\Delta b_{max} = 26$ µm	• Smaller notches get sealed in the SLM process • The minimal clearance in the hybrid SLM is defined by the radius of the used milling cutter and the finishing allowance
outer radii	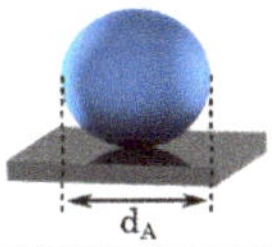	$d_A \geq 2$ mm Support structures necessary	$d_A \geq 2$ mm	• Overhanging structures have to be machined with the T-slot cutter • Upwards facing surfaces get milled by the ball end mill
inner radii	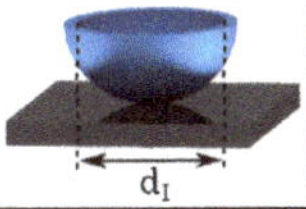	$d_I \geq 1$ mm Support structures necessary	$d_I \geq d_{\text{milling cutter}} + 2 \cdot a_{\text{finish}}$ $d_I \geq 2{,}06$ mm ($d_{\text{milling cutter}} = 2$ mm)	• Due to the geometry, an additional rounding is not necessary for an entire removal of the allowance • Powder gets expulsed by the ball endmill smoothly
external edges		$0° \leq \alpha \leq 180°$: external edges should be rounded	$0° \leq \alpha \leq 180°$: external edges are completely manufacturable	• Minimal wall thickness has to be guaranteed • The fabrication of sharp edges is possible in the hybrid SLM process

Figure A1. *Cont.*

		SLM	Hybrid SLM	Characteristics
internal edges		$0° \leq \alpha \leq 180°$: manufacturable, independent of the angle	$0° \leq \alpha \leq 90°$: Internal edges have to be rounded at least in the diameter of the milling cutter 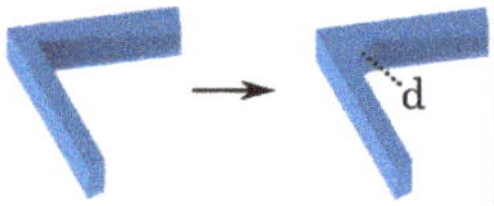 $90° \leq \alpha \leq 180°$: manufacturable	• The milling cutter cannot remove the allowance entirely at acute angles, a rounding gets necessary • At obtuse angles, the milling cutter machines the whole surface, generating a rounding itself
chamfers	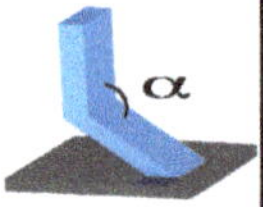	$90° \leq \alpha \leq 180°$: manufacturable	$90° \leq \alpha \leq 105°$: Transitions should be rounded minimal in the secondary radius r of the milling cutter 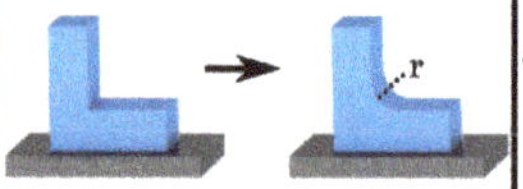 $105° \leq \alpha \leq 180°$: manufacturable	• Due to the 3-axis milling system, the secondary radius of the milling cutter defines perpendicular material transitions • From 105° inclination angle, the milling cutter removes the allowance entirely without an additional rounding

		SLM	Hybrid SLM	Characteristics
bore holes	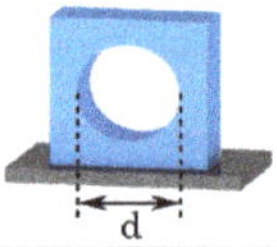	$d \leq 7.5$ mm	$d \geq d_{\text{milling cutter}} + 2 \cdot a_{\text{finish}}$ $d \geq 5{,}06$ mm ($d_{\text{milling cutter}} = 5$mm, T-slot)	• Due to insufficient heat dissipation, bore holes with a diameter greater than 7.5 mm get deformed • Self-supporting structures or elliptical bore holes can be manufactured by SLM, independent on the diameter • Dependent on the bore hole, a non millable area arises • Bottom parts have to be milled by the ball end mill
	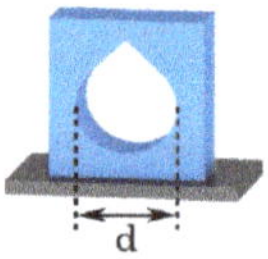	manufacturable		
	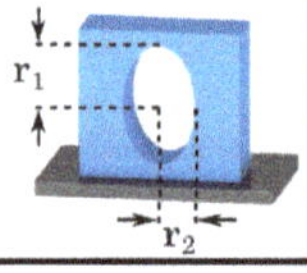	$r_1 \gg r_2$		
channels	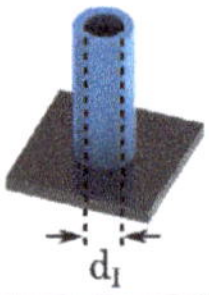	$d_I \geq 0{,}4$ mm	$d_I \geq d_{\text{milling cutter}} + 2 \cdot a_{\text{finish}}$ $d_I \geq 2{,}06$ mm ($d_{\text{milling cutter}} = 2$ mm) $d_I \geq 5{,}06$ mm ($d_{\text{milling cutter}} = 5$mm, T-slot)	• Smaller channels get sealed in the SLM process due to the heat input • Powder expulsion during the milling process is not problematic, independent on the used milling cutter

Figure A1. Design Guideline for the sole SLM and the hybrid additive Manufacturing.

References

1. Flynn, J.M.; Shokrani, A.; Newman, S.T.; Dhokia, V. Hybrid additive and subtractive machine tools—Research and industrial developments. *Int. J. Mach. Tools Manuf.* **2016**, *101*, 79–101. [CrossRef]
2. Edelmann, A.; Dubis, M.; Hellmann, R. Selective Laser Melting of Patient Individualized Osteosynthesis Plates-Digital to Physical Process Chain. *Materials* **2020**, *13*, 5786. [CrossRef] [PubMed]
3. Stoffregen, H.A.; Butterweck, K.; Abele, E. Fatigue analysis in selective laser melting: Review and investigation of thin-walled actuator housings. In Proceedings of the 25th Solid Freeform Fabrication, Austin, TX, USA, 4–6 August 2014; pp. 635–650.
4. Abele, E.; Kniepkamp, M. Analysis and optimisation of vertical surface roughness in micro selective laser melting. *Surf. Topogr. Metrol. Prop.* **2015**, *3*, 034007. [CrossRef]
5. Merklein, M.; Junker, D.; Schaub, A.; Neubauer, F. Hybrid Additive Manufacturing Technologies—An Analysis Regarding Potentials and Applications. *Phys. Procedia* **2016**, *83*, 549–559. [CrossRef]
6. Edelmann, A.; Riedel, L.; Hellmann, R. Realization of a Dental Framework by 3D Printing in Material Cobalt-Chromium with Superior Precision and Fitting Accuracy. *Materials* **2020**, *13*, 5390. [CrossRef] [PubMed]
7. Kerschbaumer, M.; Ernst, G. Hybrid manufacturing process for rapid high performance tooling combining high speed milling and laser cladding. In Proceedings of the 23th International Congress of Applicaitons of Laser & Electro Optics, San Francisco, CA, USA, 4–7 October 2004; pp. 1710–1720.
8. Cortina, M.; Arrizubieta, J.I.; Ruiz, J.E.; Ukar, E.; Lamikiz, A. Latest Developments in Industrial Hybrid Machine Tools that Combine Additive and Subtractive Operations. *Materials* **2018**, *11*, 2583. [CrossRef] [PubMed]
9. Sealy, M.; Madireddy, G.; Williams, R.E.; Rao, P.; Toursangsaraki, M. Hybrid Processes in Additive Manufacturing. *J. Manuf. Sci. Eng.* **2018**, *140*, 060801. [CrossRef]
10. Sreenathbabu, A.; Karunakaran, K.P.; Amarnath, C. Statistical process design for hybrid adaptive layer manufacturing. *Rapid Prototyp. J.* **2005**, *11*, 235–248. [CrossRef]
11. Ye, Z.P.; Zhang, Z.J.; Jin, X.; Xiao, M.Z.; Su, J.Z. Study of hybrid additive manufacturing based on pulse laser wire depositing and milling. *Int. J. Adv. Manuf. Technol.* **2017**, *88*, 2237–2248. [CrossRef]
12. Liu, C.; Yan, D.; Tan, J.; Mai, Z.; Cai, Z.; Dai, Y.; Jiang, M.; Wang, P.; Liu, Z.; Li, C.C.; et al. Development and experimental validation of a hybrid selective laser melting and CNC milling system. *Addit. Manuf.* **2020**, *36*, 101550. [CrossRef]
13. Thompson, M.K.; Moroni, G.; Vaneker, T.; Fadel, G.; Campbell, I.; Gibson, I.; Bernard, A.; Schulz, J.; Graf, P.; Ahuja, B. Design for Additive Manufacturing: Trends, opportunities, considerations, and constraints. *CIRP Ann.* **2016**, *65*, 737–760. [CrossRef]
14. Kokkonen, P.; Salonen, L.; Virta, J.; Hemming, B.; Laukkanen, P.; Savolainen, M.; Komi, E.; Junttila, J.; Ruusuvuori, K.; Varjus, S. *Design Guide for Additive Manufacturing of Metal Components by SLM Process*; VTT Research Report; VTT Technical Research Centre of Finland: Helsinki, Finland, 2016.
15. Thomas, D. The Development of Design Rules for Selective Laser Melting. Ph.D. Thesis, University of Wales and National Centre for Product Design & Development Research, Cardiff, UK, 2009; ISBN 978-3-662-54113-5.
16. Nicola, G.; Missell, F.; Zeilmann, R. Surface quality in milling of hardened H13 steel. *Int. J. Adv. Manuf. Technol.* **2010**, *49*, 53–62. [CrossRef]
17. Suresh Kumar Reddy, N.; Venkateswara Rao, P. Experimental investigation to study the effect of solid lubricants on cutting forces and surface quality in end milling. *Int. J. Mach. Tools Manuf.* **2006**, *46*, 189–198. [CrossRef]
18. Vivancos, J.; Luis, C.J.; Costa, L.; Ortiz, J.A. Optimal machining parameters selection in high speed milling of hardened steels for injection moulds. *J. Mater. Process. Technol.* **2004**, *155–156*, 1505–1512. [CrossRef]
19. Karunakaran, K.P.; Suryakumar, S.; Pushpa, V.; Akula, S. Low cost integration of additive and subtractive processes for hybrid layered manufacturing. *Robot. Comput.-Integr. Manuf.* **2010**, *26*, 490–499. [CrossRef]
20. Magana Carranza, R.; Robinson, J.; Ashton, I.; Fox, P.; Sutcliffe, C.; Patterson, E. A novel device for in-situ force measurements during laser powder bed fusion (L-PBF). *Rapid Prototyp. J.* **2021**, *27*, 1423–1431. [CrossRef]
21. Kumar, M.; Gibbons, G.J.; Das, A.; Manna, I.; Tanner, D.; Kotadia, H.R. Additive manufacturing of aluminium alloy 2024 by laser powder bed fusion: Microstructural evolution, defects and mechanical properties. *Rapid Prototyp. J.* **2021**, *27*, 1388–1397. [CrossRef]
22. Guan, J.; Wang, Q.; Chen, C.; Xiao, J. Forming feasibility and interface microstructure of Al/Cu bimetallic structure fabricated by laser powder bed fusion. *Rapid Prototyp. J.* **2021**, *27*, 1337–1345. [CrossRef]
23. Lieneke, T.; Denzer, V.; Adam, G.A.; Zimmer, D. Dimensional Tolerances for Additive Manufacturing: Experimental Investigation for Fused Deposition Modeling. *Procedia CIRP* **2016**, *43*, 286–291. [CrossRef]
24. Adam, G.A.O. Systematic Development of Design Rules for Additive Manufacturing Processes for Additive Manufacturing Processes SLS, SLM and FDM, Paderborn. Ph.D. Thesis, University Paderborn, Paderborn, Germany, 2015; ISBN 978-3-8440-3474-5.
25. Kranz, J.; Herzog, D.; Emmelmann, C. Design guidelines for laser additive manufacturing of lightweight structures in TiAl6V4. *J. Laser Appl.* **2015**, *27*, S14001. [CrossRef]
26. Wang, D.; Wu, S.; Bai, Y.; Lin, H.; Yang, Y.; Song, C. Characteristics of typical geometrical features shaped by selective laser melting. *J. Laser Appl.* **2017**, *29*, 022007. [CrossRef]
27. Calignano, F.; Lorusso, M.; Pakkanen, J.; Trevisan, F.; Ambrosio, E.P.; Manfredi, D.; Fino, P. Investigation of accuracy and dimensional limits of part produced in aluminum alloy by selective laser melting. *Int. J. Adv. Manuf. Technol.* **2016**, *88*, 451–458. [CrossRef]

28. Finazzi, V.; Demir, A.G.; Biffi, C.A.; Chiastra, C.; Migliavacca, F.; Petrini, L.; Previtali, B. Design Rules for Producing Cardiovascular Stents by Selective Laser Melting: Geometrical Constraints and Opportunities. *Procedia Struct. Integr.* **2019**, *15*, 16–23. [CrossRef]
29. Ameen, W.; Al-Ahmari, A.; Mohammed, M.K. Self-supporting overhang structures produced by additive manufacturing through electron beam melting. *Int. J. Adv. Manuf. Technol.* **2019**, *104*, 2215–2232. [CrossRef]
30. Gokuldoss, P.K.; Kolla, S.; Eckert, J. Additive Manufacturing Processes: Selective Laser Melting, Electron Beam Melting and Binder Jetting—Selection Guidelines. *Materials* **2017**, *10*, 672. [CrossRef] [PubMed]
31. Nakata, T.; Shimpo, H.; Ohkubo, C. Clasp fabrication using one-process molding by repeated laser sintering and high-speed milling. *J. Prosthodont. Res.* **2017**, *61*, 276–282. [CrossRef]
32. Torii, M.; Nakata, T.; Takahashi, K.; Kawamura, N.; Shimpo, H.; Ohkubo, C. Fitness and retentive force of cobalt-chromium alloy clasps fabricated with repeated laser sintering and milling. *J. Prosthodont. Res.* **2018**, *62*, 342–346. [CrossRef]
33. Ohkubo, C.; Sato, Y.; Nishiyama, Y.; Suzuki, Y. Titanium removable denture based on a one-metal rehabilitation concept. *Dent. Mater. J.* **2017**, *36*, 517–523. [CrossRef]
34. Wüst, P.; Edelmann, A.; Hellmann, R. Areal Surface Roughness Optimization of Maraging Steel Parts Produced by Hybrid Additive Manufacturing. *Materials* **2020**, *13*, 418. [CrossRef]
35. Maamoun, A.H.; Xue, Y.F.; Elbestawi, M.A.; Veldhuis, S.C. Effect of Selective Laser Melting Process Parameters on the Quality of Al Alloy Parts: Powder Characterization, Density, Surface Roughness, and Dimensional Accuracy. *Materials* **2018**, *11*, 2343. [CrossRef]
36. Zhang, J.; Song, B.; Wei, Q.; Bourell, D.; Shi, Y. A review of selective laser melting of aluminum alloys: Processing, microstructure, property and developing trends. *J. Mater. Sci. Technol.* **2019**, *35*, 270–284. [CrossRef]
37. Redwood, B.; Schöffer, F.; Garret, B. *The 3D Printing Handbook: Technologies, Design and Applications*; 3D HUBS: Amsterdam, The Netherlands, 2017.
38. Goindi, G.S.; Sarkar, P. Dry machining: A step towards sustainable machining—Challenges and future directions. *J. Clean. Prod.* **2017**, *165*, 1557–1571. [CrossRef]
39. Mazur, M.; Leary, M.; McMillan, M.; Elambasseril, J.; Brandt, M. SLM additive manufacture of H13 tool steel with conformal cooling and structural lattices. *Rapid Prototyp. J.* **2016**, *22*, 504–518. [CrossRef]
40. Mercelis, P.; Kruth, J.P. Residual stresses in selective laser sintering and selective laser melting. *Rapid Prototyp. J.* **2006**, *12*, 254–265. [CrossRef]
41. Casalino, G.; Campanelli, S.L.; Contuzzi, N.; Ludovico, A.D. Experimental investigation and statistical optimisation of the selective laser melting process of a maraging steel. *Opt. Laser Technol.* **2015**, *65*, 151–158. [CrossRef]
42. Koutiri, I.; Pessard, E.; Peyre, P.; Amlou, O.; de Terris, T. Influence of SLM process parameters on the surface finish, porosity rate and fatigue behavior of as-built Inconel 625 parts. *J. Mater. Process. Technol.* **2018**, *255*, 536–546. [CrossRef]
43. Adam, G.A.O.; Zimmer, D. On design for additive manufacturing: Evaluating geometrical limitations. *Rapid Prototyp. J.* **2015**, *21*, 662–670. [CrossRef]
44. Klahn, C.; Omidvarkarjan, D.; Meboldt, M. Evolution of Design Guidelines for Additive Manufacturing—Highlighting Achievements and Open Issues by Revisiting an Early SLM Aircraft Bracket. In *Industrializing Additive Manufacturing–Proceedings of Additive Manufacturing in Products and Applications, Vogel Business Media, W ürzburg*; Springer: Cham, Switzerland, 2018; pp. 3–13.
45. Mazur, M.; Leary, M.; Sun, S.; Vcelka, M.; Shidid, D.; Brandt, M. Deformation and failure behaviour of Ti-6Al-4V lattice structures manufactured by selective laser melting (SLM). *Int. J. Adv. Manuf. Technol.* **2016**, *84*, 1391–1411. [CrossRef]
46. Leary, M.; Mazur, M.; Williams, H.; Yang, E.; Alghamdi, A.; Lozanovski, B.; Zhang, X.; Shidid, D.; Farahbod-Sternahl, L.; Witt, G.; et al. Inconel 625 lattice structures manufactured by selective laser melting (SLM): Mechanical properties, deformation and failure modes. *Mater. Des.* **2018**, *157*, 179–199. [CrossRef]
47. Narasimharaju, S.R.; Liu, W.; Zeng, W.; See, T.L.; Scott, P.; Jiang, X.; Lou, S. Surface Texture Characterization of Metal Selective Laser Melted Part With Varying Surface Inclinations. *J. Tribol.* **2021**, *143*, 051106. [CrossRef]
48. Leutenecker-Twelsiek, B. Additive manufacturing in industrial series production: Component identification and design, Zürich. Ph.D. Thesis, ETH Zurich, Zurich, Switzerland, 2019; ISBN 000-3-4716-4.
49. Yamazaki, T. Development of A Hybrid Multi-tasking Machine Tool: Integration of Additive Manufacturing Technology with CNC Machining. *Procedia CIRP* **2016**, *42*, 81–86. [CrossRef]
50. Su, X.; Yang, Y.; Xiao, D.; Luo, Z. An investigation into direct fabrication of fine-structured components by selective laser melting. *Int. J. Adv. Manuf. Technol.* **2013**, *64*, 1231–1238. [CrossRef]
51. Yadroitsev, I.; Shishkovsky, I.; Bertrand, P.; Smurov, I. Manufacturing of fine-structured 3D porous filter elements by selective laser melting. *Appl. Surf. Sci.* **2009**, *255*, 5523–5527. [CrossRef]
52. Yadroitsev, I.; Gusarov, A.; Yadroitsava, I.; Smurov, I. Single track formation in selective laser melting of metal powders. *J. Mater. Process. Technol.* **2010**, *210*, 1624–1631. [CrossRef]
53. Wang, D.; Yang, Y.; Zhang, M.; Lu, J.; Liu, R.; Xiao, D. Study on SLM fabrication of precision metal parts with overhanging structures. In Proceedings of the 2013 IEEE International Symposium on Assembly and Manufacturing (ISAM), Xi'an, China, 30 July–2 August 2013; pp. 222–225.
54. Han, Q.; Gu, H.; Soe, S.; Setchi, R.; Lacan, F.; Hill, J. Manufacturability of AlSi10Mg overhang structures fabricated by laser powder bed fusion. *Mater. Des.* **2018**, *160*, 1080–1095. [CrossRef]

55. Leary, M.; Merli, L.; Torti, F.; Mazur, M.; Brandt, M. Optimal topology for additive manufacture: A method for enabling additive manufacture of support-free optimal structures. *Mater. Des.* **2014**, *63*, 678–690. [CrossRef]
56. Zhao, J.; Meng, L.; Lan, X.; Li, H.; Gao, L.; Wang, Z. Design and experimental verification of self-supporting topologies for selective laser melting. *Thin-Walled Struct.* **2021**, *161*, 107419. [CrossRef]
57. Barroqueiro, B.; Andrade-Campos, A.; Valente, R.A.F. Designing Self Supported SLM Structures via Topology Optimization. *J. Manuf. Mater. Process.* **2019**, *3*, 68. [CrossRef]
58. Metel, A.S.; Stebulyanin, M.M.; Fedorov, S.V.; Okunkova, A.A. Power Density Distribution for Laser Additive Manufacturing (SLM): Potential, Fundamentals and Advanced Applications. *Technologies* **2019**, *7*, 5. [CrossRef]
59. Asnafi, N. Application of Laser-Based Powder Bed Fusion for Direct Metal Tooling. *Metals* **2021**, *11*, 458. [CrossRef]
60. Wang, D.; Mai, S.; Xiao, D.; Yang, Y. Surface quality of the curved overhanging structure manufactured from 316-L stainless steel by SLM. *Int. J. Adv. Manuf. Technol.* **2015**, *86*, 781–792. [CrossRef]
61. Calignano, F. Design optimization of supports for overhanging structures in aluminum and titanium alloys by selective laser melting. *Mater. Des.* **2014**, *64*, 203–213. [CrossRef]
62. Zhang, K.; Fu, G.; Zhang, P.; Ma, Z.; Mao, Z.; Zhang, D.Z. Study on the Geometric Design of Supports for Overhanging Structures Fabricated by Selective Laser Melting. *Materials* **2018**, *12*, 27. [CrossRef] [PubMed]
63. Ponche, R.; Hascoet, J.Y.; Kerbrat, O.; Mognol, P. A new global approach to design for additivemanufacturing. In *Additive Manufacturing Handbook*; Badiru, A.B., Liu, D., Valencia, V.V., Eds.; Systems Innovation Series; CRC Press: Boca Raton, FL, USA, 2017.
64. Adam, G.A.; Zimmer, D. Design for Additive Manufacturing—Element transitions and aggregated structures. *CIRP J. Manuf. Sci. Technol.* **2014**, *7*, 20–28. [CrossRef]
65. Armillotta, A.; Baraggi, R.; Fasoli, S. SLM tooling for die casting with conformal cooling channels. *Int. J. Adv. Manuf. Technol.* **2014**, *71*, 573–583. [CrossRef]
66. Günther, J.; Leuders, S.; Koppa, P.; Tröster, T.; Henkel, S.; Biermann, H.; Niendorf, T. On the effect of internal channels and surface roughness on the high-cycle fatigue performance of Ti-6Al-4V processed by SLM. *Mater. Des.* **2018**, *143*, 1–11. [CrossRef]
67. Han, S.; Salvatore, F.; Rech, J.; Bajolet, J. Abrasive flow machining (AFM) finishing of conformal cooling channels created by selective laser melting (SLM). *Precis. Eng.* **2020**, *64*, 20–33. [CrossRef]
68. Samperi, M.T. Development of Design Guidelines for Metal Additive Manufacturing and Process Selection. Ph.D. Thesis, Pennsylvania State University, State College, PA, USA, 2014.

Article

Heat Treatment of NiTi Alloys Fabricated Using Laser Powder Bed Fusion (LPBF) from Elementally Blended Powders

Agnieszka Chmielewska [1,*], Bartłomiej Wysocki [2], Piotr Kwaśniak [2], Mirosław Jakub Kruszewski [1], Bartosz Michalski [1], Aleksandra Zielińska [1], Bogusława Adamczyk-Cieślak [1], Agnieszka Krawczyńska [1], Joseph Buhagiar [3] and Wojciech Święszkowski [1,*]

[1] Faculty of Material Science and Engineering, Warsaw University of Technology, Woloska 141 Str., 02-507 Warsaw, Poland; miroslaw.kruszewski@pw.edu.pl (M.J.K.); bartosz.michalski@pw.edu.pl (B.M.); ozielinska5@gmail.com (A.Z.); boguslawa.cieslak@pw.edu.pl (B.A.-C.); agnieszka.krawczynska@pw.edu.pl (A.K.)
[2] Centre of Digital Science and Technology, Cardinal Stefan Wyszynski University in Warsaw, Woycickiego 1/3, 01-938 Warsaw, Poland; b.wysocki@uksw.edu.pl (B.W.); p.kwasniak@uksw.edu.pl (P.K.)
[3] Department of Metallurgy and Materials Engineering, University of Malta, MSD 2080 Msida, Malta; joseph.p.buhagiar@um.edu.mt
* Correspondence: agnieszka.chmielewska.dokt@pw.edu.pl (A.C.); wojciech.swieszkowski@pw.edu.pl (W.Ś.)

Abstract: The use of elemental metallic powders and in situ alloying in additive manufacturing (AM) is of industrial relevance as it offers the required flexibility to tailor the batch powder composition. This solution has been applied to the AM manufacturing of nickel-titanium (NiTi) shape memory alloy components. In this work, we show that laser powder bed fusion (LPBF) can be used to create a $Ni_{55.7}Ti_{44.3}$ alloyed component, but that the chemical composition of the build has a large heterogeneity. To solve this problem three different annealing heat treatments were designed, and the resulting porosity, microstructural homogeneity, and phase formation was investigated. The heat treatments were found to improve the alloy's chemical and phase homogeneity, but the brittle $NiTi_2$ phase was found to be stabilized by the 0.54 wt.% of oxygen present in all fabricated samples. As a consequence, a Ni_2Ti_4O phase was formed and was confirmed by transmission electron microscopy (TEM) observation. This study showed that pore formation in in situ alloyed NiTi can be controlled via heat treatment. Moreover, we have shown that the two-step heat treatment is a promising method to homogenise the chemical and phase composition of in situ alloyed NiTi powder fabricated by LPBF.

Keywords: elemental powders; heat treatment; in situ alloying; laser powder bed fusion; nickel-titanium; pre-mixed powders

Citation: Chmielewska, A.; Wysocki, B.; Kwaśniak, P.; Kruszewski, M.J.; Michalski, B.; Zielińska, A.; Adamczyk-Cieślak, B.; Krawczyńska, A.; Buhagiar, J.; Święszkowski, W. Heat Treatment of NiTi Alloys Fabricated Using Laser Powder Bed Fusion (LPBF) from Elementally Blended Powders. *Materials* **2022**, *15*, 3304. https://doi.org/10.3390/ma15093304

Academic Editor: Chao Xu

Received: 25 March 2022
Accepted: 26 April 2022
Published: 5 May 2022

Publisher's Note: MDPI stays neutral with regard to jurisdictional claims in published maps and institutional affiliations.

1. Introduction

Shape memory alloys (SMAs) have generated great interest in a diverse number of engineering applications. Nickel-titanium (NiTi), also referred to as Nitinol, is the most frequently used SMA that demonstrates a stable shape memory effect and superelastic behaviour, which allows for large recoverable strains of up to 8% [1,2]. Furthermore, it has good biocompatibility, damping characteristics, corrosion resistance, and a low Young's Modulus (50–80 GPa) compared with other alloys commonly used in biomedical applications [3,4]. The specific mechanical properties of NiTi alloys render their machining a challenging task. One of the challenges of NiTi alloys is their extremely poor machinability resulting in rapid and uncontrollable tool wear and poor surface quality. This is due to the high ductility, superelasticity, low elastic modulus, and work hardening of the NiTi alloy the machined part is made of [5,6]. Demanding machining generates a high cost of the fabricated parts and limits the applicability and commercial usage of NiTi alloys. For this reason, an extremely important issue is to reduce material cost by eliminating machining, which would result in a lower cost of the fabricated elements. Thanks to additive manufacturing (AM) techniques that eliminate the need for machining, complex shapes of NiTi alloys have been produced in recent years [7–9]. Laser powder bed fusion (LPBF) is one of the powder-based AM methods allowing the fabrication of a wide variety of functional, complex three-dimensional shaped parts. In this technique the model is built directly from CAD data by melting successive layers of metal powder on top of the previous one, using thermal energy supplied by a focused and computer-controlled laser beam. It offers the opportunity to produce parts with complex geometries without resorting to cutting, pressing, grinding tools or fixtures. Since AM is a fast-growing industrial sector there is a strong need for new material development to satisfy the various engineering and medical application requirements. Despite its promise to overcome these challenges, the AM of NiTi has rarely been considered or explored for the production of NiTi devices. Current research into the AM of NiTi parts from pre-alloyed powders has been associated with difficulties concerning chemical homogeneity and chemical composition control caused by Ni evaporation during the melting process [10–12]. Since phase transition temperatures of NiTi alloys, which determine their unique properties, are highly sensitive to the alloy composition, any changes in the Ni/Ti ratio are undesirable [2,13]. One of the possible alternatives to solve the difficulties and reduce the costs of the production of a few different chemical compositions of NiTi powders (Ni/Ti ratio) is exploring the possibility of alloying Ni and Ti elemental powders in situ during the LPBF processes [14–18]. In situ alloying offers the flexibility to tailor the batch powder composition, by mixing the proper amount of pure metal powders [19,20]. The effect of in situ alloying during the LPBF process is the fabrication of the new object from a predesigned alloy composed of elemental powders [19,20]. For NiTi in situ fabrication, by knowing the amount of nickel evaporated during fabrication [2,9], the proper mixing ratio can be estimated in order to design the chemical composition of the final material [14]. Moreover, the use of nickel and titanium elemental powders is nearly three times cheaper than pre-alloyed NiTi powders currently available on the market. Thus, using elementally blended Ni and Ti powders would reduce the price of final objects.

It has been reported that for NiTi AM-built objects using both elementally blended and pre-alloyed powders, a heat treatment is essential to improve the homogeneity, mechanical properties, and thermomechanical response [8,21]. As reported by Li et al. [21] and Lee et al. [22], the formation of the $NiTi_2$ phase was observed in a NiTi alloy fabricated from pre-alloyed powder. In research work, where NiTi parts were AM-built from elementally blended powders, it was observed that the NiTi martensitic or austenitic, as well as $NiTi_2$ phases, were present [8,14–18,23–26]. Moreover, the presence of secondary phases such as Ni_3Ti and Ni_4Ti_3 were found in as-built specimens by Wang et al. [14] and Halani et al. [16], respectively. Hence, they investigated the influence of the heat treatment on the microstructure and phase composition of fabricated specimens. Based on the Ni-Ti phase diagram (Figure 1), the melting point of $NiTi_2$ is 984°, thus, Wang et al. and Halani et al. subjected the as-built samples made from elementally blended NiTi powders to a solution heat treatment at 1000 °C for 6 h and 1050° for 10 h, respectively. Nevertheless, they found that unwanted secondary Ni_3Ti and $NiTi_2$ phases were formed instead of being eliminated in the material after heat treatment.

In this study, a NiTi alloy was fabricated from elementally blended Ni and Ti powders using LPBF technique and was heat-treated to improve microstructure homogeneity. The influence of the different heat treatments on the microstructure and phase composition of the LPBF fabricated specimens was investigated. Moreover, the aim of this study was to design a heat treatment that homogenizes the microstructure and prevents porosity increase. Three heat treatments were proposed to achieve our research goals: (1) A one-step heat treatment at a temperature of 1100 °C; (2) a one-step heat treatment at a temperature of 900 °C; and (3) a two-step heat treatment, with the first step at a temperature of 900 °C and a second one at 1150 °C. Moreover, for the first time, the oxygen stabilization of the

Ni$_2$Ti phase that resulted in the formation of a thermodynamically stable Ni$_2$Ti$_4$O phase was revealed in the in situ alloyed NiTi.

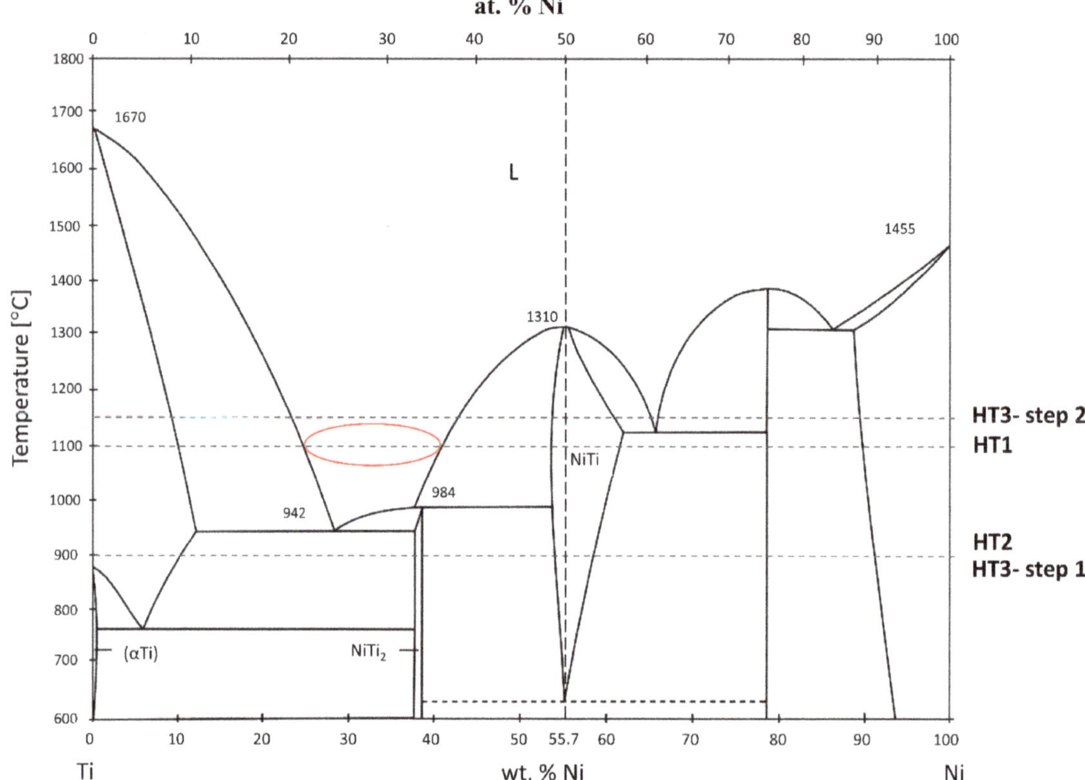

Figure 1. Phase diagram of the Ni-Ti system (based on [27]). The horizontal dashed lines represent the temperatures of the HT1–HT3 heat treatments; The vertical dashed line represents the composition of the initial Ni-Ti powder blend studied in this work; the red circle indicates the local concentrations for which there is a liquid phase at 1100 °C.

2. Materials and Methods

Powder of nickel (purity 99.9%, Eckart TLS, Bitterfeld-Wolfen, Germany) and titanium (grade 1 purity 99.7%, Eckart TLS, Bitterfeld-Wolfen, Germany) were mixed in a tumbling mixer for 2 h, without any additives, to achieve a homogenous mixture of the NiTi alloy having a composition of Ni$_{55.7}$Ti$_{44.3}$ which corresponds to pre-alloyed NiTi, used in our previous study [28]. For both powders, a particles size of <45 μm was used. A Realizer SLM 50 (Realizer GmbH, Borchen, Germany) was employed for the laser powder bed fusion (LPBF) fabrication of cylindrical samples with a diameter of 6 mm and a height of 10 mm. Argon gas was used as a protective gas and the oxygen level was kept below 0.3 vol.%. Samples were fabricated on a NiTi substrate which was maintained at 200 °C. An alternating scanning strategy including additional remelting after the first laser scanning was employed, thus each solidified layer was exposed to the laser beam two times. A laser power of 30 W and a scanning speed of 500 mm/s were used for first melting, while a slightly decreased laser power of 25 W and a scanning speed of 1000 mm/s were used for remelting. The layer thickness was set to 25 μm. The manufacturing procedure has been previously described in detail [10].

Annealing heat treatment (HT) was performed using an Elmor-1100 furnace (Falodlew, Gdańsk, Poland). This was carried out in a vacuum, using sealed quartz tubes using the parameters described in Table 1. The heating rate for each HT was fixed at 10 °C/min. After annealing, all samples were water quenched.

Table 1. Parameters of the heat treatments of in situ alloyed NiTi fabricated by LPBF.

Name	Number of Steps	Temperature [°C]	Time [h]
HT1	1	1100	10
HT2	1	900	24
HT3	2	900 + 1150	24 + 24

The samples were exposed to three different heat treatments denoted as HT1–HT3. The parameters of the HT1 heat treatment were chosen based on the literature [14,16,21] while HT2 and HT3 were chosen directly based on the NiTi phase diagram (Figure 1) to avoid liquid phase formation. The liquid metal has a bigger volume than its solid state, thus, the melt pool shrinking during solidification results in the formation of voids (pores). This phenomenon was observed for HT1 heat treatment and is discussed extensively in this article in the discussion section. Therefore, the aim behind the HT2 and HT3 heat treatments was to minimize liquid phase formation during annealing to prevent further void formation. The proposed HT1 and HT2 are single step annealing heat treatments at 1100 and 900 °C respectively, followed by immediate water quenching. The HT3 is a two-step heat treatment: firstly, samples were heated up to 900 °C and held for 24 h, then the temperature was increased to 1150 °C (based on the DTA results) and held for another 24 h. Annealing time for HT2 and HT3 was increased to 24 h based on the literature to extend the diffusion time of alloying elements according to Fick's second law [29–31]. Samples were water quenched immediately after the second step of annealing. Water quenching was performed to retain the NiTi phase and prevent the formation of brittle intermetallic phases i.e., $NiTi_2$, Ni_3Ti, Ni_4Ti_3 [32–34].

After the heat treatment samples were mounted in epoxy and mechanically polished on a Saphire 550 grinding and polishing machine (ATM Qness GmbH, Mammelzen, Germany) using SiC grinding papers (from #600 to #2000), and subsequently polished using 0.1 μm alumina oxide suspensions. Light microscopy (LM) was performed on Axio Scope Microscope (Carl Zeiss Microscopy GmbH, Kelsterbach, Germany). Porosity was determined by the Archimedes' principle. Each sample was weighed three times, non-consecutively, in air and water using an electronic balance with ±1 mg of accuracy. Scanning electron microscopy (SEM) observations using a backscattered electron (BSE) detector (SU-8000, Hitachi, Ibaraki, Japan) at 5 kV and energy dispersion spectroscopy (EDS) analysis at 10 kV were used to reveal the chemical composition. To determine the phases present, in the as-built NiTi alloy and after different heat treatments, X-ray diffraction (XRD) was performed at room temperature using a Bruker D8 Advance diffractometer (Bruker, Karlsruhe, Germany) with filtered Cu Kα (λ = 0.154056 nm) radiation. The XRD machine parameters were as follows: voltage 40 kV, current 40 mA, angular range 2Θ from 20° to 55°, step Δ2Θ 0.05° and dwell time 3 s. The XRD patterns were analysed using Bruker EVA software and a PDF-2 database (from the International Centre for Diffraction Data). A Labsys DTA/DSC (Setaram, Caluire-et-Cuire, France) differential thermal analysis (DTA) machine, with heating/cooling varying between room temperature to 1500 °C at a rate of 40°/min, was used to determine the $NiTi_2$ transformation temperature. In order to identify phases present in the HT3 sample, a cross-sectional membrane (thin foil) of the selected region was prepared by a focused ion beam (FIB) system (NB5000, Hitachi, Ibaraki, Japan). Subsequently, their microstructure was studied using a transmission electron microscope (TEM) JEOL JEM 1200 (JEOL, Tokyo, Japan) operated at 120 kV.

A TCH 600 Nitrogen/Oxygen/Hydrogen determinator (LECO, Benton Harbor, MI, USA) was used to determine the content of oxygen in all fabricated samples. The elements were converted to their oxidized form by utilizing the gas fusion method and the infrared absorption (IR) was used to measure combustion gases within a metallic sample.

3. Results

3.1. Microscopic Observation and Phase Analysis

The as-built and heat-treated samples were subjected to light microscopic observations. Microstructures of all samples are shown in Figure 2. Pores are evident in each sample and show a spherical morphology. The porosity of the samples was determined by the Archimedes principle. For HT1 heat treatment, the phenomenon of the formation of the liquid phase, described in the discussion section, appeared in the sample. This caused the material porosity to increase remarkably. The porosity of the HT1 sample is almost seven times greater than that of the as-built sample. However, the porosity measurement results of the as-built sample, as well as HT2 and HT3 samples are similar. No significant influence of heat treatment on the porosity of the sample was noticed and the slight differences in porosity are within the measurement error (Figure 2). In addition, no increase in the number of cracks was observed in any of the heat-treated samples.

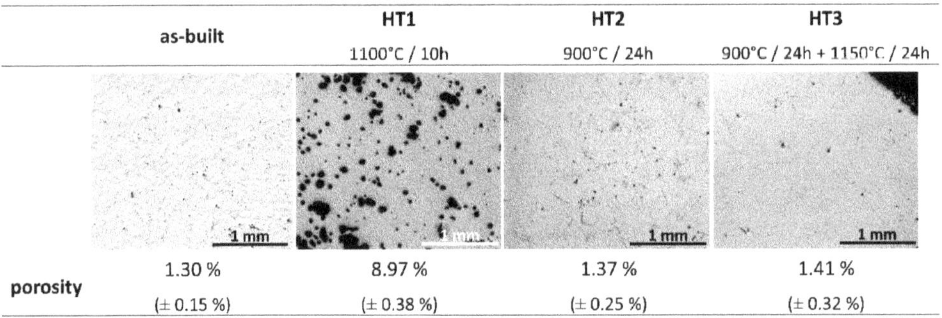

Figure 2. Light microscopic observation and porosity results of as-built sample; HT1 (1000 °C/10 h), HT2 (900 °C/24 h), and HT3 (900 °C/24 h + 1150 °C/24 h) heat-treated samples.

The XRD results of the as-built and heat-treated samples are shown in Figure 3. The XRD patterns of the as-built sample showed the presence of multiple phases, including NiTi (B2) [35] and (B19′) [36], $NiTi_2$ [37], Ni_2Ti_4O [37], Ni_3Ti [38], and Ni_4Ti_3 [39]. A distinct change in the phases was observed in the heat-treated samples. After each heat treatment only NiTi (B19′), (B2), $NiTi_2$, and Ni_2Ti_4O phases were identified. Since NiTi (B19′), $NiTi_2$ and Ni_2Ti_4O peaks overlap each other, it is not possible to distinguish them on XRD.

The phase composition homogeneity of the samples and the microstructures were captured by SEM BSE observations, as illustrated in Figure 4. The SEM BSE images of the as-built sample indicate a large heterogeneity of the chemical composition. EDS analysis (Figure 5) was performed to determine the Ni and Ti distribution in the as-built material. Distinct differences in contrast indicate the presence of titanium-rich (dark) and nickel-rich (bright) areas. In the case of heat-treated samples, only two different phases can be distinguished in the images, the dark-shaded titanium-rich phase, and the bright-shaded nickel-rich phase. EDS point analysis showed the presence of oxygen in some regions of a dark-shaded phase in samples HT1–HT3; however, the presence of oxygen was not detected in all regions of the phase in HT1 and HT2 samples. Therefore, the dark-shaded areas consist of $NiTi_2$ and/or Ni_2Ti_4O phases. A more detailed analysis of the phase composition of the dark-shaded phase in each sample is discussed in paragraph 4. Based on the SEM BSE images, the percentage amount of the dark-shaded Ti-rich phases in each of the tested samples was calculated in the Micrometer software (Micrometer, Poland) using the method described by Wejrzanowski et al. [15,16] (Table 2). The greatest amount of dark-shaded phase is found in the HT2 sample (26 ± 4%), while the least amount was found in the

HT3 sample (12.5 ± 0.5%). The amount of dark-shaded phase in the HT1 sample was 18 ± 2%. It was found that the dark-shaded phase in HT1 and HT2 samples is finer and has a more irregular shape than the phase in the HT3 sample, which is larger and has a nearly spherical shape. The presence of C, Si, and W elements detected by EDS (Figure 5) is a result of surface contamination that remains from materials used for surface polishing and preparation for TEM observations.

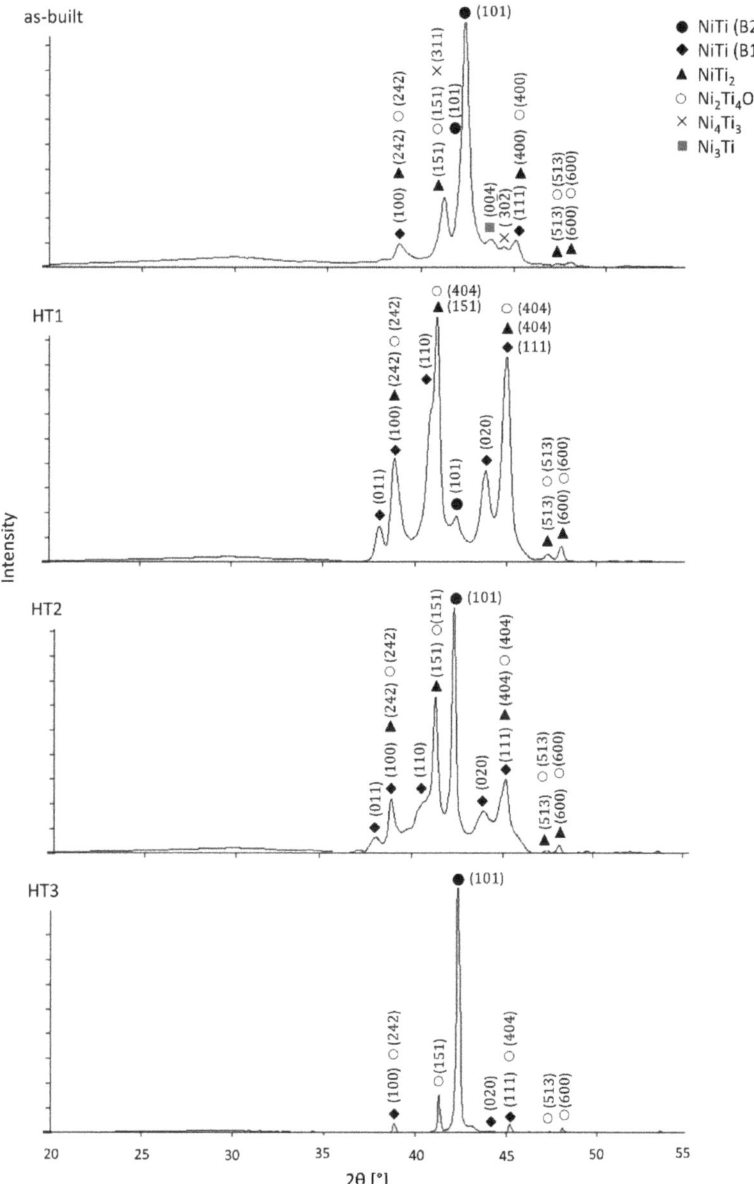

Figure 3. XRD diffractograms of as-built and HT1 (1000 °C/10 h), HT2 (900 °C/24 h), and HT3 (900 °C/24 h + 1150 °C/24 h) heat-treated samples.

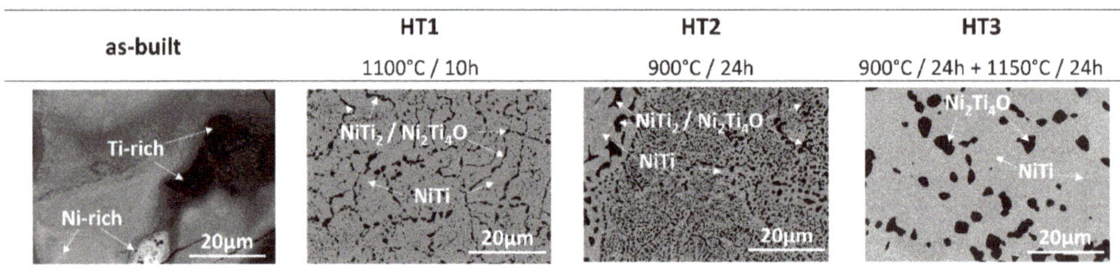

Figure 4. BSE SEM observation of the as-built sample; HT1 (1000 °C/10 h), HT2 (900 °C/24 h), and HT3 (900 °C/24 h + 1150 °C/24 h) heat-treated samples.

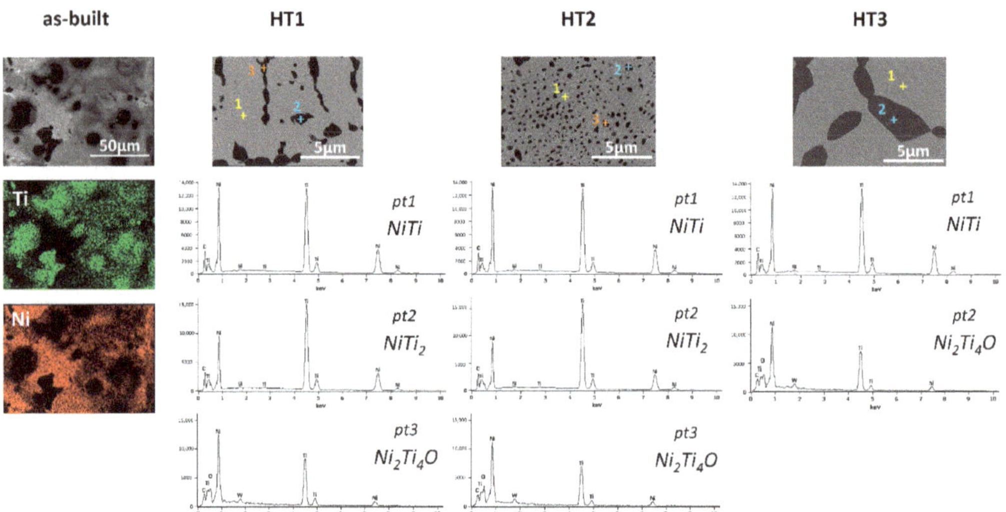

Figure 5. EDS mapping of the as-built sample indicates that the dark shaded areas are titanium-rich and bright shaded areas are nickel-rich; EDS point analysis revealed the presence of Ni, Ti, and O elements.

Table 2. The volume fraction of NiTi$_2$/Ni$_2$Ti$_4$O phases in HT1-HT3 samples.

	HT1	HT2	HT3
NiTi$_2$/Ni$_2$Ti$_4$O	18	26	12.5
volume fraction (%)	±2	±4	±0.5

3.2. Differential Thermal Analysis

The Differential Thermal Analysis (DTA) test was performed on the HT2 sample to determine the NiTi$_2$ phase transformation behaviour and indicate its melting temperature (Figure 6). The results indicate that the NiTi$_2$ melts at about 1128 °C.

3.3. Transmission Electron Microscopy

For a more detailed examination of the dark-shaded phase in the HT3 sample (Figure 4), (that was not removed after HT3, as expected) Transmission Electron Microscopy (TEM) was performed. Figure 7 shows a bright-field image of a NiTi matrix and a corresponding SAED pattern. The matrix consists of B2 cubic austenite and B19′ monoclinic martensite phases as presented by Dutkiewicz et al. [40]. The crystallographic relationship (001) B2 II (011)

B19′and [100] B2II [100] B19′ results from the SAED pattern. Figure 8 shows particles present in the NiTi matrix with a corresponding selected area electron diffraction (SAED) from a single particle. One can expect to find $NiTi_2$ or Ni_2Ti_4O (Figure 8) which are in the same crystal system and possess approximately the same lattice parameter. However, the differences between intensities of diffraction spots from {311} and {331} planes on the diffraction pattern enabled the identification of the Ni_2Ti_4O phase; a phase having XRD peaks overlapping $NiTi_2$. Moreover, in the case of the $NiTi_2$ phase, the intensity from the diffraction spot from {311} planes would be negligible contrary to the Ni_2Ti_4O phase.

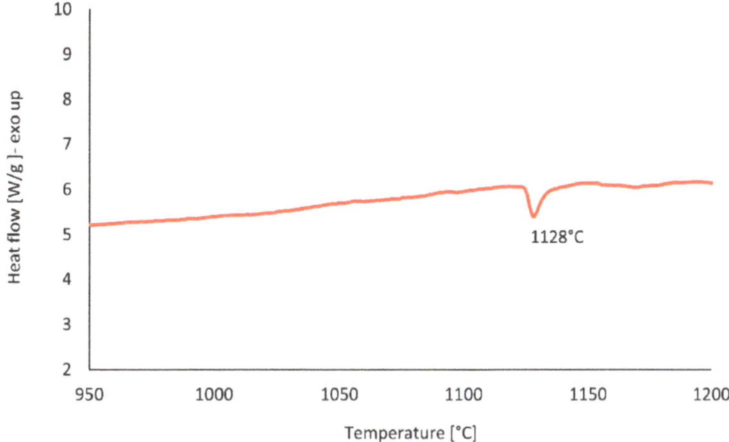

Figure 6. DTA trace of HT2 (900 °C/24 h) sample shows that $NiTi_2$ melts at about 1128 °C and solidifies during cooling at about 1113 °C.

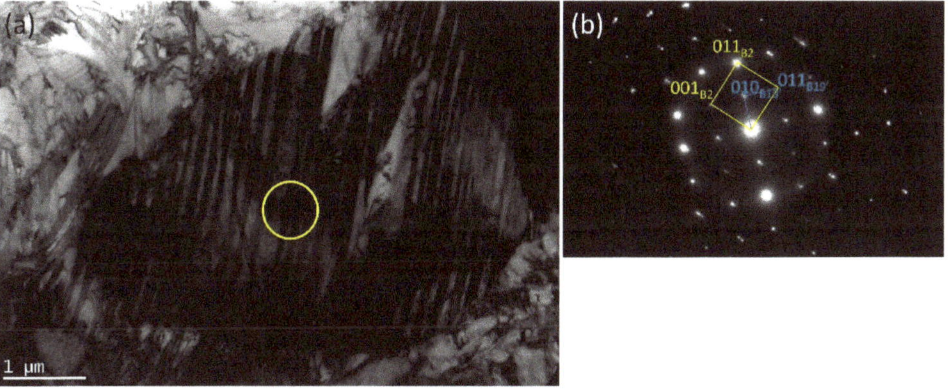

Figure 7. (**a**) A bright-field image of the NiTi matrix; (**b**) with a corresponding SAED pattern from (an area marked by a yellow circle).

3.4. Oxygen Content

The oxygen content in all samples was analysed to check whether the oxygen, which caused the formation of the Ni_2Ti_4O phase, could have dissolved in the sample during the LPBF fabrication or the heat treatment. The presence of 0.52 to 0.54 wt.% of the oxygen is observed in all samples (Table 3). There are no significant differences in the oxygen content between the as-built and heat-treated samples. Thereby, it can be assumed that oxygen was dissolved in the sample during the LPBF fabrication, and this stabilized the Ni_2Ti_4O phase.

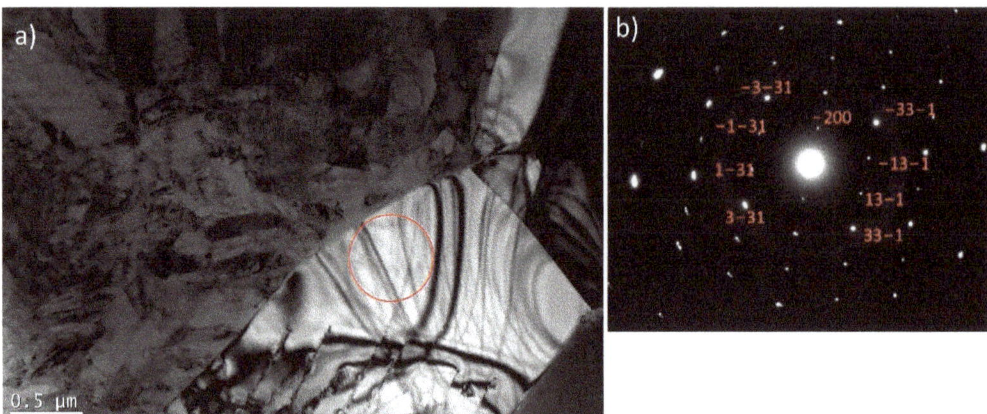

Figure 8. (**a**) TEM image of NiTi matrix with Ni_2Ti_4O particles; (**b**) SAED of Ni_2Ti_4O in the orientation [013] (an area marked by a red circle).

Table 3. Oxygen content in as-built and HT1–HT3 samples.

	As-Built	**HT1**	**HT2**	**HT3**
oxygen content (wt.%)	0.52 ±0.1	0.53 ±0.1	0.54 ±0.2	0.52 ±0.1

4. Discussion

The $Ni_{55.7}Ti_{44.3}$ alloy was fabricated by laser powder bed fusion (LPBF) from elementally blended nickel and titanium powders and the high chemical and phase composition heterogeneity suggested that heat treatment is necessary to improve homogeneity. Therefore, three heat treatments were applied to solve this technical problem.

4.1. As-Built Sample

Microscopic observation of the as-built samples is presented in Figure 2. It shows that the sample is free of cracks or delamination and has a porosity of 1.30%. SEM observations in BSE mode showed the heterogeneity of the chemical composition of the sample (Figure 4). The Ti-rich and Ni-rich areas were revealed by EDS (Figure 5), meaning that components were not fully alloyed during laser melting. XRD phase analysis (Figure 3) indicated the presence of NiTi (B2) and NiTi (B19′), $NiTi_2$, Ni_2Ti_4O, Ni_3Ti, and Ni_4Ti_3 phases. Therefore, heat treatment of the manufactured alloy is necessary to homogenise its chemical and phase composition.

4.2. First Heat Treatment—HT1

In the first heat treatment, HT1, the temperature and time were determined based on the literature. Wang et al. [14] employed LPBF to manufacture a NiTi alloy from elementally blended Ni and Ti powders and subjected the fabricated samples to a solution heat treatment at 1000 °C for 6 h and subsequently quenched in water. Halani and Shin [16] fabricated NiTi coupons using elementally blended powders by the flow-based direct energy deposition (DED) method and heat-treated them at 1050 °C for 10 h, with subsequent quenching in water. In the work presented by Li et al. [21] an additively manufactured NiTi alloy was heat-treated at temperatures that were determined based on DSC tests. It was concluded that $NiTi_2$ in the alloy studied by Li et al. [21] has a melting temperature of 1010 °C. Thus, they performed heat treatment at a temperature close to the determined $NiTi_2$ melting point (1010 °C). When analysing the NiTi phase diagram (Figure 1), it can be seen, that the widest range of NiTi phase presence (49–57 at.% of Ni) occurs at

1118 °C. The temperature of HT1 was determined based on the literature review and the NiTi phase diagram analysis. Therefore, HT1 was performed at 1100 °C for 10 h. Microscopic observation of the HT1 sample was shown in Figure 2 and the presence of large spherical pores was revealed. The porosity of the sample was 8.97%. EDS analysis showed the presence of the oxygen in some regions of a dark-shaded phase (Figure 5), however, the presence of the oxygen was not detected in all regions of the phase. SEM BSE observations (Figure 4), EDS analysis (Figure 5) and corresponding XRD analysis (Figure 3) determined the presence of NiTi (B2), NiTi (B19′), $NiTi_2$ and Ni_2Ti_4O phases. The Ni_3Ti and Ni_4Ti_3 phases on the other hand were eliminated. The amount of dark-shaded $NiTi_2$ and Ni_2Ti_4O phases in HT1 sample was $18 \pm 2\%$. Although the microstructure was significantly homogenized, the porosity of the sample increased almost seven times.

Morris and Morris [41] examined different techniques for the sintering of Ni and Ti elemental powders. They found that when samples are sintered at high temperatures, where melting of a phase occurs, high porosity is generated in the material. Moreover, they noticed a large heterogeneity in the microstructure of sintered samples. They concluded that homogeneity of the material can be increased due to solid-state interfusion, occurring at temperatures below 900 °C, when no liquid phase appears in the material. In our study, the occurrence of Ti- and Ni-rich regions were presented in the as-built sample (Figure 4). According to the NiTi phase diagram (Figure 1), the liquid phase occurs at the temperature of 1100 °C in a wide range of compositions; especially 27 to 38 wt.% of Ni. The liquid phase of a material has a larger volume than its solid counterpart, therefore, during cooling, the material will shrink, resulting in voids. These voids induced the porosity that is observed in Figure 2.

4.3. Second Heat Treatment—HT2

The second heat treatment, HT2, was aimed at homogenizing the microstructure and composition at a temperature that does not allow any phase to liquify since this could generate high porosity. Various conditions of material fabrication and, inter alia, the presence of internal stresses can influence the melting temperature (increase or decrease it) of materials [42]. To this effect, HT2 was performed at 900 °C, i.e., well below the lowest transition temperature of titanium-rich phases to the liquid state (942 °C). The HT2 time was extended to 24 h to increase the diffusion time according to Fick's second law [29–31]. Microscopic observations and porosity measurement results (Figure 2) show that the porosity has not increased and is well below to that found in the HT1 sample. Thus, the reduction of the heat treatment temperature avoided pores formation caused by the melting of the Ti-rich phases. SEM BSE images (Figure 4) revealed the presence of two phases. The presence of oxygen was revealed by EDS in some areas of the dark-shaded phase in the HT2 sample (Figure 5). XRD analysis (Figure 3) confirmed that the samples consisted of B2, B19′, $NiTi_2$, and Ni_2Ti_4O phases. It has been observed that both HT2 and HT1 reduced the number of phases when compared to the as-built samples. The amount of dark-shaded phases in the HT2 sample was $26 \pm 4\%$ (Table 2). Due to the fact that HT1 was conducted at a temperature of 1100 °C which is close to the solubility of the $NiTi_2$ phase in the fabricated material (1123 °C), more of it diffused into the surrounding phase than it was for HT2, which was carried out at a lower temperature (900 °C). Thus, the highest amount of dark-shaded $NiTi_2$ and Ni_2Ti_4O phases remained in the HT2 sample.

Since $NiTi_2$ is a brittle phase, it favours cracks formation and decreases the ductility [43] and it would be therefore beneficial to eliminate the presence of this phase. DTA analysis, shown in Figure 6, was performed to determine the melting temperature of the $NiTi_2$ phase formed in the HT2 heat-treated sample. The melting temperature of the $NiTi_2$ phase in the studied alloy was at 1128 °C. Based on this result, it was expected that the annealing heat treatment at a temperature slightly above the melting point of the $NiTi_2$ phase will allow the diffusion of the elements and the elimination of the phase from the alloy. According to the Ni-Ti phase diagram (Figure 1), the liquid phase is present for a wide range of Ti concentrations (~75 to 60 wt.%) at 1100 °C. Direct annealing conducted above this temperature leads to pronounced, local melting of the material resulting in voids formation after the heat treatment routine. Initial annealing at temperatures where only solid phases exist activates diffusional homogenization and accelerates the formation of the ordered structures which are thermodynamically favoured over solid solutions due to low enthalpies of formation [44]. In such a scenario, the temperature of a second annealing step can be adjusted to eliminate the unwanted phase by heating the material slightly above the thermal stability of the selected, ordered structure. In this case, the amount of the liquid phase is minimal, and melted elements will diffuse to neighbouring phases and/or form other, intermetallic compounds stable under given thermodynamic conditions. To conclude, the HT2 heat treatment improved the homogenization of the chemical composition, however, the $NiTi_2$ and Ni_2Ti_4O phases remained within the alloy. Therefore, the HT3 heat treatment has been designed and performed to eliminate the $NiTi_2$ and Ni_2Ti_4O phases.

4.4. Third Heat Treatment—HT3

Based on the observations made and discussed above, the third heat treatment, HT3, was designed to eliminate from the alloy the $NiTi_2$ phase through a two-step annealing treatment. Firstly, the samples were heated to 900 °C for 24 h to diffuse the elements and homogenise the chemical composition. Secondly, the samples were immediately heated to 1150 °C (no cooling between steps) and annealed for 24 h. After the second step was completed, samples were immediately water quenched. Since DTA analysis (Figure 6) showed that the $NiTi_2$ phase melts at 1128 °C, it was expected that annealing just above the melting point, thus at 1150 °C, should allow the elements to diffuse into the NiTi phase and 1150 °C temperature was chosen for our study. The motivation behind the selection of the two annealing temperatures arises from the thermal stability of intermetallic compounds occurring in NiTi mixtures. It is expected that laser melting of commercially pure Ti and Ni elemental powders can lead to local Ti- or Ni-rich regions, which favour crystallisation of $NiTi_2$, Ni_4Ti_3, or Ni_3Ti phases. The main goal behind the adopted annealing treatments is chemical homogenisation of the fabricated material, however, due to possible local inhomogeneity, the maximum temperature of heat treatment (1150 °C) was chosen slightly above the stability of $NiTi_2$ (1128 °C) to minimize the risk of formation of a large amount of liquid phase inside the samples. Microscopic observations of the HT3 sample and porosity measurement results (Figure 2) showed that the porosity did not significantly increase compared with the as-built sample. SEM BSE observation (Figure 4) and the calculation of the dark-shaded phase volume fraction content (Table 2) revealed that when compared with the HT1 (18 ± 2%) and HT2 (26 ± 4%) samples there was a reduction of dark-shaded phase and its amount in the HT3 sample was 12.5 ± 0.5%. It should be noted that the phase was not completely eliminated. Moreover, EDS analysis revealed the presence of oxygen in all regions of the dark-shaded phase (Figure 5). Recent studies on NiTi alloys have suggested that the $NiTi_2$ phase can be stabilized by oxygen. The presence of the Ni_2Ti_4O phase in NiTi alloy was reported by Kai et al. [45] and Frenzel et al. [13]. $NiTi_2$ can absorb oxygen up to 14 at.% without changing its FCC crystal structure [46]. Oxygen can dissolve in the $NiTi_2$ structure, and cause it to transform into a thermodynamically stable oxygen-rich $Ni_2Ti_4O_x$ oxide phase [45]. Both $NiTi_2$ and Ni_2Ti_4O are crystallographically very similar to each other, both possess FCC structure, thus, it is impossible to distinguish them apart using the XRD technique [47]. Hiroyuki et al. [46] have stated that the Ni_2Ti_4O phase in NiTi alloys cannot be eliminated even after two annealing heat treatments at 900 °C. In the HT3 sample NiTi (B2), NiTi (B19'), and $NiTi_2$/Ni_2Ti_4O phases were identified by XRD (Figure 3). The XRD phase analysis did not indicate the presence of TiO_2. However, it has been shown in our previous study that the thin layer of TiO_2 was found on the material's surface by X-ray photoelectron spectroscopy (XPS) [48]. Since the peaks corresponding to NiTi (B19'), $NiTi_2$ and Ni_2Ti_4O overlap each other the XRD technique cannot be used alone to determine which phase is present in the studied alloy. To this effect, TEM was performed to confirm whether the analysed dark phase, shown in Figure 4, was $NiTi_2$ or

Ni_2Ti_4O. TEM analysis of the matrix revealed the presence of both B2 and B19' phases (Figure 7). Furthermore, the differences between intensities of diffraction spots on the diffraction pattern enabled the identification of the dark phase as Ni_2Ti_4O (Figure 8). Based on this, the oxygen content in the chemical composition of the studied samples was determined and was found in all samples to be 0.52 to 0.54% (Table 3). Since the amount of oxygen in the as-built sample is comparable to all the heat-treated samples, it can be concluded that oxygen was introduced into the samples during the LPBF fabrication and not during heat treatment. The Realizer SLM50 machine used for samples fabrication does not provide a 100% pure oxygen-free atmosphere since the oxygen level in the building chamber during the fabrication process was about 0.3 vol.%. The oxygen residuals in the machine chamber caused oxygen pickup by the NiTi during the LPBF process, in a similar way to oxygen pickup by other materials reported in recent studies [28,49]. As was shown in this study, it is subsequently difficult to remove unwanted secondary phases from the material by applying heat treatments. The negative effect of oxygen residuals and other side effects of powder processing in an inert atmosphere can be eliminated during additive manufacturing processes by using a vacuum [50,51]. Vacuum LPBF processing, which is a relatively new technique for AM processes, could avoid the stabilization of the unwanted secondary phases, such as the $NiTi_2$ phase. Therefore, further research should focus on the elimination of oxygen from the building chamber. Moreover, the issue of the as-built material heterogeneity can be eliminated during AM processes using machines that give the possibility to fabricate materials at elevated temperatures. For this reason, fabrication at elevated temperatures (above 500 °C) would be implemented to perform in situ heat treatment during the LPBF process. Consequently, the transformation temperatures of the fabricated materials would be further investigated using differential scanning calorimetry, as described by Martins et al. [52].

5. Conclusions

Laser powder bed fusion (LBPF), using a remelting scanning strategy and elementally blended Ni and Ti powders, was used to additively manufacture NiTi alloys. Since the as-built components possessed a heterogeneous chemical and phase composition, three different annealing heat treatments were designed and applied. The study led to the following main conclusions:

- The selected two-step HT3 heat treatment condition (900 °C/24 h + 1150 °C/24 h) allows for the significant homogenization of the chemical and phase composition of the LPBF in situ alloyed NiTi components;
- The Ti-rich phases in the as-built material melt during the chosen HT1 temperature (1100 °C) and upon solidification shrinkage occurs resulting in pore formation;
- Oxygen pickup during the LBPF manufacturing process promoted the formation of a thermodynamically stable, oxygen-rich Ni_2Ti_4O phase that is observed even after an annealing heat treatment;
- LPBF combined with post annealing is a promising way of fabricating NiTi alloys using elemental powder blends. Elimination of the oxygen pickup during the process and decrease of the possibility for formation of the oxides like Ni_2Ti_4O could be reached by undergoing the process in vacuum conditions, however further studies proving this hypothesis should be performed.

Author Contributions: Conceptualisation, methodology and investigation, A.C.; microscopic observation, A.K.; heat treatment, P.K. and M.J.K.; DTA analysis, B.M.; XRD investigation, B.A.-C.; coupons preparation, A.Z.; writing—original draft preparation, A.C.; project administration and funding acquisition, A.C. and W.Ś.; writing—review and editing, A.C., B.W., J.B. and W.Ś.; supervision, W.Ś. All authors have read and agreed to the published version of the manuscript.

Funding: This work was supported a subsidy from the Warsaw University of Technology Faculty of Materials Science and Engineering.

Institutional Review Board Statement: Not applicable.

Informed Consent Statement: Not applicable.

Data Availability Statement: Not applicable.

Acknowledgments: The authors are thankful Łukasz Żrodowski from the Faculty of Material Science and Engineering, Warsaw University of Technology, Warsaw, Poland, for his help with the heat treatment.

Conflicts of Interest: The authors declare no conflict of interest.

References

1. Hodgson, D.E.; Wu, M.H.; Biermann, R.J. *Shape Memory Alloys*; ASM International: Novelty, OH, USA, 2013; Volume 2.
2. Elahinia, M.; Shayesteh Moghaddam, N.; Taheri Andani, M.; Amerinatanzi, A.; Bimber, B.A.; Hamilton, R.F. Fabrication of NiTi through additive manufacturing: A review. *Prog. Mater. Sci.* **2016**, *83*, 630–663. [CrossRef]
3. Greiner, C.; Oppenheimer, S.M.; Dunand, D.C. High strength, low stiffness, porous NiTi with superelastic properties. *Acta Biomater.* **2005**, *1*, 705–716. [CrossRef]
4. Duering, T.W.; Pelton, A.R. *Materials Properties Handbook: Titanium Alloys*, 1st ed.; Welsch, G., Boyer, R., Collings, E.W., Eds.; ASM International the Materials Information Society: Materials Park, OH, USA, 1994; ISBN 978-0-87170-481-8.
5. Kaynak, Y. Machining and phase transformation response of room-temperature austenitic NiTi shape memory alloy. *J. Mater. Eng. Perform.* **2014**, *23*, 3354–3360. [CrossRef]
6. Kaynak, Y.; Karaca, H.E.; Noebe, R.D.; Jawahir, I.S. Analysis of Tool-wear and Cutting Force Components in Dry, Preheated, and Cryogenic Machining of NiTi Shape Memory Alloys. *Procedia CIRP* **2013**, *8*, 498–503. [CrossRef]
7. Wang, X.; Kustov, S.; Van Humbeeck, J. A short review on the microstructure, transformation behavior and functional properties of NiTi shape memory alloys fabricated by selective laser melting. *Materials* **2018**, *11*, 1683. [CrossRef] [PubMed]
8. Farber, E.; Zhu, J.N.; Popovich, A.; Popovich, V. A review of NiTi shape memory alloy as a smart material produced by additive manufacturing. *Mater. Today Proc.* **2019**, *30*, 761–767. [CrossRef]
9. Khoo, Z.X.; Liu, Y.; An, J.; Chua, C.K.; Shen, Y.F.; Kuo, C.N. A review of selective laser melted NiTi shape memory alloy. *Materials* **2018**, *11*, 519. [CrossRef]
10. Chmielewska, A.; Wysocki, B.; Buhagiar, J.; Michalski, B.; Adamczyk-Cieślak, B.; Gloc, M.; Święszkowski, W. In situ alloying of NiTi: Influence of Laser Powder Bed Fusion (LBPF) scanning strategy on chemical composition. *Mater. Today Commun.* **2022**, *30*, 103007. [CrossRef]
11. Speirs, M.; Wang, X.; Van Baelen, S.; Ahadi, A.; Dadbakhsh, S.; Kruth, J.-P.; Van Humbeeck, J. On the Transformation Behavior of NiTi Shape-Memory Alloy Produced by SLM. *Shape Mem. Superelasticity* **2016**, *2*, 310–316. [CrossRef]
12. Bormann, T.; Müller, B.; Schinhammer, M.; Kessler, A.; Thalmann, P.; De Wild, M. Microstructure of selective laser melted nickel-titanium. *Mater. Charact.* **2014**, *94*, 189–202. [CrossRef]
13. Frenzel, J.; George, E.P.; Dlouhy, A.; Somsen, C.; Wagner, M.F.X.; Eggeler, G. Influence of Ni on martensitic phase transformations in NiTi shape memory alloys. *Acta Mater.* **2010**, *58*, 3444–3458. [CrossRef]
14. Wang, C.; Tan, X.P.; Du, Z.; Chandra, S.; Sun, Z.; Lim, C.W.J.; Tor, S.B.; Lim, C.S.; Wong, C.H. Additive manufacturing of NiTi shape memory alloys using pre-mixed powders. *J. Mater. Process. Technol.* **2019**, *271*, 152–161. [CrossRef]
15. Stoll, P.; Spierings, A.; Wegner, K. SLM processing of elementally blended NiTi shape memory alloy. *Procedia CIRP* **2020**, *95*, 121–126. [CrossRef]
16. Halani, P.R.; Shin, Y.C. In situ synthesis and characterization of shape memory alloy nitinol by laser direct deposition. *Metall. Mater. Trans. A Phys. Metall. Mater. Sci.* **2012**, *43*, 650–657. [CrossRef]
17. Zhao, C.; Liang, H.; Luo, S.; Yang, J.; Wang, Z. The effect of energy input on reaction, phase transition and shape memory effect of NiTi alloy by selective laser melting. *J. Alloys Compd.* **2020**, *817*, 153288. [CrossRef]
18. Zhang, B.; Chen, J.; Coddet, C. Microstructure and transformation behavior of in-situ shape memory alloys by selective laser melting Ti-Ni mixed powder. *J. Mater. Sci. Technol.* **2013**, *29*, 863–867. [CrossRef]
19. Mosallanejad, M.H.; Niroumand, B.; Aversa, A.; Saboori, A. In-situ alloying in laser-based additive manufacturing processes: A critical review. *J. Alloys Compd.* **2021**, *872*, 159567. [CrossRef]
20. Katz-Demyanetz, A.; Koptyug, A.; Popov, V.V. In-situ Alloying as a Novel Methodology in Additive Manufacturing. In Proceedings of the 2020 IEEE 10th International Conference Nanomaterials: Applications & Properties (NAP), Sumy, Ukraine, 9–13 November 2020; pp. 26–29. [CrossRef]
21. Li, S.; Hassanin, H.; Attallah, M.M.; Adkins, N.J.E.; Essa, K. The development of TiNi-based negative Poisson's ratio structure using selective laser melting. *Acta Mater.* **2016**, *105*, 75–83. [CrossRef]
22. Lee, J.; Shin, Y.C. Effects of Composition and Post Heat Treatment on Shape Memory Characteristics and Mechanical Properties for Laser Direct Deposited Nitinol. *Lasers Manuf. Mater. Process.* **2019**, *6*, 41–58. [CrossRef]
23. Ibrahim, H.; Dean, D.; Klarner, A.D.; Luo, A.A. The Effect of Heat-Treatment on Mechanical, Microstructural, and Corrosion Characteristics of a Magnesium Alloy with Potential Application in Resorbable Bone Fixation Hardware. In Proceedings of

the ASME 2016 11th International Manufacturing Science and Engineering Conference, MSEC2016-8822, Blacksburg, VA, USA, 27 June–1 July 2016; pp. 1–7.

24. Shiva, S.; Palani, I.A.; Mishra, S.K.; Paul, C.P.; Kukreja, L.M. Investigations on the influence of composition in the development of Ni-Ti shape memory alloy using laser based additive manufacturing. *Opt. Laser Technol.* **2015**, *69*, 44–51. [CrossRef]
25. Halani, P.R.; Kaya, I.; Shin, Y.C.; Karaca, H.E. Phase transformation characteristics and mechanical characterization of nitinol synthesized by laser direct deposition. *Mater. Sci. Eng. A* **2013**, *559*, 836–843. [CrossRef]
26. Shishkovsky, I.; Yadroitsev, I.; Smurov, I. Direct Selective Laser Melting of Nitinol Powder. *Phys. Procedia* **2012**, *39*, 447–454. [CrossRef]
27. Chekotu, J.C.; Groarke, R.; O'Toole, K.; Brabazon, D. Advances in selective laser melting of Nitinol shape memory alloy part production. *Materials* **2019**, *12*, 809. [CrossRef] [PubMed]
28. Chmielewska, A.; Jahadakbar, A.; Wysocki, B.; Elahinia, M.; Święszkowski, W.; Dean, D. Chemical Polishing of Additively Manufactured, Porous, Nickel–Titanium Skeletal Fixation Plates. *3D Print. Addit. Manuf.* **2021**. *ahead of print*. [CrossRef]
29. Li, X.; Chen, H.; Guo, W.; Guan, Y.; Wang, Z.; Zeng, Q.; Wang, X. Improved superelastic stability of NiTi shape memory alloys through surface nano-crystallization followed by low temperature aging treatment. *Intermetallics* **2021**, *131*, 8. [CrossRef]
30. Divinski, S.V.; Stloukal, I.; Kral, L.; Herzig, C. Diffusion of titanium and nickel in B2 NiTi. *Defect Diffus. Forum* **2009**, *289–292*, 377–382. [CrossRef]
31. Wilkinson, D.S. *Mass Transport in Solids and Fluids*; Cambridge University Press: Cambridge, UK, 2000; ISBN 0521624940.
32. Motemani, Y.; Nili-Ahmadabadi, M.; Tan, M.J.; Bornapour, M.; Rayagan, S. Effect of cooling rate on the phase transformation behavior and mechanical properties of Ni-rich NiTi shape memory alloy. *J. Alloys Compd.* **2009**, *469*, 164–168. [CrossRef]
33. Thomas, F. *The Effect of Various Quenchants on the Hardness and Microstructure of 60-NITINOL*; Nasa/TM-2015-218463; Glenn Research Center: Cleveland, OH, USA, 2015.
34. Chen, K.C. NiTi—Magic or phase transformations? In Proceedings of the ASEE Annual Conference & Exposition, Nashville, TN, USA, 21 June 2003; pp. 2423–2430.
35. Sitepu, H. Texture and structural refinement using neutron diffraction data from molybdite (MoO_3) and calcite ($CaCO_3$) powders and a Ni-rich $Ni_{50.7}Ti_{49.30}$ alloy. *Powder Diffr.* **2009**, *24*, 315–326. [CrossRef]
36. Kudoh, Y.; Tokonami, M.; Miyazaki, S.; Otsuka, K. Crystal structure of the martensite in Ti-49.2 at.%Ni alloy analyzed by the single crystal X-ray diffraction method. *Acta Metall.* **1985**, *33*, 2049–2056. [CrossRef]
37. Mueller, M.H.; Knott, H.W. The crystal structures of Ti_2Cu, Ti_2Ni, Ti_4Ni_2O, and Ti_4Cu_2O. *Trans. Metall. Soc. Aime* **1963**, *227*, 674–678.
38. Laves, F.; Wallbaum, J.H. Die Kristallstruktur von Ni_3Ti und Si_3Ti. *Z. Für Krist.—Cryst. Mater.* **1939**, *101*, 78–93. [CrossRef]
39. Tirry, W.; Schryvers, D.; Jorissen, K.; Lamoen, D. Electron-diffraction structure refinement of Ni_4Ti_3 precipitates in $Ni_{52}Ti_{48}$. *Acta Crystallogr. Sect. B Struct. Sci.* **2006**, *62*, 966–971. [CrossRef] [PubMed]
40. Dutkiewicz, J.; Rogal, Ł.; Kalita, D.; Węglowski, M.; Błacha, S.; Berent, K.; Czeppe, T.; Antolak-Dudka, A.; Durejko, T.; Czujko, T. Superelastic Effect in NiTi Alloys Manufactured Using Electron Beam and Focused Laser Rapid Manufacturing Methods. *J. Mater. Eng. Perform.* **2020**, *29*, 4463–4473. [CrossRef]
41. Morris, D.G.; Morris, M.A. NiTi intermetallic by mixing, milling and interdiffusing elemental components. *Mater. Sci. Eng. A* **1989**, *110*, 139–149. [CrossRef]
42. Hwang, Y.S.; Levitas, V.I. Internal stress-induced melting below melting temperature at high-rate laser heating. *Appl. Phys. Lett.* **2014**, *104*, 263106. [CrossRef]
43. Zhang, Y.; Cheng, X.; Cai, H. Fabrication, characterization and tensile property of a novel $Ti_2Ni/TiNi$ micro-laminated composite. *Mater. Des.* **2016**, *92*, 486–493. [CrossRef]
44. Povoden-Karadeniz, E.; Cirstea, D.C.; Lang, P.; Wojcik, T.; Kozeschnik, E. Thermodynamics of Ti-Ni shape memory alloys. *Calphad* **2013**, *41*, 128–139. [CrossRef]
45. Kai, W.Y.; Chang, K.C.; Wu, H.F.; Chen, S.W.; Yeh, A.C. Formation mechanism of Ni_2Ti_4Ox in NITI shape memory alloy. *Materialia* **2019**, *5*, 100194. [CrossRef]
46. Takeshita, H.T.; Tanaka, H.; Kuriyama, N.; Sakai, T.; Uehara, I.; Haruta, M. Hydrogenation characteristics of ternary alloys containing Ti_4Ni_2X (X = O, N, C). *J. Alloys Compd.* **2000**, *311*, 188–193. [CrossRef]
47. Karlík, M.; Haušild, P.; Klementová, M.; Novák, P.; Beran, P.; Perrière, L.; Kopeček, J. TEM phase analysis of NiTi shape memory alloy prepared by self-propagating high-temperature synthesis. *Adv. Mater. Process. Technol.* **2017**, *3*, 58–69. [CrossRef]
48. Chmielewska, A.; Dobkowska, A.; Kijeska-Gawrońska, E.; Jakubczak, M.; Krawczyńska, A.; Choińska, E.; Jastrzębska, A.; Dean, D.; Wysocki, B.; Święszkowski, W. Biological and Corrosion Evaluation of In Situ Alloyed NiTi Fabricated through Laser Powder Bed Fusion (LPBF). *Int. J. Mol. Sci.* **2021**, *22*, 13209. [CrossRef] [PubMed]
49. Wysocki, B.; Maj, P.; Krawczyńska, A.; Rożniatowski, K.; Zdunek, J.; Kurzydłowski, K.J.; Święszkowski, W. Microstructure and mechanical properties investigation of CP titanium processed by selective laser melting (SLM). *J. Mater. Process. Technol.* **2017**, *241*, 13–23. [CrossRef]
50. Kaserer, L.; Bergmueller, S.; Braun, J.; Leichtfried, G. Vacuum laser powder bed fusion—track consolidation, powder denudation, and future potential. *Int. J. Adv. Manuf. Technol.* **2020**, *110*, 3339–3346. [CrossRef]

51. Cooper, N.; Coles, L.A.; Everton, S.; Maskery, I.; Campion, R.P.; Madkhaly, S.; Morley, C.; O'Shea, J.; Evans, W.; Saint, R.; et al. Additively manufactured ultra-high vacuum chamber for portable quantum technologies. *Addit. Manuf.* **2021**, *40*, 101898. [CrossRef]
52. Martins, J.N.R.; Silva, E.J.N.L.; Marques, D.; Belladonna, F.; Simões-Carvalho, M.; Vieira, V.T.L.; Antunes, H.S.; Braz Fernandes, F.M.B.; Versiani, M.A. Design, metallurgical features, mechanical performance and canal preparation of six reciprocating instruments. *Int. Endod. J.* **2021**, *54*, 1623–1637. [CrossRef]

materials

Article

Additively Manufactured 316L Stainless Steel Subjected to a Duplex Peening-PVD Coating Treatment

Luana Bonnici [1], Joseph Buhagiar [1], Glenn Cassar [1], Kelsey Ann Vella [1], Jian Chen [2], Xiyu Zhang [2], Zhiquan Huang [2] and Ann Zammit [1,*]

[1] Department of Metallurgy and Materials Engineering, University of Malta, MSD 2080 Msida, Malta
[2] School of Materials Science and Engineering, SEU University, Nanjing 211189, China
* Correspondence: ann.zammit@um.edu.mt

Abstract: This research studies the individual and combined effects of mechanical shot peening and the deposition of TiAlCuN coating on additively manufactured 316L stainless steel. Shot peening has been found to induce a 40% increase in surface hardness, while the combined effect of shot peening and the coating produced an approximately three-fold increase in surface hardness when compared to the as-printed coupons. Shot peening reduced the surface roughness of printed metal coupons by 50%, showing that shot peening can also serve to improve the surface finish of as-printed 316L stainless steel components. The peening process was found to induce a compressive residual stress of 589 MPa, with a maximum affected depth of approximately 200 μm. Scratch testing of the printed and coated specimens showed complete delamination failure at a normal load of 14 N, when compared to hybrid treated samples which failed at 10 N. On the other hand, from the corrosion tests, it was found that the hybrid treated samples provided the optimal results as opposed to the other variables.

Keywords: additive manufacturing; shot peening; PVD multi-layer coating; mechanical testing; corrosion; residual compressive stresses

Citation: Bonnici, L.; Buhagiar, J.; Cassar, G.; Vella, K.A.; Chen, J.; Zhang, X.; Huang, Z.; Zammit, A. Additively Manufactured 316L Stainless Steel Subjected to a Duplex Peening-PVD Coating Treatment. *Materials* **2023**, *16*, 663. https://doi.org/10.3390/ma16020663

Academic Editor: Amir Mostafaei

Received: 9 December 2022
Revised: 30 December 2022
Accepted: 4 January 2023
Published: 10 January 2023

1. Introduction

The marine transportation industry makes up part of the global economy and trade, since primary necessities such as petroleum, foods and goods are carried by means of water. Due to their working environment, parts such as propellers and shafts are exposed to the adverse effects of corrosion, erosion, wear, and detrimental effects. The failure of the fundamental components of a ship can lead to severe consequences, mainly delays in delivery of cargo, environmental damage, financial losses and most importantly fatalities of the passengers on board [1]. Such mid-journey failure would dictate a potentially long sourcing processing and docking of the vessel. The high demand for replacement of parts could be sustained by additive manufacturing (AM).

Components suitable for the marine environment necessitate a particular combination of attributes to make them suitable for use, first and foremost hardness. Additive manufactured 316L stainless steel (SS) is known to have a higher hardness than wrought 316L SS. In a study by Yusuf et al. [1], an average microhardness value of 228 HV was obtained for AM 316L SS, while 192 HV was measured for wrought 316L SS [1]. This increase in microhardness is attributed to the finer grains within the microstructure [2]. Additionally, parts intended for engine and drive components need to possess a high tensile and yield strength to be able to withstand the high loads experienced during use. AM can produce a wide range of ultimate tensile strengths ranging from 550 to 700 MPa, while values for the yield strength vary from 300 to 600 MPa, depending on the chosen parameters used for printing the material [3–6]. Similarly, corrosion resistance is also an important characteristic for marine transportation parts due to the environment they are subjected to. Few studies on the corrosion resistance of additively manufactured 316L SS are available and no broad consensus has been reached in terms of AM performance compared to wrought equivalents. The corrosion resistance of 316L SS can be enhanced with its fine sub-granular microstructure. This is because a more uniform passive film is generated with the help of the rapid interface boundary diffusion process [7], while other work has shown that SLM components exhibited inferior corrosion resistance in comparison to wrought 316L SS, due to retained porosity [8–10]. Jung et al. [8] studied the corrosion characteristics electrochemically in seawater, where AM 316L SS emerged to have poorer qualities than wrought 316L SS.

The typically high surface roughness and residual porosity of AM parts can also cause a deterioration in the mechanical strength of critical components, weakening their performance. The application of surface treatments such as shot peening (SP) and coating deposition, is proposed to mitigate such issues. The cold work induced at the surface via SP can be beneficial in minimising pore diameter found in additively manufactured components, while also inducing work hardening, causing surface texturing and mitigation of intergranular corrosion. In addition, the characteristic resultant residual compressive stresses offset any remnant tensile stresses originating from the printing process [11]. In a study carried out by Santa-aho et al. [12], surface compressive stresses reaching up to 502 MPa were generated in the additively manufactured 316L SS, up to a depth of 80 μm. The compressive stresses generated in the AM 316L SS were quickly formed as the shots impacted the surface, removing the tensile stresses generated by the printing process [12], thereby enhancing fatigue performance [13,14]. Concurrently, shot peening produces an increase in the hardness of the surface and near-surface layers of the material [15–18]. Gundgire et al. [16] obtained a hardness of 340 HV after SP additively manufactured 316L SS with a native hardness of 230 HV. A similar result was obtained by Sugavaneswaran et al. [15], where the hardness was increased from 230.8 HV to 324.5 HV after SP additively manufactured 316L SS. Furthermore, SP is known to reduce the mean surface roughness (R_a) of additively manufactured 316L SS. Evidence from research by Rautio et al. [14] shows that after SP additively manufactured 316L SS using martensitic chromium shots, the R_a was reduced from 8.81 $\pm$ 0.06 μm to 3.93 $\pm$ 0.01 μm. Additionally, in another study by Gundgire et al. [16], it was noted that the R_a was reduced to 5.81 from 8.81 μm, after SP.

According to the author's knowledge, literature on coatings deposited by PVD on AM 316L SS is scarce—even more so in the case of novel coating systems such as TiAlCuN. Coatings deposited by PVD are mainly intended to increase the hardness of the substrate on which they have been deposited, as they have a hardness ranging between 2300 and 4079 HV [19,20], while specific additives can be used to provide specific characteristics depending on the intended application. In this case, Cu was added to the more common Ti–Al–N system for anti-biofouling purposes. Furthermore, the coating deposition enables the enhancement of tribological behaviour and corrosion resistance amongst others. Multi-layer coatings are known to have a high corrosion resistance when compared to monolayer coatings and the bare substrate [21–23]. This is consistent with work by Ananthakumar et al. [23], a positive shift from −0.837 V to −0.546 V in the corrosion potential was detected when comparing the bare substrate to the multi-layered coated substrate. On top of this, PVD coatings do not typically increase the roughness of the substrate as these conform readily with the underlying surface, making them beneficial as they do not alter the surface once deposited [24]. The coating provides corrosion protection and improvement of tribological behaviour, amongst others. In a study carried out by Tillmann et al. [25], improved adhesion was discovered after the coating deposition of CrAlN coatings on additively manufactured 316L SS.

The combination of SP and coating deposition by PVD is required to increase the hardness, to induce residual compressive stresses and to increase the corrosion resistance of the substrate. Multiple studies have been carried out on the separate application of SP and PVD on AM parts. However, according to the authors' knowledge, there is a lack of studies on the duplex combination of such surface treatments, with particular attention to those performed on AM-ed 316L SS. Therefore, the aim of this research was to study in detail, the combined effect of shot peening and PVD on AM 316L SS, mainly through the characterisation of microstructure, hardness, roughness, phase analysis, residual stress measurement, the resultant adhesion of the coating to the substrate and the corrosion resistance in a marine simulated environment.

2. Materials and Methods

2.1. Powder Processing and Manufacturing of Coupons

316L SS powder used for additive manufacturing (AM), having the composition found in Table 1(a), was provided by Jiangsu Vilory Advanced Materials Technology Company Limited (Xuzhou, China). Its particle diameter ranged from 15 to 53 µm [26]. AM was carried out by an AM Pro SP100 3D printer selective laser melting (SLM) machine (Suzhou, China) having a IPG 2000 W type laser, a laser power of 200 W and a laser speed of 950 mm/s. The beam was offset for 10 µm and each layer was 30 µm thick. Samples were printed along the y-direction and had a 20 mm diameter and a 7 mm height. Wrought 316L SS (Table 1(b)) was used as a benchmark, thereby eliminating morphological variation caused by the AM process.

Table 1. Main elements of the chemical composition of the wrought 316L SS and the powder used for AM [26,27].

Element	Cr	Ni	Mo	Si	Mn	Fe
(a) 316L SS Powder (wt.%)	17	11	2	1	1	Bal.
(b) Wrought 316L SS (wt.%)	16–18	10–14	2–3	<1	<2	Bal.

2.2. Tensile and Impact Testing

Tensile and impact tests were carried out to study the bulk tensile strength and toughness of the material. The tensile samples having dimensions shown in Figure 1 were printed according to the standard ASTM E8/E8M-16a: Standard for Tension Testing of Metallic Materials [28]. The sub-sized dimensions were chosen so that the samples could be printed within the printer's limited build volume. The tensile tests were carried out along the x-direction using an Instron Universal Testing System 5892 (Norwood, MA, USA) with the attachment of an Instron extensometer 2530 (USA). A strain rate of 1 mm per minute was used until the elastic region was exceeded and was then changed to 3 mm per minute after removing the extensometer, which strain rate was kept until fracture occurred. Six repeated tests were carried out.

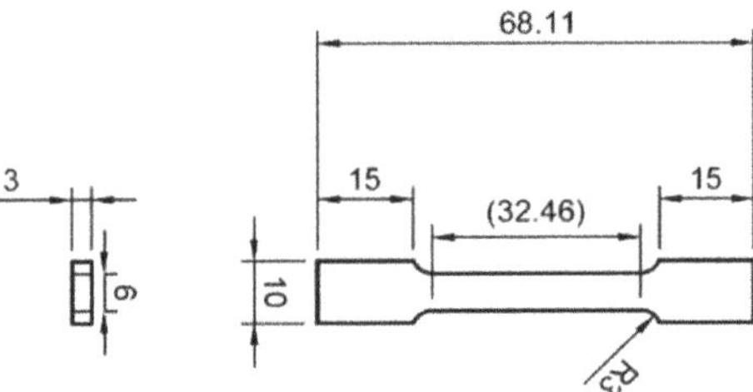

Figure 1. Schematic diagram for the tensile testing specimen. (Units are in mm).

The Charpy impact testing samples having dimensions of 55 × 10 × 10 mm, were manufactured according to the standard ASTM E23-18: Standard Test Methods for Notched Bar Impact Testing of Metallic Materials [29], as shown in Figure 2. The impact testing was performed using an Instron 450MPX-J2 (USA) motorised pendulum impact testing system and a velocity of 5.32 m/s. The reported results are the average of six measurements.

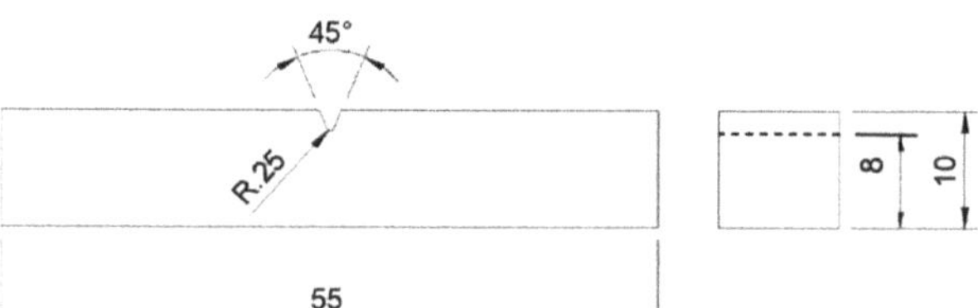

Figure 2. Schematic diagram for the impact testing specimen. (Units are in mm).

2.3. Shot Peening and Coating Deposition

Shot peening was carried out using a modified set up in an Industrial Surface Treatments Ltd. AB850 air blasting machine and S230 shots. A nozzle of 80 mm length and 7 mm diameter, nozzle to specimen distance of 100 mm, pressure of 7 bar and an Almen intensity of 0.21 mmA were used.

The coating deposition was carried out using a Teer UDP800 (Beijing, China) closed field unbalanced magnetron sputtering ion plating machine. The system was composed of 2 titanium targets, 1 aluminium and 1 copper target, with an argon and nitrogen gas inlet. Table 2 shows a summary of the coating deposition parameters.

Table 2. Coating deposition parameters.

Layer Number	Total Time (min)	Time Distribution (min)	Bias Voltage (V)	Nitrogen Flow (%)	Target Current (A) 1: Ti	2: Ti	3: Al	4: Cu
Cleaning	10	0 Ramp up to 2 8	200 400 400	0		0.5	/	/
1: Ti	15	0 Ramp up to 10 5	90	0		0.5 8	/	/
2: TiN	60	0 Ramp up to 30 30	90	100 35		8	/	/
3: TiAlN	60	0 Ramp up to 5 55	90	35		8	2 8	/
4: TiAlCuN	60	60	90	35		8	8	8

2.4. Material Characterisation

Scanning electron microscopy (SEM) was performed by a Carl Zeiss Merlin field emission scanning electron microscope (Oberkochen, Germany) having a Gemini II column. Electron dispersive spectroscopy (EDS) was performed by means of an Ametek EDAX Apollo X 2189 (Mahwah, NJ, USA) attachment located within the SEM. This was used to obtain the chemical composition and elemental maps of the studied surfaces. X-ray diffraction (XRD) phase analysis of all the specimens was carried out using a Rigaku Ultima IV X-ray diffractometer (Tokyo, Japan), equipped with crossbeam optics (CBO) set in Glancing Angle Incidence Asymmetric Bragg (GIAB) configuration, with a 3° angle of incidence and a scan range between 20° and 120°.

Residual stress measurement at the surface and near the surface up to a depth of 200 μm was carried out using the XRD, according to the standard BS EN 15305 (2008)—Non-destructive testing—Test method for residual stress analysis by X-ray diffraction [30], using the $\sin^2\psi$ method. The measurement for peak shifting for the calculation of the residual stresses was performed on the (311) austenite peak at a 2θ value of 90.68°. Each surface was tilted at seven different ψ angles between 0 and 60°. The θ-2θ scans were performed between 2θ values of 87° and 93°. An electrolyte consisting of 5.4% perchloric acid, 94% ethanol and 0.6% de-ionised water was used to carry out electropolishing on a Struers LectroPol-5 electrolytic polishing machine (Copenhagen, Denmark) to progressively remove layers of the material.

Surface micro-hardness tests were executed on a Mitutoyo MVK-H2 micro-hardness testing machine (Kawasaki, Japan), equipped with a Vickers pyramidal indenter and loaded with a 200 gf indentation load, having a dwell time of 10 s. For coating evaluation, nano-indentation tests were performed on a mirror-finished coated wrought 316L SS sample. A Nanomaterial NanoTest 600 machine (Wrexham, UK) was equipped with a 18580-a Berkovich 120° diamond tip indenter. The maximum indentation load was set to 50 mN. Thirty indentations were made on the specimen, spaced at 30 μm from each other.

The surface roughness was measured using an AEP Technology NanoMap 500 LS 3D contact profilometer (California, USA), equipped with a 1 μm stylus tip. The results presented are an average of five measurements.

2.5. Scratch Testing

A UMT Bruker TribolabTM tribometer (San Jose, CA, USA) having a 60° Rockwell type C indenter was used to execute micro-scratch testing. A ramped load starting from 0.5 N up to 40 N was used, with a force of 1.33 N/s and a scanning velocity of 0.339 mm/s. Five scratches of 10 mm each were done on the sample. The scratch morphology was then studied under the optical microscope and SEM to identify the positions and loads at which failure took place.

2.6. Corrosion Studies

Cyclic polarisation tests were performed on the cylindrical samples using a three-electrode setup to study the corrosion response of the material. The three-electrode setup was connected to a Gamry Interface 1000^{TM} potentiostat (Warminster, PA, USA), having the sample as the working electrode, a platinum coated titanium rod as the counter electrode and a saturated calomel electrode (SCE) as the reference electrode. Then, 300 mL of testing solution was prepared according to ASTM D 1141–98: Standard Practice for the Preparation of Substitute Ocean Water [31], where 1 cm^2 of surface area was exposed to the electrolyte. An initial OCP test of 2 h was performed, followed by cyclic polarisation sweeps at a rate of 0.167 mV/s, which was reversed at an apex current density of 0.5 mA/cm^2. Three repeats were performed.

2.7. Designation of Samples

Table 3 shows the samples which were tested in this study, together with respective abbreviations used throughout this article.

Table 3. Abbreviations for each sample variable.

Sample	Abbreviation
Wrought 316 LVM SS	W
As-printed 316L SS	AP
Printed and shot peened 316L SS	PSP
Wrought and Coated 316L SS	WC
Printed and coated 316L SS	PC
Printed, shot peened and coated 316L SS	PSPC

2.8. Error Calculation

The data presented in this work are the sample mean ($\bar{x}$) value obtained from the measured values, with a sample size (n), specified in each section. The quantitative data presented in graphical formats have been included with error bars, while that presented in numerical formats has been included with a ±value. This was done to ensure the correct statistical interpretation of the data. Since most of the sample sizes were smaller than 30, the error ranges were calculated using the t-distribution [32].

3. Results and Discussion

3.1. Mechanical Performance of Bulk AM 316L SS

3.1.1. Tensile Tests

Tensile properties including the Young's Modulus, yield strength, ultimate tensile strength (UTS) and elongation are recorded in Table 4.

Table 4. Tensile properties.

	Young's Modulus (GPa)	Yield Strength (MPa)	Ultimate Tensile Strength (MPa)	Elongation (%)
AP (Measured)	155 ± 7	493 ± 5	644 ± 5	40 ± 2
AP (Literature)	141–183 [33,34]	424–561 [33,35–38]	528–834 [35–38]	17–51 [33,35–38]
Wrought (Literature) [39,40]	193	205–310	515–620	30

The measured Young's Modulus, the yield strength, UTS and the elongation all fall within the ranges found in the available literature for additive manufactured 316L SS. While the UTS of the AM samples is similar to that found in literature for wrought 316L SS, some discrepancies are evident for Young's Modulus, yield strength and elongation values. A 19% decrease in Young's Modulus is evident between AM SLM and conventionally manufactured 316L SS. Similar values were obtained by Merkt [41] with a Young's Modulus of 140 GPa on AM 316L SS. This could be attributed to the parameters of the 3D printing process, specifically the build-up direction and laser power used, which provide different crystallographic orientation of the grains [33,42]. Niendorf et al. [42] report that the grains have a preferential crystallographic orientation according to the laser power utilised such that the grains were oriented in the (011) direction when a laser power of 400 W was used, while the grains were oriented in the (001) direction with a laser power of 1000 W [43]. In austenitic SS, the preferential orientation is that of (001), giving a decreased Young's Modulus [42]. In addition, the porosity present in additive manufactured materials also results in a decrease in the Young's Modulus. An increase of 60% in the yield strength, 4% in UTS and 10% in elongation can be noted for AM SLM over wrought. The significant improvement in YS can be attributed to refined microstructure obtained during the high cooling rates of the SLM process. Additionally, the high yield strength and elongation are attributed to high dislocation densities and twinning formations during the SLM process, respectively. Tensile stresses in SLM materials also lead to a high yield strength, with the high dislocation densities formed during deformation [36,44]. The small increase of UTS indicates that during testing, SLM materials do not exhibit the same amount of work hardening as the wrought.

3.1.2. Impact Tests

A total impact energy, resulting in impact toughness, of 75 ± 2 J was obtained (Table 5) after fracturing the sample (Figure 2). This value falls within the range found in literature for AM SLM 316L SS. This value is slightly lower than that of wrought 316L stainless steel.

Table 5. Impact properties.

	Maximum Load (kN)	Total Energy (J)
AM SLM (Measured)	15 ± 0.10	75 ± 2
AM SLM (Literature) [45–48]	/	60–100
Wrought (Literature) [45–48]		120–180

A micrographic analysis of the fractured surface was carried out, as shown in Figure 3a–c. The material which has experienced compressive loading during impact testing shows a distinctively flattened morphology, whereby the material texture both that formed by plastic deformation and features characteristic of AM were compressed against each other, as shown in Figure 3a. In fact, cross-sectional evaluation has revealed some residual porosity of less than 0.2% in volume. Ductile deformation is evidenced by the cup-and-cone structure shown in Figure 3c and pores (Figure 3b), cracks and dimples. The pores, a characteristic of ductile fracture, are formed due to insufficient bonding of melt pools which are next to each other, during the solidification processing [49,50].

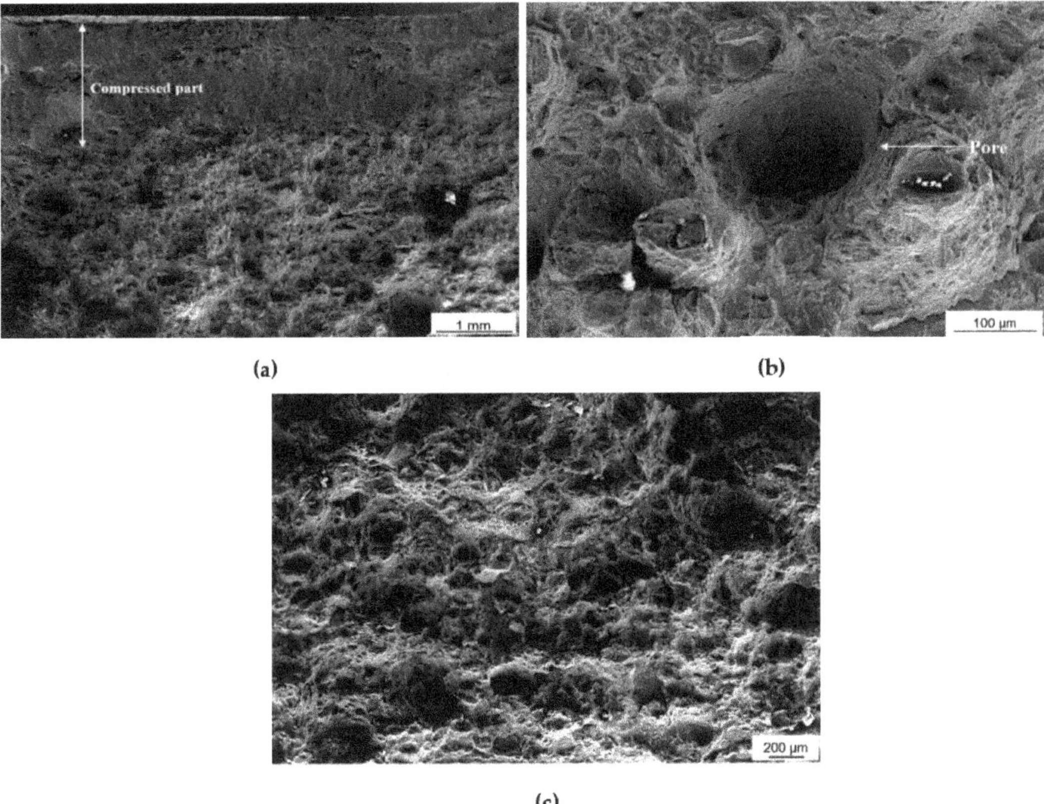

Figure 3. Images portraying the fractured surface of the notched sample (**a**) Compressed part after fracture, showing brittle behaviour. (**b**,**c**) Characteristics of ductile behaviour.

3.2. Surface and Microstructure Analysis

Figure 4 shows the microstructure of the AP 316L SS composed of an austenitic matrix. At higher magnifications, shown in Figure 5, columnar and cellular dendritic structures were observed. Such structures are formed following molten metal solidification. Additionally, Figure 6a–d respectively show optical micrographs of the surface of AP, PSP, PC and PSPC specimens. These micrographs show the 100% coverage produced by the shot peening. Figure 6b,d show the individual SP dimple characteristics. SP generated a less rough surface than the as-printed, as will be observed later in Section 3.3.

Figure 4. Austenitic microstructure of as-printed 316L SS.

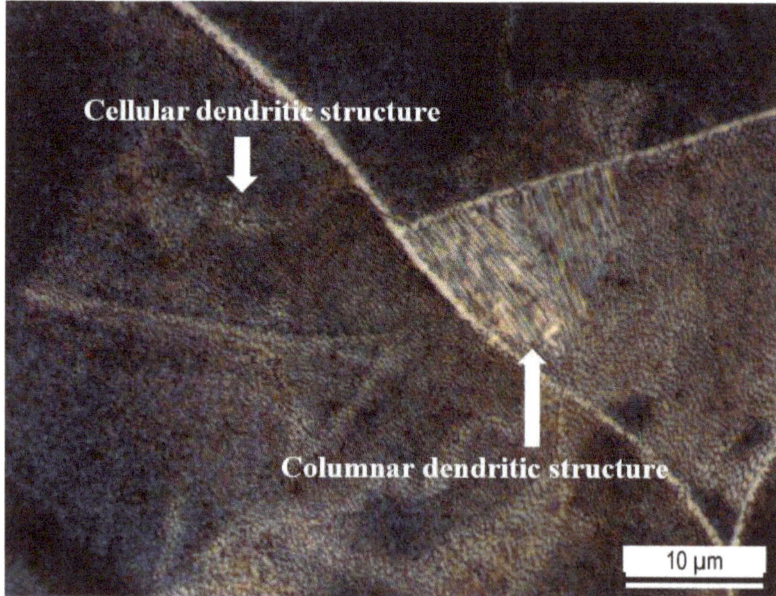

Figure 5. Columnar and cellular dendritic structures in the AP microstructure.

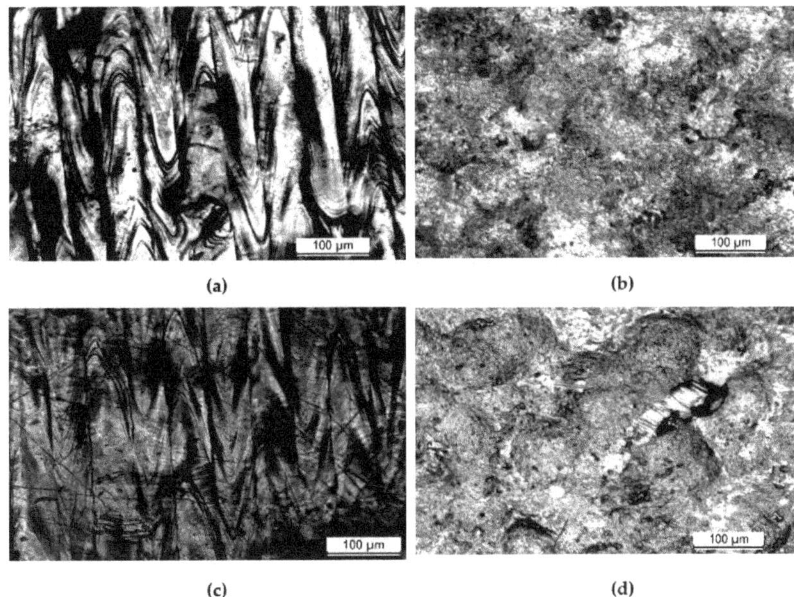

(a) **(b)**

(c) **(d)**

Figure 6. Micrographs for the surface: (**a**) AP (**b**) PSP (**c**) PC and (**d**) PSPC.

3.3. Roughness Analysis

Figure 7 shows a comparison between the AP, shot peened, coated and hybrid treated coupons. The peening treatment resulted in a 50% reduction for Ra and 80% for Rz and can be attributed to the fact that as the shots impinge the surface, the rough crests resulting from the printing process are compressed, the protrusions on the surface are deformed radially, forming individual dimples which flatten the surface and thus, reduces the roughness. The surface roughness reduction by SP was also reported by Sugavaneswaran et al. [15] where a 50% deduction in the average surface roughness was discovered after SP AM 316L SS, using S390 shots with a 1 mm diameter, for 15 min, and with a 200% coverage.

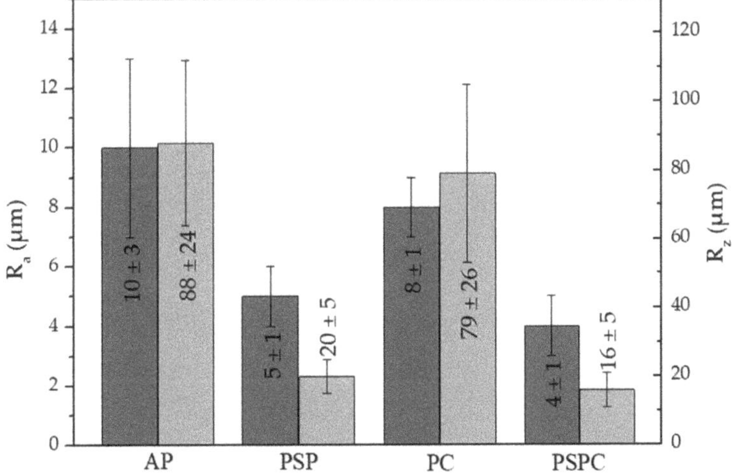

Figure 7. R_a and R_z roughness values for as-printed, printed and shot peened, printed and coated and printed, shot peened and coated 316L SS.

3.4. XRD Phase Analysis

Figure 8 portrays the XRD diffractographs for the AP, shot peened, coated and hybrid treated coupons. The main differences identified between the AP and shot peened diffractographs were: (i) change in relative intensity at the (111) and (200) peaks, (ii) a poorer definition of the (110) ferrite peak, (iii) broadening of XRD peaks in the shot peened sample and (iv) a slight peak shift for the (200) peak. The broadening and shifting of the XRD peaks are attributed to the macro and micro residual stresses induced by SP [51,52]. Similar differences between the printed and coated, and the hybrid samples were obtained, including: (i) change in relative intensity in the (111) and (200) peaks and (ii) XRD peaks broadening due to the induction of macro and micro-residual stresses.

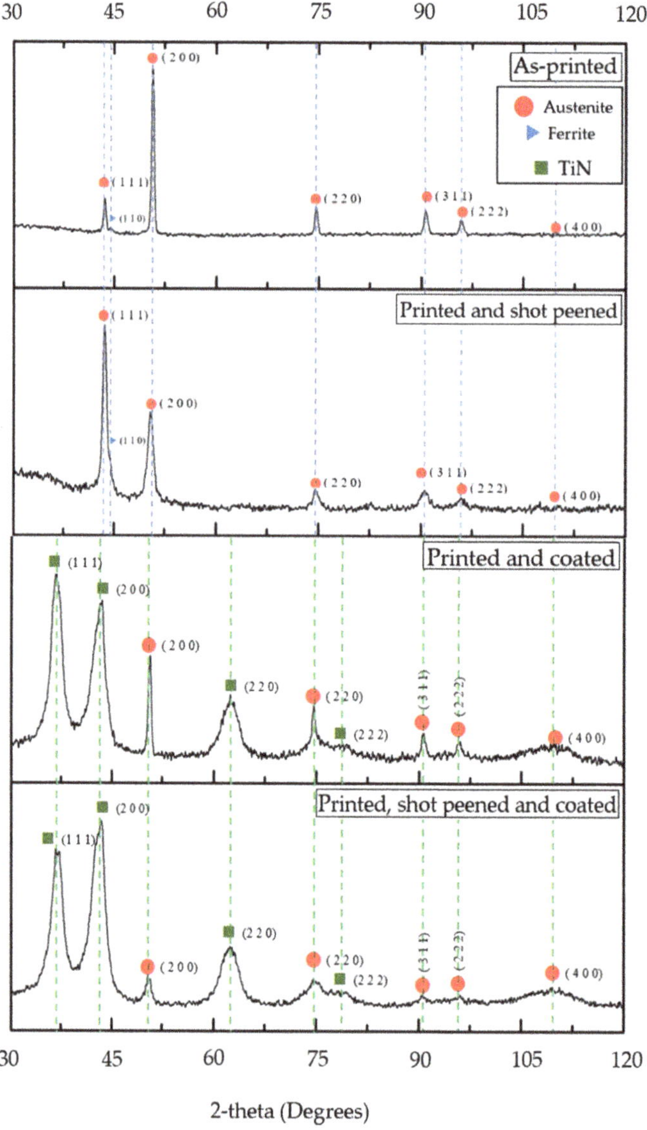

Figure 8. XRD for the as-printed (AP), printed and shot peened (PSP), printed and coated (PC) and printed, shot peened and coated (PSPC).

When analysing the diffractographs of the coated and hybrid treated, the peaks obtained agree with those obtained by Man et al. [53] when studying TiAlN thin films. The two peaks of (111) and (200) for TiN are in the same position as austenite. The XRD technique did not detect any Al and Cu crystalline compounds with the two elements having diffused to form a solid solution. The XRD patterns of the coated surfaces are different from those of the substrate, confirming that there are distinct phases of the coating, even though some of the austenite peaks were still present.

3.5. XRD Stress Evaluation

A surface residual stress of 61 ± 4 MPa, -589 ± 6 MPa and -693 ± 8 MPa was obtained for the as-printed, printed and shot peened and printed, shot peened and coated respectively. Figure 9 portrays the residual stresses developed along the depth for the as-printed (AP) and shot peened (PSP) samples.

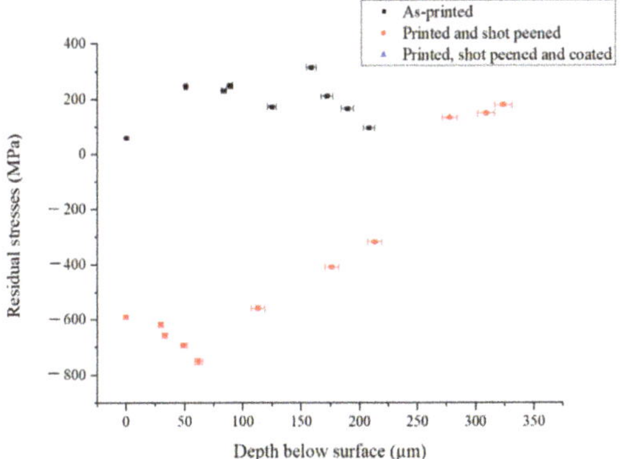

Figure 9. Residual stresses for the AP and PSP against the material depth removed.

The AP samples exhibited tensile stresses of around 61 MPa, both at the surface and the sub-surface. Tensile stress values have been associated with several mechanisms including: (i) temperature gradient, (ii) re-melting and solidification of layers and (iii) inhomogeneous lattice spacing [54]. On the other hand, the SP treatment induced a maximum compressive residual stress of around 589 MPa. Figure 10 shows that the depth of the shot peened layer is around 250 μm. This is in line with other work on AM 316L SS carried out by Gundgire et al. [16], where the affected depth was in the range of 225 to 275 μm.

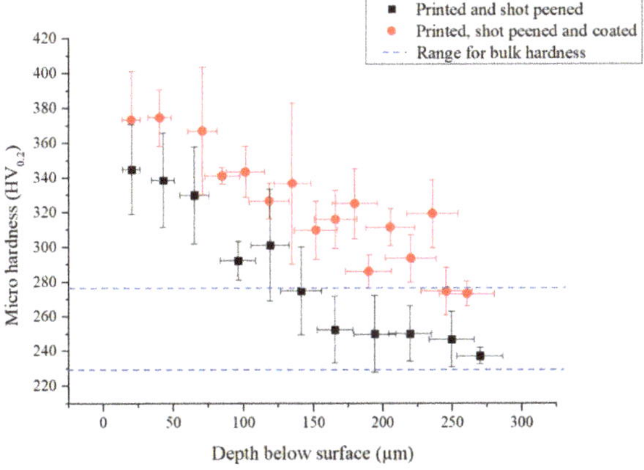

Figure 10. Microhardness-depth profile for printed and shot peened and printed, shot peened and coated 316L SS.

Compressive residual stresses for the hybrid treated specimens were similar to those obtained on the shot peened coupon. A surface stress of -693 ± 7 MPa was achieved which reached -561 ± 5 MPa at an affected depth of 54 µm. This outcome shows that the coating deposition on the peened AM specimen did not remove the beneficial compressive stresses induced by the peening process.

3.6. Hardness Studies

Table 6 shows that shot peening treatment improves the hardness of the as-printed material by 40%, from 238 HV to 334 HV. This is in line with the study by Gundgire et al. [16], where a hardness of 340 to 360 HV was achieved after SP AM 316L SS. This increase is attributed to plastic deformation taking place during SP. The intrinsic hardness of the coating, which was measured by nanohardness, shows that it further improves the characteristics of the surface of the material. Then, the combination of the shot peening and the coating treatment provides an even superior compound value of hardness, giving 2.9 times increase in hardness over the as-printed samples.

Table 6. Results obtained from impact testing.

Material	Surface Hardness
AP	$238 \pm 4\ HV_{0.2}$
PSP	$334 \pm 16\ HV_{0.2}$
PSPC	$691 \pm 23\ HV_{0.2}$
TiAlCuN coating *	$3022 \pm 54\ HV$

* This measurement was obtained after carrying out a nanohardness test. The other three measurments were obtained via microhardness.

Figure 10 showcases the microhardness depth profile of the shot peened and hybrid sample. It can be noted that the affected depth is also around 250 µm, which is in line with the affected depth obtained in the residual stress measurement (Section 3.5). The affected depth is comparable with that of 189 µm and 225–275 µm obtained by Maamoun et al. [55] and Gundgire et al. [16], respectively.

3.7. Material Testing

3.7.1. Adhesion Tests

The coating and the hybrid treatment showed a similar behaviour of coating characteristics following scratch testing. Figure 11 shows that L_{C2} and L_{C3} were detected along the wear track of coated, while for the hybrid treated L_{C3} only was identified. L_{C1} could not be identified for both variables. The L_{C2} characteristic was made from initial delamination of the coating, while the L_{C3} characteristic was made from interfacial shell-shaped spallation. The earliest sign of adhesive failure and substrate exposure was noted on the hybrid at a distance of 2.6 ± 0.34 mm, and as shown in Figure 11, this corresponds to a scratching load of 10 ± 1 N. This is in contrast with the results obtained in a study by Tillmann et al. [54], in which a CrAlN coating was deposited on AM 316L SS. Both L_{C2} and L_{C3} were obtained at a force of 7 ± 1.7 N and 38.4 ± 3.5 N, respectively. This shows that the CrAlN coating provided better adhesion to the substrate as it failed at higher loads.

Further mixture of cohesive and adhesive failure was identified throughout the wear track of coated and the hybrid treated, showing more delamination and interfacial shell spallation, as shown in Figures 12c and 13c. These were formed as the scratch load increased. Such characteristics were elevated since the soft material was not able to properly support the coating from cracking and forming such defects. As observed in Figure 13c, the perforation of the scar did not result in total coating delamination, even as the load was increased. These were replaced by interval delamination because of residual stress relaxation during coating spallation.

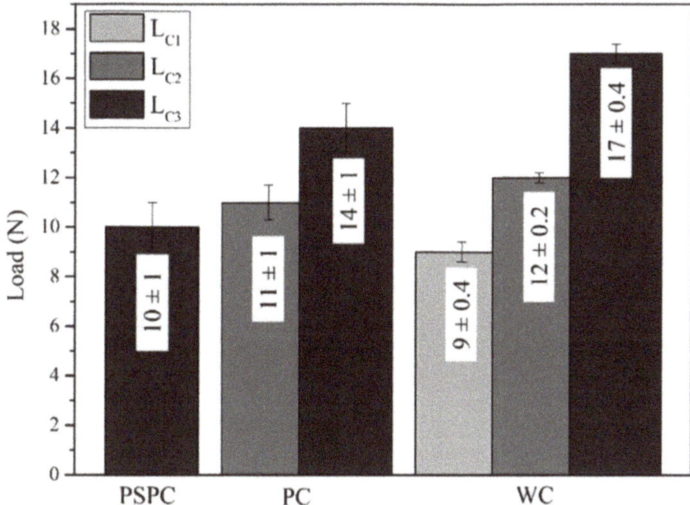

Figure 11. Critical loads achieved at ramped loads for the hybrid, printed and coated and wrought and coated.

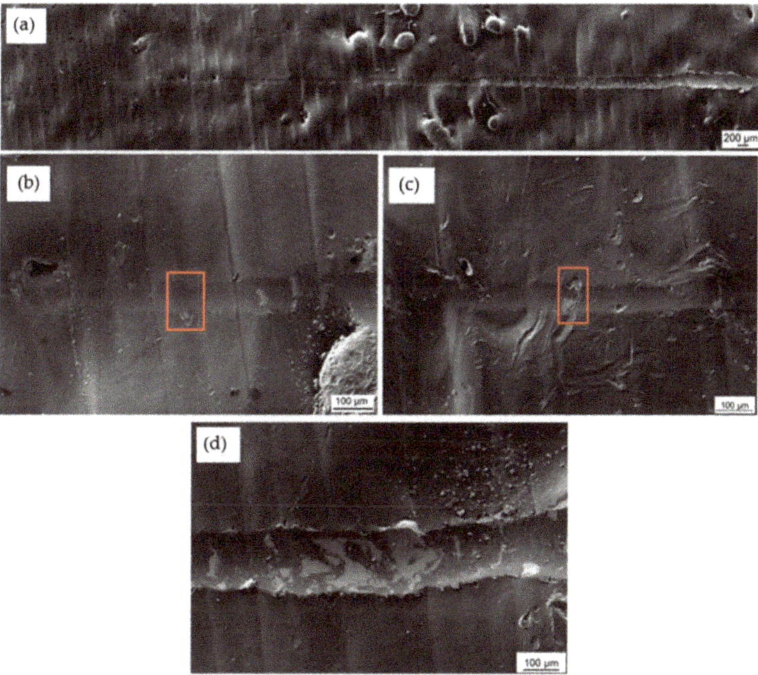

Figure 12. SEM topographical images of the wear scar morphology on the coated specimen: (a) Whole length of the wear scar (b) Red box showing location of L_{C2} (c) Red box showing location of L_{C3} and (d) Further delamination along the wear track.

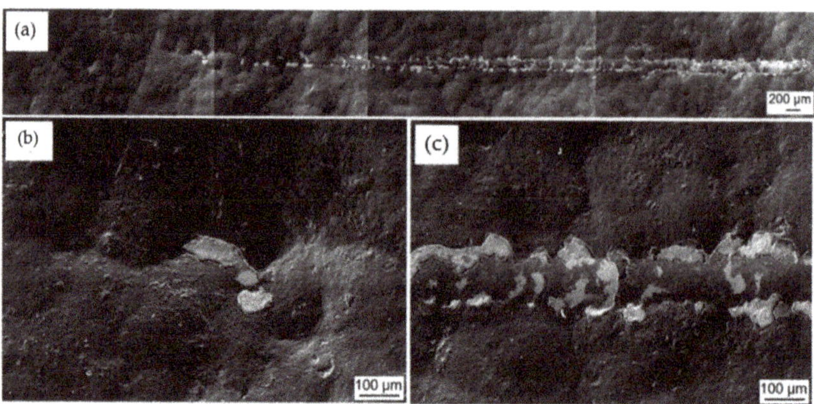

Figure 13. SEM topographical images of the wear scar morphology on the printed and hybrid treated 316L SS: (**a**) Whole length of the wear scar (**b**) Location of L_{C3} and (**c**) Further delamination along the wear track.

Additionally, the scratch testing on the wrought and coated was performed to serve as a control to the coated and the hybrid. During this testing, all the three adhesion characteristics were identified. At low loads, L_{C1} was shown, which was characterised by forward chevron cracks, longitudinal to the scratch track, showing cohesive failure (Figure 14b). Adhesive failure was then identified at a load of 12 ± 0.2 N at which load the coating delaminated along the scratch track (Figure 14c). At further higher loads, L_{C3} (Figure 14d) took place at a load of 17 ± 0.4 N. This characteristic consisted of full interfacial shell spallation, with full delamination of the coating taking place longitudinally along the scratch track. Chevron cracks and localised chipping were found along the track.

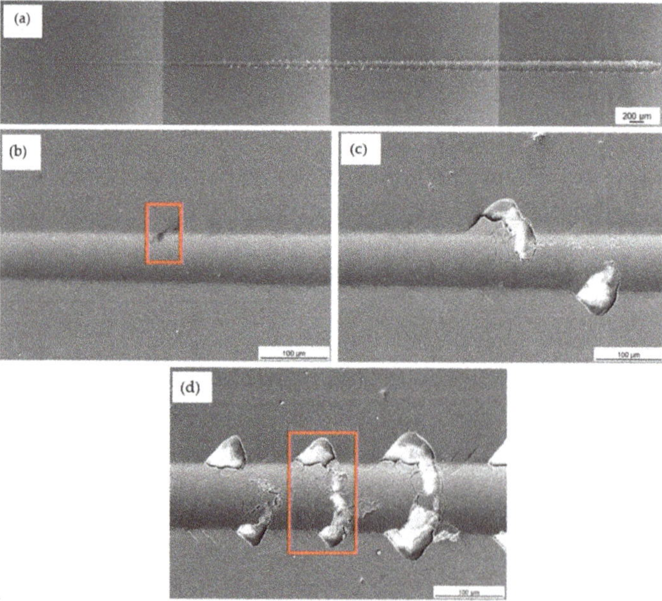

Figure 14. SEM topographical images of the wear scar morphology on the printed coated 316L SS: (**a**) Whole length of the wear scar (**b**) Red box showing location of L_{C1} (**c**) Location of L_{C2} and (**d**) Red box showing location of L_{C3}.

When comparing the values in Figure 11, the PC performed better than the PSPC evidenced by the first failure mode detected at a higher load than that measured on the hybrid equivalent. The similar nature of the coated chemical makeup at the surface suggests that the difference in performance can be attributed both to the test mechanics, where the tip interaction changes with the roughness of the sample being measured and the improved load support provided by the harder and stiffer coating. The results of the PC are also superior to the wrought and coated, since the first failure on the wrought and coated was seen at an earlier load. The wrought and coated tests were performed to analyse all the three characteristics synonymous with scratch testing.

3.7.2. Corrosion Tests
OCP Curves

The OCP curves for the wrought, as-printed, polished, shot peened and hybrid treated followed a similar behaviour, until they stabilised for the rest of the curve's duration. This shows that the setup had stabilised and was ready to carry out the cyclic polarisation test. However, the same cannot be said for the coated curve. This is due to the fluctuations in the curve taking place. The reason for these fluctuations could be due to metastable pits growing but their growth is stopped abruptly and repassivated. This repeatable behaviour obtained from a set of repeats for each set, provided an early indication of how the sample was going to perform in the cyclic polarisation test, providing poor corrosion results, performing the worst with respect to the six samples tested.

Cyclic Polarisation Curves

Figure 15 shows the representative curves for the cyclic polarisation tests, while Table 7 provides a summary of important numerical values extracted from the plot and repeats. The most noble E_{corr} was obtained by the AP, whilst the most negative was achieved by the PC sample. The rest of the samples had very similar E_{corr} ranging from -210 to -180 mV. The more noble an E_{corr} is, the lower the corrosion susceptibility [56] and the higher the stability of the passive film [7].

The highest E_{break} was obtained for PP at 776 ± 127 mV, whilst the least was achieved by the PC sample at 241 ± 80 mV. The range for the other samples is between 330 and 690 mV. The E_{break}, also known as the pitting potential, is the lowest potential at which the material will succumb to pitting corrosion. Above the E_{break}, new pits will form [57]. Therefore, the higher the E_{break}, the more resistance to pitting and the further improved stability of the oxide film [56,58].

A fluctuating current density in the passive region of the anodic scans shows the formation of metastable pits [7]. This was showed by the wrought, as-printed, shot peened and coated samples. Their growth is stopped rapidly and repassivated [59]. This repassivation takes place with the aid of the salt films formed by the electrolyte. They suppress the transfer of cations and more growth of the pits. From the curves in Figure 15b, it can be noted that the polished sample showed the lowest metastable pits formation, while the as-printed and the coated showed the most metastable pits formation, with the most fluctuations below E_{break}. From this analysis, it can be concluded that the surface roughness impacts the metastable pit formation, the smaller the surface roughness, the less formation of metastable pits. The values of the passive current density go hand in hand with those of the E_{break}. The smaller the anodic current, (at E_{break}), a denser passive oxide film is formed. Therefore, a smaller current density is preferred. All of the samples had a current density of around 0.4–3 $\mu A/cm^2$, except for the PC, which had a larger current density of around 12 $\mu A/cm^2$, at E_{break}.

The E_{prot} is the point of intersection of the forward and reverse scans. The higher the E_{prot}, the least prone to corrosion the material is [56]. Therefore, from Table 7, the smallest E_{prot} was identified for the PSPC sample, while the largest was that of the wrought. This shows that the wrought has a stable passive film and pit growth is restricted at an earlier potential value.

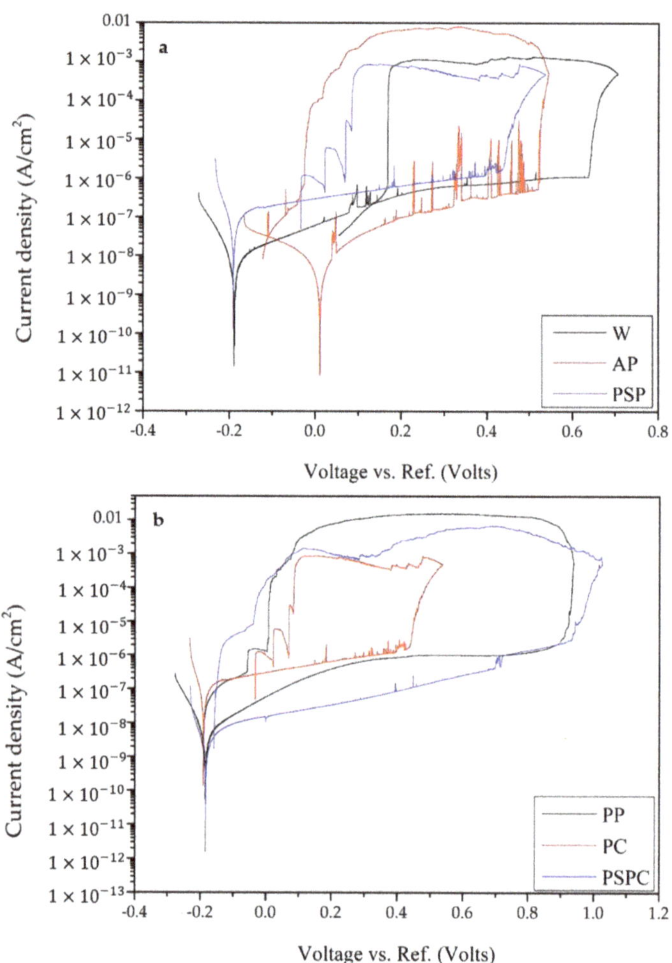

Figure 15. Representative cyclic curves for the six studied surfaces for: (**a**) Wrought, as-printed and shot peened and (**b**) polished, coated and hybrid.

Table 7. Average corrosion testing results for each variable.

	Passive Current Density ($\mu A/cm^2$)	E_{corr} (mV)	E_{break} (mV)	E_{prot} (mV)	$E_{break-Eprot}$ (mV)	Presence of Pits and Delamination
W	1 ± 0.1	-197 ± 10	686 ± 24	60 ± 36	626 ± 60	No
AP	0.4 ± 0.2	-2 ± 14	500 ± 36	-133 ± 17	633 ± 53	No
PP	3 ± 0.6	-181 ± 14	776 ± 127	-199 ± 37	975 ± 164	No
PSP	1 ± 0.6	-198 ± 15	334 ± 92	-53 ± 17	387 ± 109	No
PC	12 ± 1	-500 ± 20	241 ± 80	-138 ± 25	379 ± 105	Yes
PSPC	2 ± 0.8	-210 ± 20	595 ± 168	-208 ± 25	803 ± 193	Yes

All of the six curves show a positive hysteresis loop, which is linked to pitting. The larger the hysteresis loop, the more location for pitting to occur and the less pitting corrosion resistant the material is. Therefore, the bigger the hysteresis loop, the more damage that is occurring on the passive oxide film and the more difficulty to restore it. All the curves show a positive hysteresis loop, with the biggest hysteresis loop identified for the PP. This was determined by finding the difference between E_{break} and E_{prot}, as shown in Table 7.

Figure 15a shows the three curves with similar behaviour, that of the W, AP and SP. The results in Figure 15b show that the PC had a poor corrosion resistance when comparing it to the PP and the PSPC, which performed the best. The testing solution damaged the coating and formulated pits which reach the substrate, damaging the sample. As already mentioned, the surface roughness plays an important part in the formation of pits. The PC has a surface roughness of 8 ± 1 μm, which is similar to the 10 ± 3 μm of the AP. This shows that the higher the surface roughness the more surface area for pits to form.

Surface Analysis after Testing

Figure 16 shows SEM micrographs of each sample after corrosion testing. Severe damage to the coupons was not evident. However, pits and pores of different sizes were visible on all the samples. In the PC micrographs, the printing striations and directions were revealed, while the PSP and the PSPC samples both show the dimples on the surface which are a characteristic of SP. Additionally, in the PSPC sample, a part of the TiAlCuN coating was delaminated following corrosion testing. Figure 17a,b shows that, upon inspecting the PC sample at a higher magnification, multiple pits were found, as opposed to those found on the shot peened (Figure 17c) and hybrid treated (Figure 17d). On the PSPC less pits were identified, whose benefit will be explained later.

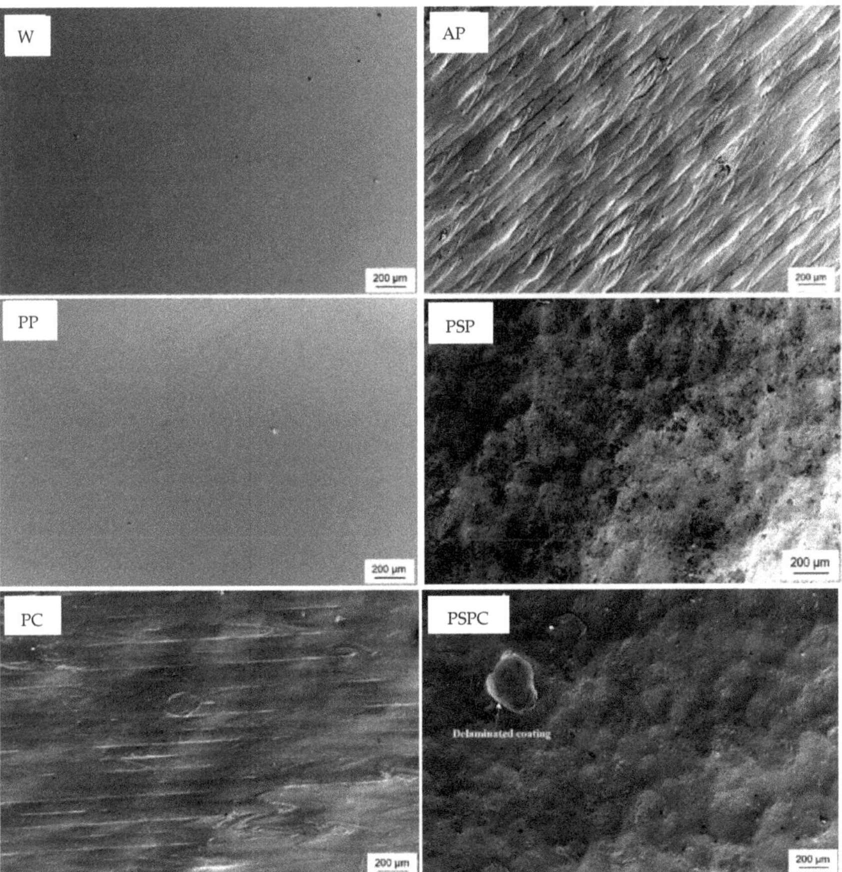

Figure 16. Micrographs of surfaces following corrosion testing.

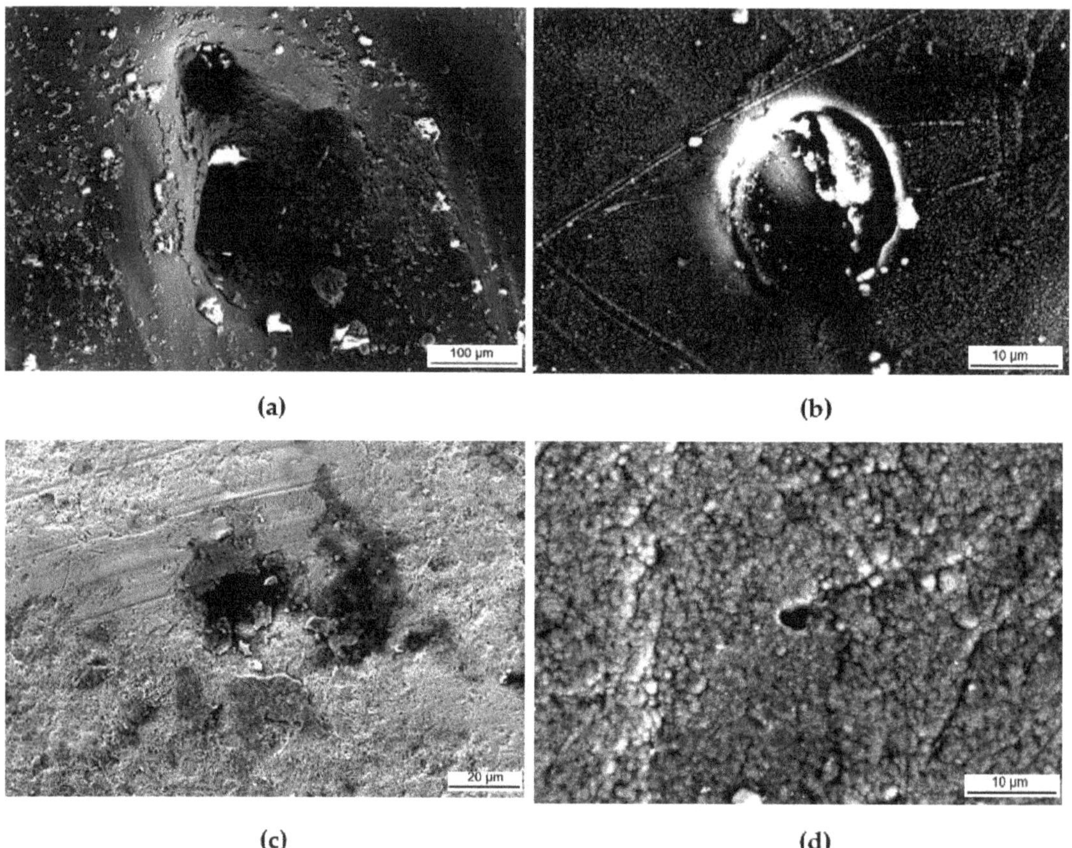

Figure 17. Pits identified on (**a**,**b**) PC, (**c**) PSP and (**d**) PSPC.

The above demonstrates that the PC has the most corrosion susceptibility, while the PSPC has the least corrosion susceptibility, indicating that the combined effect of the surface treatments of shot peening and PVD provided superior corrosion qualities.

4. Conclusions

This study was carried out to analyse the effect of the surface treatments of shot peening and TiAlCuN coating on AP additive manufactured 316L SS, on the surface and sub-surface of the material. The main conclusions from this study include:

- Microscopy and XRD phase analysis showed that the as-printed 316L SS was composed of an austenitic matrix, characterised with columnar and cellular dendritic together, together with the presence of some ferrite.
- XRD stress measurement highlighted tensile residual stresses in the as-printed samples and compressive residual stresses in the shot peened and hybrid treated samples. Compressive residual stresses of 589 MPa for an approximate depth of 250 μm were generated by the cold working achieved by shot peening.
- A 40% increase in surface hardness was obtained on the printed and shot peened specimens, while a 2.9 times increase was achieved following the application of the combined surface treatments. This shows that the coating possesses a high hardness, which when combined with shot peening, improves the material characteristics.

- A 50% decrease for R_a and an 80% decrease for R_z were found following the application of shot peening on the as-printed specimens. This shows that shot peening has the added advantage of improving the surface finish of additive manufactured components.
- The application of the TiAlCuN coating on the as-printed provided better adhesion characteristics of the additive manufactured 316L stainless steel, than on the hybrid counterpart. This could be attributed to the test mechanics, where the tip interaction is changing with the roughness and the improved load support provided by the harder and stiffer coating.
- The printed and coated combination had the worst corrosion behaviour, while the printed and hybrid treated specimens exhibited the best corrosion behaviour showing that the combined effect of the surface treatments of shot peening and PVD provided optimal corrosion qualities.

The results achieved in this study show optimal qualities for applying a shot peening treatment combined with the deposition of a coating on additive manufactured 316L stainless steel, making this combination of material processing ideal for a range of demanding applications involving bulk mechanical loading, susceptibility to wear under contact loads and corrosion damage, including many such instances found in the maritime transportation industry.

Author Contributions: Conceptualization, A.Z., J.C. and J.B.; methodology, L.B., X.Z. and Z.H.; formal analysis, L.B.; investigation, L.B., A.Z., J.B. and G.C.; writing—original draft preparation, L.B.; writing—review and editing, L.B., K.A.V., A.Z., J.B., G.C. and J.C.; supervision, A.Z., J.B., G.C. and J.C.; project administration, A.Z. and J.C.; funding acquisition, A.Z. and J.C. All authors have read and agreed to the published version of the manuscript.

Funding: This study was part of the *SEAM—Surface Engineering for Additive Manufactured parts used in marine transportation* project, funded by the Malta Council for Science and Technology (MCST) as part of the Sino-Malta fund 2019-02 and the Ministry of Science and Technology (MOST) of the People's Republic of China (2019YFE0191500).

Institutional Review Board Statement: Not applicable.

Informed Consent Statement: Not applicable.

Data Availability Statement: Not applicable.

Conflicts of Interest: The authors declare no conflict of interest.

References

1. Yusuf, S.M.; Chen, Y.; Boardman, R.; Yang, S.; Gao, N. Investigation on porosity and microhardness of 316L stainless steel fabricated by selective laser melting. *Metals* **2017**, *7*, 64. [CrossRef]
2. Saeidi, K.; Gao, X.; Zhong, Y.; Shen, Z.J. Hardened austenite steel with columnar sub-grain structure formed by laser melting. *Mater. Sci. Eng. A* **2015**, *625*, 221–229. [CrossRef]
3. DebRoy, T.; Wei, H.L.; Zuback, J.S.; Mukherjee, T.; Elmer, J.W.; Milewski, J.O.; Beese, A.M.; Wilson-Heid, A.D.; De, A.; Zhang, W. Additive manufacturing of metallic components—Process, structure and properties. *Prog. Mater. Sci.* **2018**, *92*, 112–224. [CrossRef]
4. Zhang, D.; Sun, S.; Qiu, D.; Gibson, M.A.; Dargusch, M.S.; Brandt, M.; Qian, M.; Easton, M. Metal Alloys for Fusion-Based Additive Manufacturing. *Adv. Eng. Mater.* **2018**, *20*, 1700952. [CrossRef]
5. Wang, D.; Song, C.; Yang, Y.; Bai, Y. Investigation of crystal growth mechanism during selective laser melting and mechanical property characterization of 316L stainless steel parts. *Mater. Des.* **2016**, *100*, 291–299. [CrossRef]
6. Sun, S.; Brandt, M.; Easton, M. *Powder Bed Fusion Processes: An Overview*; Elsevier Ltd.: Amsterdam, The Netherlands, 2017. [CrossRef]
7. Lodhi, M.J.K.; Deen, K.M.; Haider, W. Corrosion behavior of additively manufactured 316L stainless steel in acidic media. *Materialia* **2018**, *2*, 111–121. [CrossRef]
8. Jung, G.S.; Park, Y.H.; Kim, D.J.; Lim, C.S. Study on corrosion properties of additive manufactured 316L stainless steel and alloy 625 in seawater. *Corros. Sci. Technol.* **2019**, *18*, 258–266. [CrossRef]
9. Trelewicz, J.R.; Halada, G.P.; Donaldson, O.K.; Manogharan, G. Microstructure and Corrosion Resistance of Laser Additively Manufactured 316L Stainless Steel. *J. Mater.* **2016**, *68*, 850–859. [CrossRef]
10. Sun, Y.; Moroz, A.; Alrbaey, K. Sliding wear characteristics and corrosion behaviour of selective laser melted 316L stainless steel. *J. Mater. Eng. Perform.* **2014**, *23*, 518–526. [CrossRef]

11. Wycisk, E.; Solbach, A.; Siddique, S.; Herzog, D.; Walther, F.; Emmelmann, C. Effects of defects in laser additive manufactured Ti-6Al-4V on fatigue properties. *Phys. Procedia* **2014**, *56*, 371–378. [CrossRef]

12. Santa-Aho, S.; Kiviluoma, M.; Jokiaho, T.; Gundgire, T.; Honkanen, M.; Lindgren, M.; Vippola, M. Additive manufactured 316L stainless-steel samples: Microstructure, residual stress and corrosion characteristics after post-processing. *Metals* **2021**, *11*, 182. [CrossRef]

13. Kumar, M.D.B.; Aravindan, K.M.; Jebaraj, A.V.; Kumar, T.S. Effect of Post-Fabrication Treatments on Surface Residual Stresses of Additive Manufactured Stainless Steel 316L. *FME Trans.* **2020**, *49*, 87–94. [CrossRef]

14. Rautio, T.; Jaskari, M.; Gundgire, T.; Iso-Junno, T.; Vippola, M.; Järvenpää, A. The Effect of Severe Shot Peening on Fatigue Life of Laser Powder Bed Fusion Manufactured 316L Stainless Steel. *Materials* **2022**, *15*, 3517. [CrossRef] [PubMed]

15. Sugavaneswaran, M.; Jebaraj, A.V.; Kumar, M.D.B.; Lokesh, K.; Rajan, A.J. Enhancement of surface characteristics of direct metal laser sintered stainless steel 316L by shot peening. *Surf. Interfaces* **2018**, *12*, 31–40. [CrossRef]

16. Gundgire, T.; Jokiaho, T.; Santa-aho, S.; Rautio, T.; Järvenpää, A.; Vippola, M. Comparative study of additively manufactured and reference 316 L stainless steel samples—Effect of severe shot peening on microstructure and residual stresses. *Mater. Charact.* **2022**, *191*, 112162. [CrossRef]

17. Iswanto, P.T.; Akhyar, H.; Faqihudin, A. Effect of shot peening on microstructure, hardness, and corrosion resistance of AISI 316l. *J. Achiev. Mater. Manuf. Eng.* **2018**, *89*, 19–26. [CrossRef]

18. Moradi, A.; Heidari, A.; Amini, K.; Aghadavoudi, F.; Abedinzadeh, R. The effect of shot peening time on mechanical properties and residual stress in Ti-6Al-4V alloy. *Metall. Res. Technol.* **2022**, *119*, 401. [CrossRef]

19. Björk, T.; Westergård, R.; Hogmark, S.; Bergström, J.; Hedenqvist, P. Physical vapour deposition duplex coatings for aluminium extrusion dies. *Wear* **1999**, *225–229*, 1123–1130. [CrossRef]

20. Vogli, E.; Tillmann, W.; Selvadurai-Lassl, U.; Fischer, G.; Herper, J. Influence of Ti/TiAlN-multilayer designs on their residual stresses and mechanical properties. *Appl. Surf. Sci.* **2011**, *257*, 8550–8557. [CrossRef]

21. Mani, S.P.; Kalaiarasan, M.; Ravichandran, K.; Rajendran, N.; Meng, Y. Corrosion resistant and conductive TiN/TiAlN multilayer coating on 316L SS: A promising metallic bipolar plate for proton exchange membrane fuel cell. *J. Mater. Sci.* **2021**, *56*, 10575–10596. [CrossRef]

22. Vladescu, A.; Cotrut, C.M.; Kiss, A.; Balaceanu, M.; Braic, V.; Zamfir, S.; Braic, M. Corrosion resistance of the TiN, TiAlN and TiN/TiAlN nanostructured hard coatings. *UPB Sci. Bull. Ser. B Chem. Mater. Sci.* **2006**, *68*, 57–64.

23. Ananthakumar, R.; Subramanian, B.; Kobayashi, A.; Jayachandran, M. Electrochemical corrosion and materials properties of reactively sputtered TiN/TiAlN multilayer coatings. *Ceram. Int.* **2012**, *38*, 477–485. [CrossRef]

24. Bouzakis, K.D.; Asimakopoulos, A.; Skordaris, G.; Pavlidou, E.; Erkens, G. The inclined impact test: A novel method for the quantification of the adhesion properties of PVD films. *Wear* **2007**, *262*, 1471–1478. [CrossRef]

25. Tillmann, W.; Hagen, L.; Stangier, D.; Dias, N.F.L.; Görtz, J.; Kensy, M.D. Lapping and polishing of additively manufactured 316L substrates and their effects on the microstructural evolution and adhesion of PVD CrAlN coatings. *Surf. Coat. Technol.* **2021**, *428*, 127905. [CrossRef]

26. Jiangsu Vilory Advanced Materials Technology Company Limited. *Certificate of Quality—316LVM Stainless Steel*; Jiangsu Vilory Advanced Materials Technology Company Limited: Xuzhou, China, 2021.

27. L. Klein SA. *Inspection Certificate of 316L SS*; L. Klein SA: Biel, Switzerland, 2018.

28. *ASTM E8*; Standard Test Methods for Tension Testing of Metallic Materials. ASTM-International: West Conshohocken, PA, USA, 2022.

29. *ASTM E23-18a*; Standard Test Methods for Notched Bar Impact Testing of Metallic Materials. ASTM-International: West Conshohocken, PA, USA, 2018.

30. *BS EN 15305:2008*; Non-Destructive Testing—Test Method for Residual Stress Analysis by X-ray Diffraction. B. S. Institution: London, UK, 2008.

31. *ASTM D1141-98*; Standard for the Preparation of Substitute Ocean Water. ASTM-International: West Conshohocken, PA, USA, 1999; pp. 98–100.

32. *BS ISO 2602:1980*; Statistical Interpretation of Test Results—Estimation of the Mean—Confidence Interval. B. S. Institution: London, UK, 1981.

33. Röttger, A.; Boes, J.; Theisen, W.; Thiele, M.; Esen, C.; Edelmann, A.; Hellmann, R. Microstructure and mechanical properties of 316L austenitic stainless steel processed by different SLM devices. *Int. J. Adv. Manuf. Technol.* **2020**, *108*, 769–783. [CrossRef]

34. Rosnitschek, T.; Seefeldt, A.; Alber-Laukant, B.; Neumeyer, T.; Altstädt, V.; Tremmel, S. Correlations of geometry and infill degree of extrusion additively manufactured 316l stainless steel components. *Materials* **2021**, *14*, 5173. [CrossRef]

35. Suryawanshi, J.; Prashanth, K.G.; Ramamurty, U. Mechanical behavior of selective laser melted 316L stainless steel. *Mater. Sci. Eng. A* **2017**, *696*, 113–121. [CrossRef]

36. Qiu, C.; Al Kindi, M.; Aladawi, A.S.; Al Hatmi, I. A comprehensive study on microstructure and tensile behaviour of a selectively laser melted stainless steel. *Sci. Rep.* **2018**, *8*, 7785. [CrossRef]

37. Mertens, A.; Reginster, S.; Contrepois, Q.; Dormal, T.; Lemaire, O.; Lecomte-Beckers, J. Microstructures and mechanical properties of stainless steel AISI 316L processed by selective laser melting. *Mater. Sci. Forum* **2014**, *783–786*, 898–903. [CrossRef]

38. Jaskari, M.; Ghosh, S.; Miettunen, I.; Karjalainen, P.; Järvenpää, A. Tensile Properties and Deformation of AISI 316L Additively Manufactured with Various Energy Densities. *Materials* **2021**, *14*, 5809. [CrossRef]

39. Callister, W.D.; Retwisch, D.G. *Materials Science and Engineering*, 9th ed.; John Wiley & Sons Ltd.: Hoboken, NJ, USA, 2015.

40. A. S. Metals. AISI Type 316L Stainless Steel, Annealed Bar. 2014. Available online: https://asm.matweb.com/search/SpecificMaterial.asp?bassnum=mq316q (accessed on 3 December 2022).

41. S. Merkt. *Qualification of Additively Manufactured Lattice Structures for Customized Component Functions*; RWTH Aachen University: Aachen, Germany, 2015.

42. Niendorf, T.; Leuders, S.; Riemer, A.; Richard, H.A.; Tröster, T.; Schwarze, D. Highly anisotropic steel processed by selective laser melting. *Metall. Mater. Trans. B Process Metall. Mater. Process. Sci.* **2013**, *44*, 794–796. [CrossRef]

43. Sun, Z.; Tan, X.; Tor, S.B.; Chua, C.K. Simultaneously enhanced strength and ductility for 3D-printed stainless steel 316L by selective laser melting. *NPG Asia Mater.* **2018**, *10*, 127–136. [CrossRef]

44. Wang, Y.M.; Voisin, T.; McKeown, J.T.; Ye, J.; Calta, N.P.; Li, Z.; Zeng, Z.; Zhang, Y.; Chen, W.; Roehling, T.T.; et al. Additively manufactured hierarchical stainless steel with high strength and ductility. *Nat. Mater.* **2017**, *17*, 63–70. [CrossRef] [PubMed]

45. Cherry, J.A.; Davies, H.M.; Mehmood, S.; Lavery, N.P.; Brown, S.G.R.; Sienz, J. Investigation into the effect of process parameters on microstructural and physical properties of 316L stainless steel parts by selective laser melting. *Int. J. Adv. Manuf. Technol.* **2015**, *76*, 869–879. [CrossRef]

46. Yasa, E.; Deckers, J.; Kruth, J.P.; Rombouts, M.; Luyten, J. Charpy impact testing of metallic selective laser melting parts. *Virtual Phys. Prototyp.* **2010**, *5*, 89–98. [CrossRef]

47. Wang, C.; Lin, X.; Wang, L.; Zhang, S.; Huang, W. Cryogenic mechanical properties of 316L stainless steel fabricated by selective laser melting. *Mater. Sci. Eng. A* **2021**, *815*, 141317. [CrossRef]

48. Zhong, Y.; Liu, L.; Wikman, S.; Cui, D.; Shen, Z. Intragranular cellular segregation network structure strengthening 316L stainless steel prepared by selective laser melting. *J. Nucl. Mater.* **2016**, *470*, 170–178. [CrossRef]

49. Kurgan, N.; Varol, R. Mechanical properties of P/M 316L stainless steel materials. *Powder Technol.* **2010**, *201*, 242–247. [CrossRef]

50. Zhong, Y.; Rännar, L.E.; Liu, L.; Koptyug, A.; Wikman, S.; Olsen, J.; Cui, D.; Shen, Z. Additive manufacturing of 316L stainless steel by electron beam melting for nuclear fusion applications. *J. Nucl. Mater.* **2017**, *486*, 234–245. [CrossRef]

51. Astaraee, A.H.; Miresmaeili, R.; Bagherifard, S.; Guagliano, M.; Aliofkhazraei, M. Incorporating the principles of shot peening for a better understanding of surface mechanical attrition treatment (SMAT) by simulations and experiments. *Mater. Des.* **2017**, *116*, 365–373. [CrossRef]

52. Sun, Q.; Liu, X.; Han, Q.; Li, J.; Xu, R.; Zhao, K. A comparison of AA2024 and AA7150 subjected to ultrasonic shot peening: Microstructure, surface segregation and corrosion. *Surf. Coat. Technol.* **2018**, *337*, 552–560. [CrossRef]

53. Man, B.Y.; Guzman, L.; Miotello, A.; Adami, M. Microstructure, oxidation and H2-permeation resistance of TiAlN films deposited by DC magnetron sputtering technique. *Surf. Coat. Technol.* **2004**, *180–181*, 9–14. [CrossRef]

54. Carpenter, K.; Tabei, A. On residual stress development, prevention, and compensation in metal additive manufacturing. *Materials* **2020**, *13*, 255. [CrossRef] [PubMed]

55. Maamoun, A.H.; Elbestawi, M.A.; Veldhuis, S.C. Influence of shot peening on alsi10mg parts fabricated by additive manufacturing. *J. Manuf. Mater. Process.* **2018**, *2*, 40. [CrossRef]

56. Al-Mamun, N.S.; Deen, K.M.; Haider, W.; Asselin, E.; Shabib, I. Corrosion behavior and biocompatibility of additively manufactured 316L stainless steel in a physiological environment: The effect of citrate ions. *Addit. Manuf.* **2020**, *34*, 101237. [CrossRef]

57. Khamaj, J.A. Cyclic polarization analysis of corrosion behavior of ceramic coating on 6061 Al/SiCp composite for marine applications. *Prot. Met. Phys. Chem. Surf.* **2016**, *52*, 886–893. [CrossRef]

58. Lodhi, M.J.K.; Deen, K.M.; Greenlee-Wacker, M.C.; Haider, W. Additively manufactured 316L stainless steel with improved corrosion resistance and biological response for biomedical applications. *Addit. Manuf.* **2019**, *27*, 8–19. [CrossRef]

59. Esmailzadeh, S.; Aliofkhazraei, M.; Sarlak, H. Interpretation of Cyclic Potentiodynamic Polarization Test Results for Study of Corrosion Behavior of Metals: A Review. *Prot. Met. Phys. Chem. Surf.* **2018**, *54*, 976–989. [CrossRef]

 *materials*

Review

Wire Arc Additive Manufacturing (WAAM) for Aluminum-Lithium Alloys: A Review

Paula Rodríguez-González *, Elisa María Ruiz-Navas and Elena Gordo

Departamento de Ciencia e Ingeniería de Materiales e Ingeniería Química, IAAB, Universidad Carlos III de Madrid, Avda. De la Universidad 30, 38911 Leganés, Spain
* Correspondence: paularod@ing.uc3m.es

Abstract: Out of all the metal additive manufacturing (AM) techniques, the directed energy deposition (DED) technique, and particularly the wire-based one, are of great interest due to their rapid production. In addition, they are recognized as being the fastest technique capable of producing fully functional structural parts, near-net-shape products with complex geometry and almost unlimited size. There are several wire-based systems, such as plasma arc welding and laser melting deposition, depending on the heat source. The main drawback is the lack of commercially available wire; for instance, the absence of high-strength aluminum alloy wires. Therefore, this review covers conventional and innovative processes of wire production and includes a summary of the Al-Cu-Li alloys with the most industrial interest in order to foment and promote the selection of the most suitable wire compositions. The role of each alloying element is key for specific wire design in WAAM; this review describes the role of each element (typically strengthening by age hardening, solid solution and grain size reduction) with special attention to lithium. At the same time, the defects in the WAAM part limit its applicability. For this reason, all the defects related to the WAAM process, together with those related to the chemical composition of the alloy, are mentioned. Finally, future developments are summarized, encompassing the most suitable techniques for Al-Cu-Li alloys, such as PMC (pulse multicontrol) and CMT (cold metal transfer).

Keywords: WAAM; Al-Li alloys; wire production; DED (directed energy deposition)

Citation: Rodríguez-González, P.; Ruiz-Navas, E.M.; Gordo, E. Wire Arc Additive Manufacturing (WAAM) for Aluminum-Lithium Alloys: A Review. *Materials* **2023**, *16*, 1375. https://doi.org/10.3390/ma16041375

Academic Editors: Bartłomiej Wysocki, Joseph Buhagiar and Tomasz Durejko

Received: 21 December 2022
Revised: 13 January 2023
Accepted: 29 January 2023
Published: 6 February 2023

1. Metal Additive Manufacturing Techniques

Additive manufacturing (AM) is defined as "a process of joining materials to make objects from 3D model data; it is usually layer upon layer, as opposed to subtractive and formative methods of manufacturing" [1]. The techniques in metal AM are divided into direct and indirect processes. The indirect processes require post-forming procedures, while the techniques included in direct AM methods use a high-power laser or electron beam as a heat source and the bonding mechanism is completely melted [2]. In addition, only two direct methods can produce metallic parts: directed energy deposition (DED) and powder bed fusion (PBF), and just one process can create an additively manufactured component from wire feedstock, direct energy deposition [3].

The aerospace industry is one of the major industries and essential players in the AM market [4] and is an impetus for innovation in aircraft materials and design structures. For instance, new aluminum-lithium alloys are being studied due to their low density, high specific modulus, and excellent fatigue, which help reduce aircraft weight and improve performance [5,6]. Therefore, the need for new production processes to obtain complex and lightweight parts combined with the application of new Al alloys has led to a growing interest in the design of new Al parts produced by AM.

Among all the AM processes, directed energy deposition (DED) and powder bed fusion are those most used in the industries for aluminum alloys. Powder-fusing systems are ideal for small and intricate parts with sophisticated and complex features, while wire-fusing systems with DED technology generally have high deposition rates but poor-quality surface finishes [7]. The advantages and disadvantages of using powder and wire in additive manufacturing are described below.

The main advantages of metallic wire are the full use of raw materials free of waste, and the easy handling and storage without any special conditions or requirements, except for a closed packing to keep the surface clean since the surface of metallic wire must be free of impurities. The main disadvantage is the lack of commercial wire with different compositions, as is the case with many high-strength aluminum alloys that are not commercially available as wire. Additionally, the chemical composition of the wire used cannot be modified, in contrast to powder feedstock, that can be mixed with alloying elements. Consequently, only a few compositions can be employed in the final part.

When powder is used in DED techniques, a high-quality one is required to avoid defects in the final part. Thus, high-quality powder is commonly used in order to achieve the nominal composition and reduce the concentration of interstitial elements, such as oxygen and nitrogen, especially those reactive with Ti and Al. However, some current studies use powder blending to tailor the final alloy composition, which provides great flexibility in alloy design [8]. Critical powder features are the packing density and flowability, which are related to the shape and size of the powder particles. The powder should spread evenly across a bed and form a gapless layer. Smooth and spherical particles flow more easily than irregular ones with a rough surface. The most commonly used methods for obtaining powders with these characteristics are gas atomization and plasma atomization.

Furthermore, the powder particles must melt completely since partial fusion of the particles could produce defects in the final product. Powder handling poses a health risk, and its storage must also be adequate, requiring specific conditions free of humidity and exposure to extreme weather. All these powder requirements are common for all bed fusion techniques.

2. Processes to Produce Metal Wires for AM

Production of wires is achieved by different techniques, such as casting, extrusion, and drawing. The most conventional process is casting, which uses ingots of metallic alloys. It is an elementary method that has been used for many years.

Drawing produces wires from rods, bars or plates. By means of a pulling force, the material goes through a die (a rigid tool with a wear-resistant surface), changing and reducing the cross-section. The process requires cleaning and lubrication before going through a die. Kabayama and Taguchi [9,10] defined the most important features of the drawing process as follows:

1. Lubricant (friction coefficient, viscosity, surface treatment)
2. Wire properties (yield stress, elastic modulus, strain rate, strain-hardening)
3. Die geometry (reduction angle, bearing region length, reduction area, material).

The extrusion process is also an alternative for producing wires. Friction extrusion uses a heat source generated by the rotating friction between the raw material and the dies used in the process under load. When the material exhibits plastic behavior, it is forced to flow through the die.

Researchers at the University of South Carolina (USC) obtained 2050 and 2195 aluminum wires. They used friction extrusion to obtain the preliminary rod, followed by drawing, giving the wires a length range between 1.7 and 2.3 m [11]. They also showed that 15 drawing steps can be needed to obtain the final wire with a 1.6 mm diameter. The 0.1 mm step size was used to reduce 2.7 mm starting diameters to 1.6 mm. Annealing and reannealing can alleviate work hardening caused by drawing to prevent wire breaking in the posterior drawing [12]. The purpose of this work was to design wires by WAAM and explore the possibility of modifying the characteristics and initial compositions to obtain a specific final wire. This research concluded that post-extrusion drawing could improve the applicability of extruded wires in the following ways:

Obtaining the desired diameter
Improving surface finish
Increasing the total length

Moreover, powder consolidation in the wires is possible. The direct extrusion of powders, which is a simple metal-forming process, is being developed to obtain preliminary rods. In the subsequent step, these rods can be drawn to obtain wires. Some studies have already been carried out for this purpose, obtaining high-strength Al alloy rods by direct extrusion processes [13,14].

Another process to produce wires from powders is the Conform™ process (Figure 1). Continuous extrusion or Conform™ is used to transform powders, particulates, or waste products, such as machining swarf, into a rod/wire, by means of severe plastic deformation processes. It is possible to obtain a diameter wire of < 5 mm by means of cold drawing with 100% of the material used.

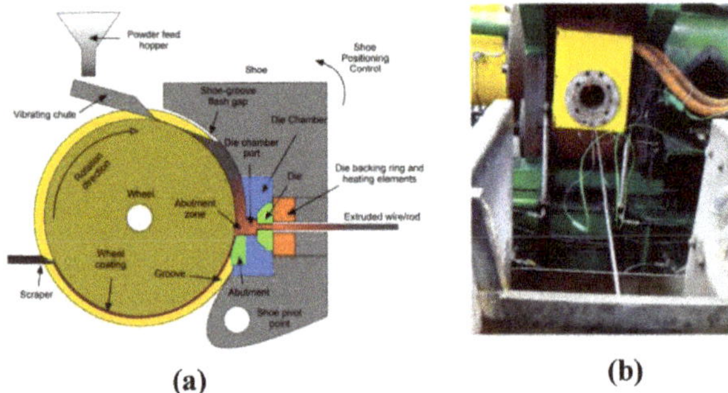

Figure 1. (**a**) Scheme of the Conform machine. (**b**) Extruded wire exiting from the Conform machine and entering the water trough for quenching [15].

Katsas et al. [16] compared the microstructure, texture and superplastic properties developed during Conform™ with a conventional extrusion for a particulate Al–4Mg–1Zr alloy. In this case, the microstructural differences of conventional extrusion from the center to the surface are observed: from a bimodal distribution of coarsely deformed and finely equiaxed grains in the center to a standard distribution of refined equiaxed grains on the surface. In the Conform™, a uniform distribution of refined grains was observed throughout the cross-section. The central region is extruded without further recrystallization, whereas in the surface regions where the strain and strain rate are higher, secondary recrystallization occurs. This work demonstrated that a fully consolidated product of aluminum alloys produced from a particulate feedstock with a uniform refined grain structure could be obtained with good superplastic properties. Conform™ has been used since the 1970s for the production of copper and aluminum rods [17].

At present, there are commercial wires of some aluminum alloys sold, such as 1350, 1100, 1199, 5056, and 6061. New processes and strategies to obtain the wires are being developed to address current demands. The design and selection of the wire are essential to the performance and quality of the deposited metal to obtain a final part with the required properties. For this purpose, the properties of the wires must be studied, in which it is necessary to define the most critical ones, such as good weldability, among others. Gu et al. [18] studied the qualities of the external surface, microhardness, porosity, composition, and microstructures of 4043 aluminum wires. The starting material must be free of defects and imperfections for the application of WAAM. Alloying elements are critical for part porosity, microstructure, and final properties. Controlling small additions of alloying elements can considerably modify the welding behavior of the alloy. Apart from the development of specific wires, the selection of a suitable WAAM technique is key to building functional parts.

WAAM, whose classification is explained below, has been established as a potential technology for the large-scale production of aluminum alloy parts; however, its application is currently limited by the porosity and low mechanical properties attained.

3. Classification of the WAAM Techniques and Their Characteristics

The classification of the WAAM techniques based on the type of welding technique used is shown in Figure 2: gas metal arc welding (GMAW), plasma arc welding (PAW), and non-consumable tungsten electrode welding (gas tungsten arc welding, GTAW) [19].

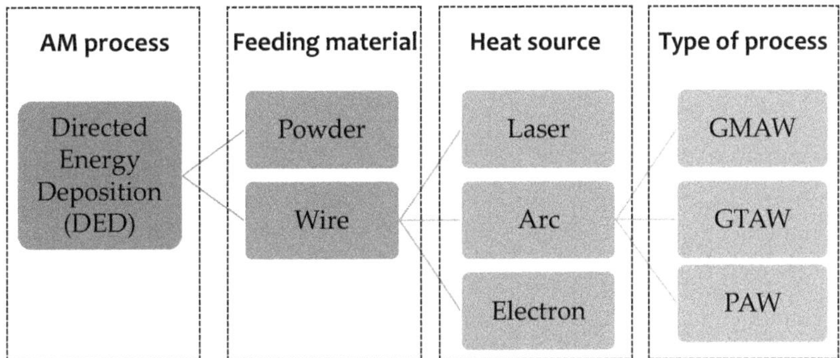

Figure 2. Classification of the DED process [3].

GMAW generates an electric arc as the heat source between the consumable metal electrode and the workpiece. GMAW has two variants: metal inert gas (MIG) and metal active gas (MAG). It is ideal for producing parts on a large scale in short periods of time. These techniques basically consist of an electric arc established between the tip of a consumable wire and the part under the protection of an inert or active gas that protects the weld pool and the adjacent material. The deposition rate is 15–160 g/min [20], which is higher compared to GTAW and PAW.

GTAW and PAW generate an electric arc between the nonconsumable tungsten electrode and the workpiece. The difference with GMAW is that GTAW and PAW require a wire feed that is externally provided. The orientation and the wire feeding direction determine the characteristics and quality of the deposited material [7]. PAW and GTAW have features in common, such as a nonconsumable electrode to establish the electric arc and using an inert shielding gas without filler material [19,20]. However, PAW is a higher energy density process, where excellent stability is achieved due to the arc passing through an orifice between the cathode and anode. Consequently, the weld bead obtained by PAW tends to be narrow since it allows the welding speed to be controlled. The deposition rates are low, approximately 1 g/min, achieving a high-quality surface finish.

On the other hand, the deposition rate for the GTAW technique can be as high as 30 g/min, and the total wall width is thicker (4–15 mm) than the one obtained by PAW (2 mm). The beam's penetration is higher with PAW, causing the fusion of the previously deposited layers and compromising the wall's stability [20].

New techniques have been developed to improve the method according to the material used. A variation of the GMAW process, also known as a modified metal inert gas (MIG), is a cold metal transfer (CMT). CMT, known as the freezing process, compared to the other welding techniques [3], alternates cooling and heating based on a short circuit with high and low current and voltage. It applies noticeably less heat input than traditional GMAW and differs from the latter in its excellent control over penetration, a high wire melting efficiency, and a high deposition rate. Another variant is GMAW in tandem. A tandem uses two independent welding systems, synchrony, and two wires. The energy input and deposition rates are higher.

4. Overview of WAAM

In 1925, Baker [21] carried out studies using an electric arc as a heat source and metal wire as feedstock, but the earliest research into WAAM dates back to 1926, when Baker patented "the use of an electric arc as a heat source to generate 3D objects depositing molten metal in superimposed layers" [22]. Baker used a new technique to build a 3D object using welding, the oldest known attempt to use welding technology in additive manufacturing. The same year, Eschholz [23] used an electric arc to deposit metal to create a variety of ornamentation using only a single layer and identified the primary process variables, such as arc current, depth of penetration, travel speed, substrate material, bead width, and height.

In 1935, an electric arc was covered beneath a bed of granulated flux known as SAW (submerged arc welding). The process patented by Jones, Kennedy, and Rothermund [24] required a continuously fed consumable solid or tubular (metal-cored) electrode. In 1947, 20 years later, Carpenter et al. [25] patented the method for metal coating of metal pipes by electric fusion. This invention was used in the production of magnesium retorts, having a carbon steel base clad with a high chromium, high nickel steel alloy coating of one-half inch and a base metal thickness of one inch.

Numerous patents were accepted as the first steps in wire and arc additive manufacturing. In 1950, a process for additive manufacturing with wire deposition was described by Muller Albert et al. [26]. In 1971, Ujiie (Mitsubishi) fabricated a pressure vessel using SAW, electroslag, and TIG, and also employed different wires to provide functionally graded walls. In 1983, Kussmaul [27] used shape welding to manufacture high-quality, large, nuclear structural steel parts with a deposition rate of 80 kg/h and a total weight of 79 tons [28].

In 1990, Acheson [29] designed automatic welding equipment for weld build-up. In 1997, Dickers et al. emphasized the importance of the weld bead geometry and conducted numerous trials of singular weld beads [30], modifying parameters, such as voltage, wire feed rate, wire stick out, distance from the nozzle, etc. In addition, a feedback loop between the welder and the robot controller was shown to be essential for increasing process uniformity, according to Dickers' software-focused research [31]. During the same year, Spencer et al. [32] focused their studies on the manufacture of thicker walls. However, there was incomplete penetration of the material, and tilting the torch to deposit several adjacent strands was unsuccessful. Finally, thanks to the combination of the GMAW technique of three axes with a Siemens controller, it was possible to build layers on a platform that could tilt and rotate.

In 2002, GMAW-WAAM was improved by using the CAD/CAM integrated system [33]. Song et al. (2007) designed the hybrid manufacturing, integrating gas metal arc welding (GMAW) and milling, reducing the dimensional accuracy from ± 0.5 mm of pure WAAM to ± 0.01 mm [34].

In 2010, a breakthrough was achieved by Almeida et al. [35]. They fabricated a 1-meter-long, fully dense structure of Ti6Al4V with a cold metal transfer (CMT) welding process with a high deposition rate of 3 Kg/h, and achieved a deposition efficiency of over 80%.

In 2015, Gaddes [30] used constant current/constant voltage (CC/CV), GMAW short circuit transfer (off-the-shelf unit), called CMT-WAAM, to control the weld bead geometry of the process.

In recent years, WAAM techniques have progressed, incorporating post-processing such as heat treatment of parts and research into corrosion behavior. In addition, the integration of rapid prototyping using CMI-WAAM allows bimetallic materials to be obtained and memory alloys to be shaped.

Recent studies have also focused on increasing the deposition rate. Project development by Stewart Williams at Cranfield University has made considerable contributions to the fields of aluminum and WAAM [36–38]. The researchers developed the WAAM process capable of creating the most significant part, a six-meter-long, 300-kg, double-sided spar created from aerospace-grade aluminum [39].

5. Overview of Al-Cu-Li Alloys

5.1. Introduction to Al-Li Alloys

The most widely used aluminum alloys in the aerospace industry are the precipitation-hardening Al-Cu alloys (2xxx series) and Al-Zn alloys (7xxx series) due to their high strength-to-weight ratio. Al-Cu-Li alloys are lightweight and ideal for reducing weight and achieving lighter and stronger parts [40]. However, Al-Cu-Li alloys are expensive compared to other aluminum alloys, with a cost that is three to five times higher than the others.

Research into Al-Cu-Li alloys began in the U.S. and Germany in the early 1920s. Although this research was interrupted for many years, in 1980, researchers developed the second generation of Al-Cu-Li alloys [41]. These alloys clearly have improved mechanical properties compared to the first-generation ones. However, the properties still could not meet most aircraft specifications in terms of thermal stability, corrosion resistance, isotropy, and weldability. The third generation of Al-Li alloys was then developed to overcome these issues [40], and some of them were finally used in the aerospace industry. For instance, AA2060 and AA2050 alloys present excellent properties, such as thermal stability, corrosion resistance, and high specific strength [42].

5.2. Al-Cu-Li Alloys and Their Applications

Despite the fact that the development of third-generation alloys has led to new applications in the aerospace industry, their high cost has not allowed them to be fully exploited and implemented in new applications.

Conventional aluminum alloys, such as 2024 and 7055, are still widely used. The most common aluminum alloys for aerospace are 2014, 2024, 5052, 6061, 7050, 7068, 7075, 2219, 6063, and 7475. However, the alloys that are being employed for additive manufacturing techniques are mainly 2524, 7055, 7150, and AA2024, which are currently employed in the Boeing 777.

Nevertheless, some Al-Li alloys, such as 8090, have been proposed to replace AA2014, AA7010, AA2024, and AA7075 in various locations of the EH101 helicopters for structural and nonstructural applications. In particular, in floor installations, brackets, stiffeners, longerons, bulkheads, tail cone skins, flying control structures, door rails, and seat tracks [43]. AA8090 has also been evaluated for cryogenic fuel tanks.

According to Al-Cu-Li alloys, Figure 3 shows the most used Al-Cu-Li alloys. The compositions can be found in the Aluminum Association, revised in 2018 [44]. These alloys can be divided into two main groups: high-copper and low-lithium, and low-copper and high-lithium.

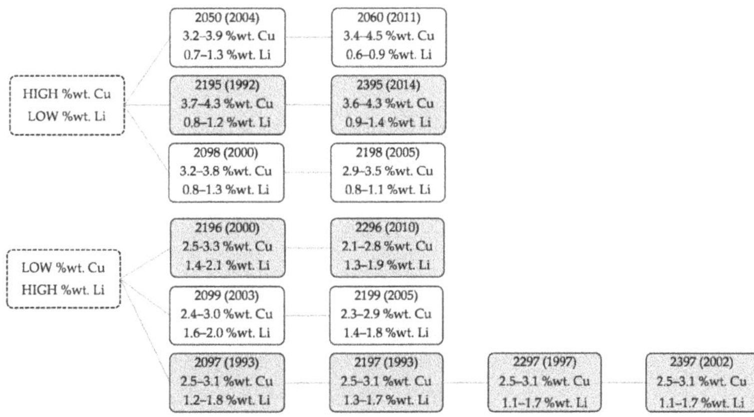

Figure 3. Scheme of the most used Al-Cu-Li alloys.

First, 2050 and 2060 alloys present excellent properties, such as thermal stability, corrosion resistance, high strength, good weldability, good toughness, and lightweight [42]. A 2050 alloy has been recently evaluated for cryogenic tank applications in space launch vehicles based on its excellent fracture toughness and stability at cryogenic temperatures. In particular, 2050 is used for wing spars and ribs in commercial aircraft/launch vehicle structures. It is an alternative to legacy 7xxx-series alloys, such as 7050 and 7075, due to its higher short-transverse strength and stress corrosion cracking resistance, in addition to a significant density reduction [45,46].

A 2060 alloy was launched in 2011 by Alcao Inc. [43] and is a relatively new alloy. This alloy is used in the aerospace industry, principally in the fuselage panel [47], particularly in the fuselage skin and lower wing structures [47]. When it is used instead of the 2524-T3 or 2024-T351, it may save 7% and 14% of weight, respectively [48].

NASA has employed the 2195 (Al-Li-Cu-Zr) alloy for space components [45,49]. This alloy provides good weldability, ultra-high strength, in particular, excellent resistance to fracture at shallow temperatures, and reduced density. A 2195 alloy was developed in 1994 by NASA for the cryogenic sections of the super light weight external tank (SLWT), an integral component of the early space shuttle launch systems; mainly used in the form of a plate [50]. This alloy reduced density and increased strength, thanks to a new structural design that reduced 3400 kg of the 27,000-kg tank (a 12.5% weight savings). There were savings of millions of dollars due to this weight reduction and increased payload capacity for the shuttle [45,51].

The successive 2395 alloy provides a higher strength, and it is used as drop-in replacements for 7x5x-T77511 for fuselage frames, floor beams, and upper wing stringers [45].

Both 2198 and 2199 alloys are also used for fuselage skin application [45]. A 2198 alloy was developed in 2005 and chosen afterwards by Airbus for the fuselage skin in the A350 aircraft [52]. It is used in commercial aircraft as a sheet product.

A 2196 alloy is typically used to replace conventional alloys, such as AA7075, when weight savings are needed. This alloy is used for well-suited stringer, fuselage frame, and floor structure applications where a combination of high strength, high toughness, and low density is required. A 2196 alloy was originally developed for the Hubble solar panel frame. 2196 alloy content Ag, which increases the cost of the raw material. However, it is very promising since NASA uses it in extrusions for commercial aircraft [49] and it is also part of the Airbus A380 plane [53] due to its properties.

Both 2099 and 2199 alloys were developed for space and aircraft applications, such as aircraft fuselages and lower-wing applications; their precipitates, dispersoids, and elements have shown attractive properties [6,40]. A 2099 alloy was developed by the U.S. Air Force for its potential applications with laser beam welding [54].

Finally, 2097 alloy is used as a replacement for 2124. This alloy is used for the B.L. 19 longeron in the F16 fighter jet and the F16 bulkhead area due to the improved combination of density, modulus, corrosion resistance, high-temperature stability, and resistance to fatigue crack growth propagation compared to conventional alloys [45]. This alloy presents three new versions: 2197 (1993), 2297 (1997), and 2397 (2002). This last one has been used for Lockheed Martin F-16 bulkheads and other parts for military installations [53].

5.3. Influence of the Main Alloying Elements

The addition of lithium reduces the density of the alloy substantially while increasing its strength more than any other element added. Each increment of 1 wt.% of lithium decreases the density by 3% and increases the elastic modulus (E) by approximately 6% [51,55,56]. Approximately 14–16 at.% (4.7 wt.%) of Li can be wholly dissolved in solid Al at 600 °C [57].

In addition, the low atomic weight of Li (6.94 g/mol) gives it the highest heat capacity compared to any other metal (in terms of J/kg·K). These properties and its relatively good thermal conductivity make Li an appealing high-temperature heat transfer material [57]. It can be a disadvantage because aluminum also presents high thermal conductivity, making it challenging in some undercooling processes.

Al-Li precipitates tend to nucleate heterogeneously on grain boundaries in slowly cooled and overheated alloys. The precipitates sequence can be described as a solid solution $\alpha + \delta' + \delta \rightarrow \alpha + \delta$ where δ is the equilibrium phase (AlLi) and δ' metastable phase (Al$_3$Li). They can also easily nucleate homogeneously in the matrix, forming spherical precipitates, thanks to their low interfacial energy with the matrix (approximately 14 MJ/m^2) and low precipitation activation energy [58].

The addition of copper improves strength and hardness and reduces corrosion resistance. During aging, the precipitates, solute-rich domains produce the strengthening effect in the alloy. These areas are fully coherent with the matrix, but the atomic spacings are different enough to distort the crystal lattice without discontinuity in the matrix. When the movement of dislocations is obstructed, an increase in strength is attained. Al-Cu systems follow this sequence during heating, where GP corresponds to Guinier–Preston zones [59]:

Super saturated solid solution $\rightarrow$ Cu clustering $\rightarrow$ G.P.1 $\rightarrow$ G.P. 2 (θ'') $\rightarrow$ θ' $\rightarrow$ θ

θ'' is an intermediate precipitate, with a tetragonal structure, which maintains coherency with the Al matrix. When θ' appears, the strengthening is reduced. Further heating causes the transformation from θ' to θ, an equilibrium precipitate, with the composition Al$_2$Cu.

When copper and lithium interact, they form part of the main hardening mechanism in aluminum alloys, specifically Al-Cu-Li alloys, where the most critical strengthening phases are: T1 (Al$_2$CuLi), T2 (Al$_6$CuLi$_3$), and TB (Al$_{15}$Cu$_8$Li$_3$) [53]. Some studies show that ternary T1 (Al$_2$CuLi) is the primary strengthening phase, and binary phases, such as δ' (Al$_3$Li) and θ' (Al$_2$Cu), despite also being present, they contribute less to strengthening.

The precipitates found in Al-Cu-Li alloys depend on the Li relative content. θ' and θ (Al$_2$Cu) phases are formed with low Li contents (<0.6%), while the main strengthening phase, T1(Al$_2$CuLi) precipitates with medium Li content (<1.4–1.5%). The δ' precipitates appear when higher Li contents (>1.4–1.5%) are employed [60]. B. Cai et al. [61], among other studies, confirmed the presence of T1 (Al2CuLi), θ' (Al2Cu) in the 2060 alloy.

The addition of magnesium increases strength due to solid solution strengthening. Additions greater than 1.6 wt.% Mg in Al-Cu alloys promote the formation of the metastable, incoherent S' (Al$_2$CuMg) phase near grain boundaries [62].

The addition of silver and Mg stimulates the nucleation of a fine and uniform dispersion of T1 phase [63]. However, it is essential to take into account that adding elements, such as Ag, makes the alloy more expensive.

The addition of manganese promotes the formation of precipitates of the Al$_{20}$Cu$_2$Mn$_3$ phase, which controls grain size and texture during the thermomechanical processing [64]. This helps improve creep resistance and damage tolerance in fracture toughness and fatigue.

Titanium interacts with the Al matrix to form Al$_3$Ti intermetallic, strengthening in addition to conventional precipitation hardening (such as in the θ' (Al$_2$Cu) phase). A study shows that adding 0.6% Ti to an Al-Cu-Mg-Ag alloy promotes the precipitation of finer and denser θ' (Al$_2$Cu) phase. It also promotes the formation of the intermetallic phase Al$_3$Ti. In contrast, if the addition of Ti is increased to 1.1%, no improvements are observed; the results reveal that a high volume of Al$_3$Ti phase is likely detrimental to strengthening [65].

Furthermore, it is widely used as a grain refiner. In particular, and regarding the WAAM techniques, Wang, L. et al. [66] studied the Al-Mg alloy for WAAM, using ER5356 as the filling wire and adding Ti powder between layers as a grain refiner. This study reports that the Al3Ti phase provides heterogeneous nucleation cores. The addition of Ti powder during WAAM can promote effective transformation from columnar to equiaxed grains at the interlayer interface. The ultimate tensile strength and elongation increase by 20.25 MPa and 3.13% in the horizontal direction, and by 25.89 MPa and 6.97% in the vertical direction. This study highlights the addition of Ti as a grain refiner as an excellent strategy to obtain isotropic and improved mechanical properties in the WAAM technique.

The addition of zirconium promotes the formation of coherent dispersoid β' (Al_3Zr) and θ' and T1 phases and reduces the solubilities of lithium and magnesium in Al alloys [67]. Zr inhibits recrystallization and grain growth at elevated temperatures [68].

The small additions of Scandium act as a core for the formation of very refined grain microstructures. At the same time, its influence has led to the development of a new alloy, Scalmalloy (AlMgSc), developed by the Airbus Group Innovations [69]. The addition of Sc results in high mechanical properties, good ductility, and specific resistance.

Sales et al. [70] demonstrated that adding Sc in AA 5183 and AA 5356 alloys promoted the formation of Al_3Sc intermetallic particles, with an increase of nearly 60 MPa in the ultimate tensile strength and yield stress.

The systems become more complicated when more than one alloying element is added, for instance, Al-Cu-Li, Al-Li-Mg, Al-Li-Zr, Al-Cu-Mg, or Al-Cu-Li-Mg-(Ag) systems. The effects of alloying elements in Al alloys are summarized as shown in Table 1. It should be noted that promoting the homogeneous grain and fine microstructures is ideal for obtaining good properties in the final piece by WAAM.

Table 1. Effects of alloying elements.

Elements	wt.%	Physical Properties	Mechanical Properties	Chemical Properties	Others
Li	0.6–2.1	↓ density ↓ weight	↑ strength ↑ hardness by age hardening		δ' (Al_3Li) [57] promote T1, T2, θ'
Cu	2.4–4.5		↑ strength ↑ hardness by age hardening	↓ corrosion resistance	θ (Al_2Cu), T1 (Al_2CuLi), T2 (Al_6CuLi_3), TB ($Al_{15}Cu_8Li_3$) [53,71]
Mg	0.1–1.1		↑ strength ↑ hardness by solid solution	↑ corrosion resistance and weldability	Al_2 (Cu, Li-Mg) S' (Al_2CuMg) [62]
Ag	0.05–0.6		↑ strength ↑ hardness by age hardening	↑ resistance to stress corrosion	promote T1 [63]
Mn	0.1–0.5		↑ strength ↑ hardness by age hardening ↑ creep resistance and damage tolerance		$Al_{20}Cu_2Mn_3$ [64]
Ti	0.1–0.12		↑ strength by grain size reduction		Al_3Ti [65,66]
Zr	0.04–0.2		↑ strength by grain size reduction	↓ resistance to stress corrosion cracking	Al_3Zr [67] promote θ' and T1
Sc			↑ strength by grain size reduction		Al_3Sc

6. General Welding Problems for Aluminum Alloys

The most relevant concerns related to aluminum alloys applied to additive manufacturing techniques are: the oxidation of the material surface due to the formation of Al_2O_3, the solidification shrinkage (compared to ferrous metallic materials) due to the wide solidification temperature range, the high coefficient of thermal expansion (CTE), which leads to cracking phenomena due to high solidification stresses and shrinkage, high reflectivity, high thermal conductivity (which leads to rapid dissipation of heat from the scanned area and requires greater source heat), and the high solubility of hydrogen in liquid aluminum (which leads to pore formation) [72].

6.1. Porosity

One of the challenges for aluminum parts is producing porosity-free pieces through welding processes. The leading cause is hydrogen, which has a high solubility in molten aluminum but a poor solubility in solids. The gas dissolved in the molten metal weld is trapped during the solidification, forming pores in the solidified weld. The hydrogen content over 1 mL/100 g of aluminum produces excessive fusion zone porosity [56]. The dispersion-strengthened aluminum alloys produced via rapid solidification by powder metallurgy processing often exhibit a residual hydrogen content of over 1–5 mL/100 g of aluminum, showing an increased tendency to form a fusion zone porosity [56].

In addition, another cause of the formation of porosity comes from the aluminum oxide layer, which is rapidly formed on the surface due to its high Al affinity for oxygen. Its melting point is approximately 2050 °C, while aluminum melts at 660 °C [73]. When aluminum is melted, the oxide film is entrapped in the aluminum melting pool. This film has to be removed before welding in order to reduce the risk of porosity during welding.

The chemical composition is likewise critical in the formation of porosity. Elements such as magnesium have a beneficial effect. It has been demonstrated that magnesium at 6% raises the solubility in solids and there is up to a two-fold reduction in hydrogen absorption [74]. However, the rest of the alloying elements, such as copper and silicon, do not have the same effect, and other elements, such as zinc and lithium, which have high vapor pressures, evaporate at high temperatures and leave porosity in the material.

Different attempts have been made to reduce porosity. The combination of the interlayer rolling process and WAAM has been studied for straight walls with 2319 and 5087 aluminum alloy wires. In this study, interlayer rolling was employed between each deposited layer, and different rolling loads were applied. It was observed that pores bigger than 5 μm in diameter were eliminated at a rolling load of 45 kN [75]. However, interlayer rolling, and post-deposition heat treatments were needed for lower applied loads, showing a reduction in the number and the percentage of the pore area.

6.2. Cracking

Cracking is a high-temperature phenomenon that usually takes place in alloys in the liquid melt pool. When this defect occurs during WAAM in aluminum alloys, it is known as solidification cracking; it is not the common defect known as hot-cracking, which occurs in arc welding and takes place in the partial melted zone [76].

The addition of alloying elements changes the freezing temperature of the pure metal, modifying the solidification range. Pores may also promote the generation of cracks during solidification.

When the material solidifies, the liquid with the lowest solidification point is retained in the interdendritic spaces. In this mushy zone, solidification, and thermal shrinkage exhibit stress on the solid network. This was observed in 1950; investigations such as Novikov [77], Sigworth [78], and Eskin et al. [79] described how a tear (hot-crack) initiates above the solidus temperature and propagates in the interdendritic liquid film. The fracture surface is usually smooth and sometimes shows solid bridges connecting both sides of the crack [80–84]. Indeed, some studies show that cracking occurs in the late stages of solidification. This happens when solid volume fractions are above 85–95% [82] and the solid phase is organized in a continuous network of grains. Different mechanisms can be found in the literature describing this phenomenon [84,85].

The addition of alloying elements to Al always leads to Al alloys with different solidification ranges, which means that Al alloys can be susceptible to cracking. The Varestraint test, developed in the 1960s by Savage and Lundin at Rensselaer Polytechnic Institute [86], is a simple test to isolate the metallurgical variables that cause cracking [87,88]. In the 1990s, Lin and Lippold [89] advanced further in the research and used Varestraints testing. Finally, they determined the magnitude of the crack-susceptible region through the measure of the temperature range in which cracking occurs. Samples were tested over a range of growing strains, and the maximum crack length in the fusion zone as a function of temperature was measured.

Singer and Jennings [90] reported the first graphs for the Al-Si system, which show crack length (inches) vs. percentage of alloying elements.

Later, Pulphrey et al. [91] investigated the cracking of many binary aluminum alloys, using the same theory as Singer et al. A maximum cracking is reached in each alloy, where the temperature interval is observed between solidus and liquidus in the solidification range [92].

The alloys can be characterized by plotting relative to the theoretical peak in cracking susceptibility under equilibrium and temperature, using the appropriate conditions (Figure 4).

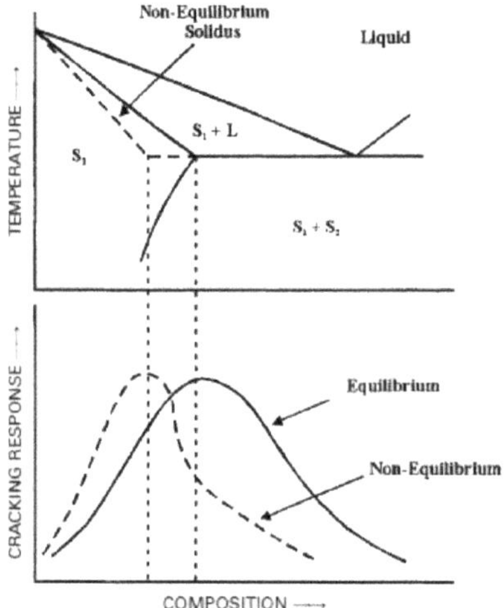

Figure 4. Weld metal cracking susceptibility as a function of composition in a simple binary alloy system: equilibrium and nonequilibrium solidification conditions are considered [93].

These graphs are known as cracking tests; they determine the composition range within which the alloy presents a high risk of cracking. The test consists of applying a load to the weld transversely to measure the length of the crack. This measure reports the specific range of sensitivity to cracking. Some examples related to the main alloying elements of aluminum alloys are seen in Figure 5.

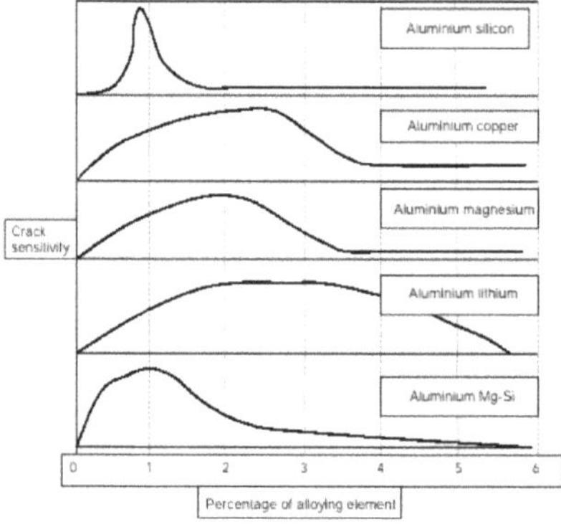

Figure 5. Graphs of crack sensitivity for different alloying elements in an Al matrix [74].

For instance, the 2040 and 7075 alloys containing a significant amount of Cu and Mg are susceptible to cracking, as reflected in their high values of total crack length, and thus poor weldability is expected.

Cracking susceptibility is expected for aluminum alloyed with Cu and Li content of approximately 2.5 wt.%, in particular in alloys such as 2090 and 2091. In Al-Cu-Li systems, a reduction of Li content below 2% and an increase of Cu content above 3% should reduce susceptibility, as in the case of 2060 and 2195 alloys.

6.3. Humping Phenomenon

Some studies, such as that by Adebayo et al. [94], have focused on the factors that promote the phenomenon of humping in WAAM. According to their research, humps are a consequence of the heat sink formed due to an accumulation of material, which reduces the penetration of the arc. To avoid this phenomenon, they suggest removing the undulation of the weld bead and using low travel speeds [94]. There are three factors that must be controlled. First, the high momentum of the backward fluid flow causes the initiation and growth of swelling. Second, the large variation of the capillary pressure of the liquid channel in the welding direction can promote the shrinkage of a liquid channel. Finally, the capillary instability makes the weld pool unstable and susceptible to collapse [95].

Some studies focused on the modification of variables in the process of reducing humps. Cold metal transfer (CMT) technology for Al-Mg alloy shows that a one-way continuous-arc trajectory presents better regularity in the part geometry, owing to the unique starting point imposed in this case; on the contrary, when using other ways, humps are formed, affecting the frequency of the deposits [96].

6.4. Lack of Fusion, Delamination, Residual Stresses, and Discontinuity in Weld Bead

The lack of fusion is caused by insufficient overlap between passes. Some causes include low energy input, an improper torch angle, an inappropriate weld position, and insufficient filler wire material, among others [97].

Delamination is produced when the underlying material is not completely melted, which causes delamination or separation between neighboring layers. This is due to low energy input. However, other causes of delamination by WAAM are residual stresses and contamination of parts or substrate surfaces [97,98].

Residual stress is the stress that remains in a material after all external loading has been removed. If the residual stress is large enough, it will have a dangerously negative impact on the component's mechanical properties and fatigue life. The region between the substrate and the buildup wall experiences the highest stresses in WAAM.

Discontinuity and deviation in the weld bead are other defects that are visually observed on the surface of the weld bead and occur when high energy input is used. When the arc is unstable, there are deviations in deposition volume, and the position or orientation is inadequate [98].

6.5. Particular Welding Problems for Al-Li Alloys

Welding issues in Al-Cu-Li alloys are presented as two types: cracking and porosity. The first one is related to high susceptibility to cracking when the process causes high levels of thermal stress and solidification shrinkage. The Al-Cu-Li alloys were designed to optimize strength and compositions that promote good mechanical properties; they were not prepared for optimum conditions in weld crack resistance. The peak tearing susceptibility of high-purity Al-Li binary alloys appears at 2.6 wt.% lithium [99,100] (Figure 6). Therefore, Al-Li alloys with this content would be more likely to promote significant tearing during the welding process.

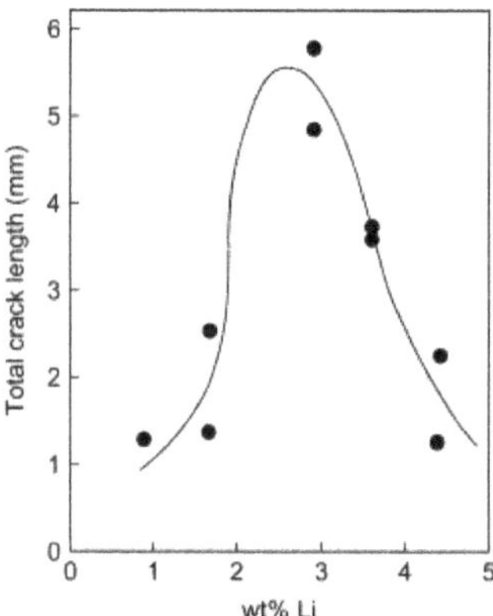

Figure 6. Independent peak of cracking susceptibility of Al-Li binary alloys [100].

However, not only can lithium be the cause of the welding defects in these alloys, but it can also promote porosity. This is due to the activity of lithium during the welding process. Li atoms tend to interact with H atoms, increasing the hydrogen solubility of the liquid aluminum. Some of the components that can be formed are lithium oxide (LiO_2), lithium hydroxide (LiOH), lithium carbonate (Li_2CO_3), and lithium nitride (LiN) [93,101].

Despite these problems, it is known that there are already some alloys with lithium contents close to the maximum cracking point that are used for welding, such as 2099-T83. Al-Li alloy is being used for the construction of lower-wing stringers by welding processes [102]. However, at present, the knowledge base related to the weldability of Al-Li alloys is still limited.

7. Suitable WAAM Techniques for Al-Cu-Li Alloys

There is scant use of WAAM in Al-Cu-Li alloys. The first research study carried out focused on post-WAAM treatments to improve properties, such as removing porosity and increasing strength. However, other problems, such as the surface oxide film (alumina), which has a higher melting point than Al, cannot be addressed by those post-treatments. For this purpose, the most recent studies have focused on a variant of the WAAM technique that works with alternating current (AC).

AC mode allows the current waveform (frequency and balance) to be modified. In one half of the cycle, the electrode will be negative (with the base plate positive), and in the next half, the electrode tip will be positive (with the base plate negative). The balance represents the relationship between the penetration (EN, electrode negative) and cleaning action (EP, electrode positive) in the percentage of the cycle. The AC mode is necessary for Al-Li alloys since it allows both actions to be worked on: cleaning the oxide layer and heating a weld bead [103]. During the cleaning, the electrons remove the oxide layer from the aluminum surface, and the shielding gas prevents new oxide from being formed during the welding process.

VP-GTAW (variable polarity gas tungsten arc welding) was applied successfully on an Al-Cu-Li alloy, in particular 2050 wire, and a thin straight wall was deposited [104]. The results showed that the inner layers consisted of refined, equiaxed grains, compared to the interlayers, which consisted of coarse, columnar grains. The secondary phases were θ (Al_2Cu) and δ' (Al_3Li) phases, which were dispersed along the grain boundaries after post-deposited heat treatment. The mechanical properties, such as microhardness, in the heat-treated sample were 141HV, which showed an increase of 98.6% compared to that of the deposited one (71HV), and 55% compared to that of the wire (91HV).

Other variants of GMAW, such as cold metal transfer (CMT), are of particular interest for Al-Cu-Li alloys: CMT-P, CMT-ADV, and CMT-PADV. The CMT-P refers to pulsing recurrent, in which the high pulse current results in a higher heat input compared to the conventional CMT-ADV. The advanced path involves a polarity reversal in the short circuit; and the combination of both is CMT-PADV and pulse-advanced. CMT-PADV, developed by Fronius [20], greatly improves porosity due to the control of the polarity and the pulse cycles [105]. The most significant characteristics are the low thermal heat input, the high deposition rate, and the low spatter, which have been shown to eliminate gas pores due to an oxide cleaning effect.

Derekar et al. [106] showed results using CMT and pulsed-MIG techniques with ER5183 wire on a wrought plate substrate of Al–Mg–Mn alloy. Despite the fact that ER5183 does not contain lithium, the magnesium range makes it susceptible to common welding problems, similar to Al-Cu-Li alloys. Deposited material by pulsed-MIG showed a higher percentage of medium-size (0.21–0.30 mm) and large-size pores (>0.31 mm); and a higher pore volume for medium-sized pores (0.21–0.3 mm) than those obtained by CMT. However, small-size pore volume was more significant in CMT. The picked-up hydrogen was higher for pulsed-MIG than for CMT due to the higher arc energy and the more extensive melt at higher temperatures.

To evaluate the applicability of the alloy to WAAM techniques, it is necessary to carry out numerous weldability tests. No alloy should be discarded due to its chemical composition or the possibility of the evaporation of elements with low vapor pressure, such as Li or Zn.

The most studied 2xxx alloy is the 2024. However, new Al-Cu alloys, such as 2219 and 2319, have also been incorporated into recent studies, where they appear more susceptible to experiencing problems during the welding process. However, when the 2319 wire was deposited on 2219-T851 aluminum plates by CMT-PADV [107], no crack was observed. The process was successful because it had a significant oxide cleaning effect on the weld bead surface. Both variant processes (CMT-ADV and CMT-PADV) allow the thermal profile to be modified, resulting in a refined, equiaxed microstructure and the elimination of porosity. Gu et al. [107] showed a reduction in porosity, fine equiaxed grain structures, and uniformly distributed θ-Al$_2$Cu phases for 2319 filler wire by the CMT-PADV process, which not only implies a reduction in porosity similar to the studies mentioned above, but that microstructure with good conditions is also possible for aluminum alloys.

Cong et al. [108,109] used Al-Cu wires with CMT techniques. CMT produces a high heat input and has excellent penetration; consequently, coarse columnar grains are formed, preventing the hydrogen from escaping. As a result, many pores with pore sizes varying from 10 to >100 μm were obtained.

They observed that a lower arc penetration by the variable method CMT-P decreased the escape distance for hydrogen and promoted a smaller number of pores. However, their research found that CMT-ADV mode efficiently eliminated the porosity, obtaining fine equiaxed grain structures, and concluded that the three key factors are the low heat input, a shallower penetration, and alternating polarities that produce a cleaning oxide effect (pore sizes > 50 μm). The CMT-PADV showed impressive results with no pores over 10 μm.

Cong et al. also reported the differences between wall and block structures. A block structure significantly reduces porosity due to the formation of refined microstructures using the CMT-P and CMT-ADV processes. In a block structure, the material available in the surroundings extracts heat, increasing the cooling rate, while the dissipation of heat by conduction is only possible through the underlying layers in the wall structure.

Zhang et al. [110] found that VP-CMT produced a high ultimate strength because the columnar grains transformed into equiaxed grains. As the process consists of a pulsed arc mode to produce an oscillation in combination with alternating changes in arc polarity, it promotes heterogeneous core points by breaking the dendrite arms. The range of the tensile strength anisotropy of the transverse and longitudinal tensile samples was found to be between 8 and 27%.

The combined effect of CMT, interlayer rolling, and heat treatment on the porosity and oxide cleaning surfaces of aluminum alloy by WAAM is an area of interest for many researchers. The interlayer rolling contributes to reducing porosity and greatly influences the grain structure. When rolling is applied, depending on the load, precipitates break into smaller sizes. After heat treatment, a uniform distribution of refined, smaller grains is attained. For the 2024 alloy, a preliminary study by Fixter et al. [111] showed that, with adequate control of porosity and subsequent heat treatment, the tensile properties could be improved through the WAAM process to levels comparable to those of the standard wrought products. By rolling each added layer, the interpass deformation was found to lead to further refinement in grain size and improved ductility.

New pulse technologies, such as PMC (pulse multicontrol) and PMC mix, are based on pulse-controlled spray arcs optimized by tight control algorithms. The company Fronius has managed to modify the power source platform (TPS) to obtain better welding results. Process stability is improved, heat input is reduced compared to MIG, and there is almost no spatter. Consistently, good penetration is guaranteed, there is less undercutting, and it is possible to weld more quickly and more cost-effectively. PMC Mix technology combines this pulse-controlled transfer to cycles controlled short-circuit, generating a colder phase and reducing the heat input even more [96,112].

Gomes et al. [96] also compared the PMC and CMT techniques for Al alloys. The porosity was smaller than 175 μm in diameter and was dispersed in both cases. Samples by PMC and PMC mix showed a pore fraction of 0.21% and 0.80%, respectively, much lower compared to those by CMT and CMT-P techniques, which showed 0.54% and 1.16%, respectively. These results are particularly interesting for Al-Cu-Li alloys. Regarding mechanical properties, nearly isotropic properties were obtained with a difference of 13 MPa higher in the longitudinal direction of the deposition. Therefore, the small pore fractions and the regularity of the deposits obtained with PMC and PMC Mix, compared with CMT and Pulsed-MIG, point to the benefits of these techniques.

Synchro-feed welding [113] is the most advanced technology and produces a high-quality and high-speed weld without requiring a post-weld cleanup. It incorporates a driven wire feeder within the torch body, advancing the welding wire forward to create an arc. It then retracts the wire while synchronizing with a specialized weld current waveform that extinguishes the arc to make consistent droplet transfer with virtually zero weld spatter. This way, it combines a speedy wire feed control and ultra-low spatters with ultra-low heat input and ultra-low smut. Finally, it results in a very neat, precise weld laid down at a rate of up to 100 inches (254 cm) per minute using a welding current of up to 300 amps [113].

Another improvement was found by Zhang et al. [114] y means of the workpiece vibrating in combination with the VP-CMT technique. The most conclusive results were the refined grain and the homogenized grain distribution with increased vibration. The average grain size decreased due to the over-threshold bending stresses induced by workpiece vibration, breaking the dendrite arms and evolving into more nuclei. In this way, the workpiece vibration was able to significantly reduce the porosity from 6.66% to 1.52% and improve the mechanical properties; the workpiece vibration induced the molten pool stirring, which removed the fine grain zone of the interlayers and the pore defects [114]. Therefore, the combination of vibration and WAAM techniques will help to significantly reduce porosity in Al-Cu-Li alloys.

8. Conclusions

WAAM is a promising manufacturing alternative to conventional manufacturing. The high deposition rates make it an ideal technology for manufacturing large metal components.

However, its use is constrained by the limited availability of commercial wire. Meanwhile, although WAAM techniques are under continuous development, as previously explained for the latest advances, the material and desired wire composition remain unaddressed by researchers. WAAM technology and material should have a parallel development in order to achieve potential breakthroughs for new aerospace components.

This review summarizes the features of conventional and innovative processes of wire production, such as the extrusion and conform processes, and seeks to encourage further research into the development of new metallic wires for WAAM.

In particular, this review provides the features of the main Al-Cu-Li alloys that are of great interest, along with their specific industrial applications, aimed at fostering and promoting the selection of the most suitable compositions. The addition of different alloying elements to aluminum has been discussed to find the most suitable wires for WAAM, focusing on lithium because of its particular strengthening benefit compared to other alloying elements. Additionally, alloying elements, such as Cu and Mn strengthen the alloy by age hardening, while Ti and Zr directly influence grain size reduction.

Finally, this review addresses specific issues for aluminum, such as porosity, cracking, and humping phenomena, and the specific defects in WAAM, such as lack of fusion and delamination, with particular interest in the role that lithium plays in porosity and cracking problems. Future strategies and developments to improve the results obtained in WAAM have been presented. The possibility to design and develop new metallic wires for the most precise and sensitive WAAM technologies, such as CMT, is open and attainable through future research.

Author Contributions: P.R.-G.: writing—original draft, conceptualization, methodology, investigation, and formal analysis. E.G.: review and editing, conceptualization, methodology, investigation, formal analysis, and resources. E.M.R.-N.: review and editing, conceptualization, methodology, investigation, formal analysis, and resources. All authors have read and agreed to the published version of the manuscript.

Funding: The authors would like to thank the funding provided for this research by the Regional Government of Madrid (Dra. Gral. Universidades e Investigación) through the project S2018/NMT4411 (ADITIMAT-CM).

Institutional Review Board Statement: Not applicable.

Informed Consent Statement: Not applicable.

Data Availability Statement: Not applicable.

References

1. *ASTM F2792-1*; Standard Terminology for Additive Manufacturing Technologies. ASTM International: West Conshohocken, PA, USA, 2012.
2. Guo, N.; Leu, M.C. Additive manufacturing: Technology, applications and research needs. *Front. Mech. Eng.* **2013**, *8*, 215–243. [CrossRef]
3. Derekar, K.S. A review of wire arc additive manufacturing and advances in wire arc additive manufacturing of aluminium. *Mater. Sci. Technol.* **2018**, *34*, 895–916. [CrossRef]
4. Altıparmak, S.; Yardley, V.; Shi, Z.; Lin, J. Challenges in Additive Manufacturing of High-Strength Aluminium Alloys and Current Developments in Hybrid Additive Manufacturing. *Int. J. Light. Mater. Manuf.* **2020**, *4*, 246–261. [CrossRef]
5. Zhang, X.; Liang, E. Metal additive manufacturing in aircraft: Current application, opportunities and challenges. *IOP Conf. Series Mater. Sci. Eng.* **2019**, *493*, 012032. [CrossRef]
6. Boţilă, L. Considerations regarding aluminum alloys used in the aeronautic/aerospace industry and use of wire arc additive manufacturing WAAM for their industrial applications. *Weld. Mater. Test.* **2020**, *4*, 1–16.
7. Ding, D.; Pan, Z.; Cuiuri, D.; Li, H. Wire-feed additive manufacturing of metal components: Technologies, developments and future interests. *Int. J. Adv. Manuf. Technol.* **2015**, *81*, 465–481. [CrossRef]
8. Ramosena, L.A.; Dzogbewu, T.C.; Preez, W. Direct Metal Laser Sintering of the Ti6Al4V Alloy from a Powder Blend. *Materials* **2022**, *15*, 8193. [CrossRef]
9. Kabayama, L.K.; Taguchi, S.P.; Martínez, G.A.S. The influence of die geometry on stress distribution by experimental and FEM simulation on electrolytic copper wiredrawing. *Mater. Res.* **2009**, *12*, 281–285. [CrossRef]
10. Ikumapayi, O.M.; Ojolo, S.J.; Afolalu, S.A. Experimental and Theoretical Investigation of Tensile Stress Distribution During Aluminium Wire Drawing. *Eur. Sci. J.* **2015**, *11*, 86–102. [CrossRef]
11. Tang, W.; Reynolds, A.P. Production of wire via friction extrusion of aluminum alloy machining chips. *J. Mater. Process. Technol.* **2010**, *210*, 2231–2237. [CrossRef]

12. Li, X. Study of Friction Extrusion and Consolidation. Ph.D. Thesis, University of South Carolina, Columbia, SC, USA, 2016.

13. Taleghani, M.A.J.; Navas, E.M.R.; Torralba, J.M. Microstructural and mechanical characterisation of 7075 aluminium alloy consolidated from a premixed powder by cold compaction and hot extrusion. *Mater. Des.* **2014**, *55*, 674–682. [CrossRef]

14. Rodríguez-González, P.; Ruiz-Navas, E.M.; Gordo, E. Effect of heat treatment prior to direct hot-extrusion processing of Al–Cu–Li alloy. *Metals* **2022**, *12*, 1046. [CrossRef]

15. BThomas, M.; Derguti, F.; Jackson, M. Continuous extrusion of a commercially pure titanium powder via the Conform process. *Mater. Sci. Technol.* **2016**, *33*, 899–903. [CrossRef]

16. Katsas, S.; Dashwood, R.; Todd, G.; Jackson, M.; Grimes, R. Characterisation of Conform^{TM} and conventionally extruded Al–4Mg–1Zr. Effect of extrusion route on superplasticity. *J. Mater. Sci.* **2010**, *45*, 4188–4195. [CrossRef]

17. Pardoe, J.A. Conform continuous extrusion of metal powders into products for electrical industry: Development experience. *Powder Met.* **1979**, *22*, 22–28. [CrossRef]

18. Gu, J.L.; Ding, J.L.; Cong, B.Q.; Bai, J.; Gu, H.M.; Williams, S.W.; Zhai, Y.C. The Influence of Wire Properties on the Quality and Performance of Wire+Arc Additive Manufactured Aluminium Parts. *Adv. Mater. Res.* **2014**, *1081*, 210–214. [CrossRef]

19. Li, J.L.Z.; Alkahari, M.R.; Rosli, N.A.B.; Hasan, R.; Sudin, M.N.; bin Ramli, F.R. Review of Wire Arc Additive Manufacturing for 3D Metal Printing. *Int. J. Autom. Technol.* **2019**, *13*, 346–353. [CrossRef]

20. Rodrigues, T.A.; Duarte, V.; Miranda, R.M.; Santos, T.G.; Oliveira, J.P. Current Status and Perspectives on Wire and Arc Additive Manufacturing (WAAM). *Materials* **2019**, *12*, 1121. [CrossRef]

21. *US423647A*; Method of Making Decorative Articles. Baker Ralph: Perth, Australia, 1925.

22. Baker. *The Use of an Electric Arc as a Heat Source to Generate 3D Objects Depositing Molten Metal in Superimposed Layers*; Baker: Chicago, IL, USA, 1926.

23. *US1533239A*; Ornamental Arc Welding. CBS Corp.: New York, NY, USA, 1925.

24. *US44142A*; Electric Welding. Union Carbide Corp.: Houston, TX, USA, 1935.

25. Carpenter, O.; Kerr, H. Method and apparatus for metal coating metal pipes by electric fusion. United States Patent US2427350A, 1943.

26. *US2504868A*; Electric Arc Welding. Airco Inc.: Punta Gorda, FL, USA, 1950. [CrossRef]

27. Kussmaul, K.; Schoch, F.W.; Luckow, H. High-quality large components shape welded by a SAW process. *Weld. J.* **1983**, *62*, 17–24.

28. Yan, L. Wire and Arc Additive Manufacture (WAAM) Reusable Tooling Investigation. Ph.D. Thesis, Cranfield University, Bedford, UK, 2012.

29. *US4952769A*; Automatic Welding Apparatus for Weld Build-Up and Method of Achieving Weld Build-Up. Bow Mills Bank and Trust: South Street Bow, NH, USA, 1990.

30. Gaddes, J.S. *Parametric Development of Wire 3D Printing*; Auburn University: Auburn, AL, USA, 2015.

31. Dickens, P.M.; Pridham, M.S.; Cobb, R.C.; Gibson, I.; Dixon, G. Rapid Prototyping Using 3-D Welding. In *Annual International Solid Freeform Fabrication Symposium*; The University of Texas at Austin: Austin, TX, USA, 1992.

32. Spencer, J.D.; Dickens, P.M.; Wykes, C.M. Rapid Prototyping of Metal Parts by Three-Dimensional Welding. *Proc. Inst. Mech. Eng. B J. Eng. Manuf.* **1998**, *212*, 175–182. [CrossRef]

33. Wang, H.F.; Zhang, Y.L. CAD/CAM integrated system in collaborative development environment. *Robot. Comput. Integr. Manuf.* **2002**, *18*, 135–145. [CrossRef]

34. Song, Y.-A.; Park, S.; Choi, D.; Jee, H. 3D welding and milling: Part I–a direct approach for freeform fabrication of metallic prototypes. *Int. J. Mach. Tools Manuf.* **2005**, *45*, 1057–1062. [CrossRef]

35. Almeida, P.; Williams, S. Innovative process model of Ti–6Al–4V additive layer manufacturing using cold metal transfer (CMT). In *21st Annual International Solid Freeform Fabrication Symposium—An Additive Manufacturing Conference, SFF*; University of Texas at Austin: Austin, TX, USA, 2010.

36. Gornyakov, V.; Sun, Y.; Ding, J.; Williams, S. Modelling and optimising hybrid process of wire arc additive manufacturing and high-pressure rolling. *Mater. Des.* **2022**, *223*, 111121. [CrossRef]

37. Chen, X.; Wang, C.; Ding, J.; Bridgeman, P.; Williams, S. A three-dimensional wire-feeding model for heat and metal transfer, fluid flow, and bead shape in wire plasma arc additive manufacturing. *J. Manuf. Process.* **2022**, *83*, 300–312. [CrossRef]

38. Eimer, E.; Williams, S.; Ding, J.; Ganguly, S.; Chehab, B. Mechanical performances of the interface between the substrate and deposited material in aluminium wire Direct Energy Deposition. *Mater. Des.* **2023**, *225*, 111594. [CrossRef]

39. Cranfield University. Is This the Largest Metal 3D Part Ever Made? 2016. Available online: https://www.cranfield.ac.uk/press/news-2016/is-this-the-largest-metal--3d-part-ever-made (accessed on 28 December 2022).

40. Rioja, R.J.; Liu, J. The evolution of Al-Li base products for aerospace and space applications. *Metall. Mater. Trans. A Phys. Metall. Mater. Sci.* **2012**, *43*, 3325–3337. [CrossRef]

41. Skrabec, Q.R. *Aluminum in America: A History*; McFarland, Incorporated, Publishers: Jefferson, North Carolina, 2017.

42. Jawalkar, C.S.; Kant, S. A Review on use of Aluminium Alloys in Aircraft Components. *I-Manag. J. Mater. Sci.* **2015**, *3*, 33–38.

43. El-Aty, A.A.; Xu, Y.; Zhang, S.; Ma, Y.; Chen, D. Abd, Experimental investigation of tensile properties and anisotropy of 1420, 8090 and 2060 Al-Li alloys sheet undergoing different strain rates and fibre orientation: A comparative study. *Procedia Eng.* **2017**, *207*, 13–18. [CrossRef]

44. Alloys, W.A. *International Alloy Designations and Chemical Composition Limits for Wrought Aluminum and Wrought Aluminum Alloys Use of the Information*; The Aluminum Association, Inc.: Arlington County, VA, USA, 2018. Available online: www.aluminum.org (accessed on 28 December 2022).

45. Lumley, R. *Fundamentals of Aluminium Metallurgy: Recent Advances*; Elsevier Science: Amsterdam, The Netherlands, 2018.

46. Dorin, T.; Vahid, A.; Lamb, J. Aluminium Lithium Alloys. In *Woodhead Publishing Series in Metals and Surface Engineering*; Woodhead Publishing: Sawston, UK, 2018; pp. 387–438. [CrossRef]

47. Zhang, X.; Huang, T.; Yang, W.; Xiao, R.; Liu, Z.; Li, L. Microstructure and mechanical properties of laser beam-welded AA2060 Al-Li alloy. *J. Mater. Process. Technol.* **2016**, *237*, 301–308. [CrossRef]

48. Abd El-Aty, A.; Xu, Y.; Guo, X.; Zhang, S.-H.; Ma, Y.; Chen, D. Strengthening mechanisms, deformation behavior, and anisotropic mechanical properties of Al-Li alloys: A review. *J. Adv. Res.* **2018**, *10*, 49–67. [CrossRef] [PubMed]

49. Hales, S.; Hafley, R. Aluminum-Lithium Alloy Thick Plate for Launch Vehicles. In Proceedings of the National Space and Missiles Materials Symposium, Scottsdale, AZ, USA, 27 June–2 July 2010.

50. Niedzinski, M. The evolution of constellium Al-Li alloys for space launch and crew module applications. *Light Metal Age* **2019**, *77*, 36–42.

51. Schlatter, S. *Improvements of Mechanical Properties in Aluminum-Lithium Alloys*; Sagina Valley State University: University Center, MI, USA, 2013; pp. 31–46.

52. Djukanovic, G. *Aluminium-Lithium Alloys Fight Back*; Aluminum Insider: Colmar, France, 2017; Available online: https://aluminiuminsider.com/aluminium-lithium-alloys-fight-back/ (accessed on 28 December 2022).

53. Prasad, N.E.; Gokhale, A.; Wanhill, R.J.H. *Aluminum-Lithium Alloys: Processing, Properties, and Applications*; Elsevier Science: Amsterdam, The Netherlands, 2013.

54. Molian, P.A.; Srivatsan, T.S. Laser-beam weld microstructures and properties of aluminum-lithium alloy 2090. *Mater. Lett.* **1990**, *9*, 245–251. [CrossRef]

55. Molian, P.A.; Srivatsan, T.S. Weldability of aluminium-lithium alloy 2090 using laser welding. *J. Mater. Sci.* **1990**, *25*, 3347–3358. [CrossRef]

56. Davis, J.R. *Aluminum and Aluminum Alloys*; ASM International: Almere, The Netherlands, 1993.

57. Russell, A.M.; Lee, K.L. *Structure-Property Relations in Nonferrous Metals*; John Wiley & Sons, Inc.: Hoboken, NJ, USA, 2005. [CrossRef]

58. Sanders, T.H. *Aluminum-Lithium Alloys*; Metallurgical Society of AIME: Warrendale, PA, USA, 1981; p. 388.

59. Campbell, F.C. *Manufacturing Technology for Aerospace Structural Materials*; Elsevier Science: Amsterdam, The Netherlands, 2011.

60. Prasad, N.E.; Ramachandran, T. Phase Diagrams and Phase Reactions in Al–Li Alloys. In *Aluminum-lithium Alloys*; Elsevier BV: Amsterdam, The Netherlands, 2014; pp. 61–97. [CrossRef]

61. Cai, B.; Zheng, Z.; He, D.; Li, S.; Li, H. Friction stir weld of 2060 Al-Cu-Li alloy: Microstructure and mechanical properties. *J. Alloys Compd.* **2015**, *649*, 19–27. [CrossRef]

62. Ion, J. *Laser Processing of Engineering Materials: Principles, Procedure and Industrial Application*; Elsevier Science: Amsterdam, The Netherlands, 2005.

63. Polmear, I.; Chester, R. Abnormal age hardening in an Al-Cu-Mg alloy containing silver and lithium. *Scr. Met.* **1989**, *23*, 1213–1221. [CrossRef]

64. Giummarra, C.; Thomas, B.; Rioja, R. New Aluminum alloys for aerospace applications. In Proceedings of the Light Metals Technology Conference, 2007; Volume 2007.

65. Xiao, D.; Wang, J.; Ding, D. Effect of titanium additions on mechanical properties of Al-Cu-Mg-Ag alloy. *Mater. Sci. Technol.* **2004**, *20*, 1199–1204. [CrossRef]

66. Wang, L.; Suo, Y.; Liang, Z.; Wang, D.; Wang, Q. Effect of titanium powder on microstructure and mechanical properties of wire + arc additively manufactured Al-Mg alloy. *Mater. Lett.* **2019**, *241*, 231–234. [CrossRef]

67. Abbaschian, R.; Reed-Hill, R.E. *Physical Metallurgy Principles*; Cengage Learning: Boston, MA, USA, 2008.

68. Safyari, M.; Moshtaghi, M.; Hojo, T.; Akiyama, E. Mechanisms of hydrogen embrittlement in high-strength aluminum alloys containing coherent or incoherent dispersoids. *Corros. Sci.* **2021**, *194*, 109895. [CrossRef]

69. Airbus-Group, Scalmalloy-RP. 2014. Available online: http://www.technology-licensing.com/etl/int/en/What-we-offer/Technologies-for-licensing/Metallics-and-related-manufacturing-technologies/Scalmalloy-RP.html (accessed on 28 December 2022).

70. Sales, A.; Ricketts, N.J. Effect of scandium on wire arc additive manufacturing of 5 series aluminium alloys. In *Minerals, Metals and Materials Series*; Springer: Berlin/Heidelberg, Germany, 2019. [CrossRef]

71. Mishra, R.S.; Sidhar, H. *Friction Stir Welding of 2XXX Aluminum Alloys Including Al-Li Alloys*; Elsevier Inc.: Amsterdam, The Netherlands, 2016. [CrossRef]

72. Ding, Y.; Muñiz-Lerma, J.; Trask, M.; Chou, S.; Walker, A.; Brochu, M. Microstructure and mechanical property considerations in additive manufacturing of aluminum alloys. *MRS Bull.* **2016**, *41*, 745–751. [CrossRef]

73. Olabode, M.; Kah, P.; Martikainen, J. Aluminium alloys welding processes: Challenges, joint types and process selection. *Proc. Inst. Mech. Eng. Part B J. Eng. Manuf.* **2013**, *227*, 1129–1137. [CrossRef]

74. Mathers, G. *The Welding of Aluminium and Its Alloys*; Woodhead Publishing: Sawston, UK, 2002. [CrossRef]

75. Gu, J.; Ding, J.; Williams, S.W.; Gu, H.; Ma, P.; Zhai, Y. The effect of inter-layer cold working and post-deposition heat treatment on porosity in additively manufactured aluminum alloys. *J. Mater. Process. Technol.* **2016**, *230*, 26–34. [CrossRef]
76. Langelandsvik, G.; Akselsen, O.M.; Furu, T.; Roven, H.J. Review of aluminum alloy development for wire arc additive manufacturing. *Materials* **2021**, *14*, 5370. [CrossRef] [PubMed]
77. Novikov, I. *Goryachelomkost Tsvetnykh metallov i Splavov (Hot Shortness Non-Ferrous Metals and Alloys)*; Nauka: Moscow, Russia, 1966.
78. Sigworth, G. Hot Tearing of Metals. *AFS Trans.* **1999**, *106*, 1053–1069.
79. Eskin, D.; Suyitno; Katgerman, L. Mechanical properties in the semi-solid state and hot tearing of aluminium alloys. *Prog. Mater. Sci.* **2004**, *49*, 629–711. [CrossRef]
80. Eskin, D.G.; Katgerman, L. A Quest for a New Hot Tearing Criterion. *Met. Mater. Trans. A* **2007**, *38*, 1511–1519. [CrossRef]
81. Grandfield, J.; Eskin, D.G.; Bainbridge, I. *Direct-Chill Casting of Light Alloys: Science and Technology*; Wiley: Hoboken, NJ, USA, 2013.
82. Eskin, D.G. *Physical Metallurgy of Direct Chill Casting of Aluminum Alloys*; CRC Press: Boca Raton, FL, USA, 2008.
83. Chesonis, C. *Light Metals 2019*; Springer International Publishing: Berlin/Heidelberg, Germany, 2019.
84. Katgerman, D.L.; Eskin, D.G. In search of the prediction of hot cracking in aluminium alloys. In *Hot Cracking Phenomena in Welds II*; Springer Science & Business Media: Berlin/Heidelberg, Germany, 2008. [CrossRef]
85. Böllinghaus, T.; Herold, H.; Cross, C.E.; Lippold, J.C. *Hot Cracking Phenomena in Welds II*; Springer: Berlin/Heidelberg, Germany, 2008.
86. Lundin, C.D.; Savage, W.F. The Varestraint test(Varestraint test used to evaluate base metal weldability and influence of particular welding process and welding variables on hot cracking). Weld. J. Res. Suppl. 1965; 44, pp. 433-s–442-s.
87. Lippold, J.C. *Welding Metallurgy and Weldability*; Wiley: Hoboken, NJ, USA, 2014.
88. Böllinghaus, T.; Herold, H. *Hot Cracking Phenomena in Welds*; Springer: Berlin/Heidelberg, Germany, 2005.
89. Lippold, J.C.; Lin, W. Weldability of commercial Al-Cu-Li alloys. *Mater. Sci. Forum* **1996**, *217*, 1685–1690. [CrossRef]
90. Singer AR, E.; Jennings, P.H. HotShortness of the Aluminum-Silicon Alloys of Commercial Purity. *J. Inst. Met.* **1947**, *73*, 197–212.
91. Pumhrey, W.I.; Lyons, J.V. Cracking During the Casting and Welding of the More Common Binary aluminum alloys. *J. Inst. Met.* **1948**, *74*, 439.
92. Meister, R.P.; Martin, D.C. Welding of Aluminum and Aluminum Alloys. Defense Metals Information Center, Battelle Memorial Institute: Columbus, OH, USA, 1967.
93. Kostrivas, T.; Lippold, J.C. Weldability of Li-bearing aluminum alloys. *Int. Mater. Rev.* **1999**, *44*, 217–237. [CrossRef]
94. Adebayo, A.; Mehnen, J.; Tonnellier, X. Limiting travel speed in additive layer manufacturing. In Proceedings of the ASM International Conference: Trends in Welding Research, Chicago, IL, USA, 4–8 June 2012.
95. Omiyale, B.O.; Olugbade, T.O.; Abioye, T.E.; Farayibi, P.K. Wire arc additive manufacturing of aluminium alloys for aerospace and automotive applications: A review. *Mater. Sci. Technol.* **2022**, *38*, 391–408. [CrossRef]
96. Gomes, B.F.; Morais, P.J.; Ferreira, V.; Pinto, M.; De Almeida, L.H. Wire-arc additive manufacturing of Al-Mg alloy using CMT and PMC technologies. *MATEC Web Conf.* **2018**, *233*, 00031. [CrossRef]
97. Albannai, A.I. A Brief Review on The Common Defects in Wire Arc Additive Manufacturing. *Int. J. Curr. Sci. Res. Rev.* **2022**, *05*, 4556–4576. [CrossRef]
98. Reisch, R.; Hauser, T.; Kamps, T.; Knoll, A. Robot Based Wire Arc Additive Manufacturing System with Context-Sensitive Multivariate Monitoring Framework. *Procedia Manuf.* **2020**, *51*, 732–739. [CrossRef]
99. Sanders, T.H.; Starke, E.A. (Eds.) *Aluminum-Lithium Alloys II*; The Metallurgical Society of AIME: Warrendale, PA, USA, 1984; Volume 31, p. 313.
100. Srivatsan, T.; Sudarshan, T. Welding of Lightweight Aluminum-Lithium Alloys. *Weld. Res. Suppl.* **1991**, *70*, 173–183.
101. Xiao, R.; Zhang, X. Problems and issues in laser beam welding of aluminum-lithium alloys. *J. Manuf. Process.* **2014**, *16*, 166–175. [CrossRef]
102. Bárta, J.; Simeková, B.; Marônek, M.; Dománková, M. Electron Beam Welding of 2099-T83 Aluminium-lithium Alloy Thick Plates. *MATEC Web Conf.* **2019**, *269*, 02010. [CrossRef]
103. Sarrafi, R.; Kovacevic, R. Cathodic cleaning of oxides from aluminum surface by variable-polarity arc. 2010; Volume 89, pp. 1S–10S.
104. Zhong, H.; Qi, B.; Cong, B.; Qi, Z.; Sun, H. Microstructure and Mechanical Properties of Wire + Arc Additively Manufactured 2050 Al–Li Alloy Wall Deposits. *Chin. J. Mech. Eng.* **2019**, *32*, 92. [CrossRef]
105. Wang, H.; Jiang, W.; Ouyang, J.-H.; Kovacevic, R. Rapid prototyping of 4043 Al-alloy parts by VP-GTAW. *J. Mater. Process. Technol.* **2004**, *148*, 93–102. [CrossRef]
106. Derekar, K.S.; Addison, A.; Joshi, S.S.; Zhang, X.; Lawrence, J.; Xu, L.; Melton, G.; Griffiths, D. Effect of pulsed metal inert gas (pulsed-MIG) and cold metal transfer (CMT) techniques on hydrogen dissolution in wire arc additive manufacturing (WAAM) of aluminium. *Int. J. Adv. Manuf. Technol.* **2020**, *107*, 311–331. [CrossRef]
107. Gu, J.; Cong, B.; Ding, J.; Williams, S.W.; Zhai, Y. *Wire + Arc Additive Manufacturing of Aluminium*; SFF Symposium: Austin, TX, USA, 2014; pp. 451–458.
108. Cong, B.; Ding, J.; Williams, S. Effect of arc mode in cold metal transfer process on porosity of additively manufactured Al-6.3%Cu alloy. *Int. J. Adv. Manuf. Technol.* **2014**, *76*, 1593–1606. [CrossRef]

109. Cong, B.; Ouyang, R.; Qi, B.; Ding, J. Influence of Cold Metal Transfer Process and Its Heat Input on Weld Bead Geometry and Porosity of Aluminum-Copper Alloy Welds. *Rare Met. Mater. Eng.* **2016**, *45*, 606–611. [CrossRef]
110. Zhang, C.; Li, Y.; Gao, M.; Zeng, X. Wire arc additive manufacturing of Al-6Mg alloy using variable polarity cold metal transfer arc as power source. *Mater. Sci. Eng. A* **2018**, *711*, 415–423. [CrossRef]
111. Fixter, J.; Gu, J.; Ding, J.; Williams, S.W.; Prangnell, P.B. Preliminary investigation into the suitability of 2xxx alloys for Wire-Arc Additive Manufacturing. *Mater. Sci. Forum* **2016**, *877*, 611–616. [CrossRef]
112. Fronius. Pulse Multi Control: Fronius Has Its Finger on the Pulse os Controlled and Fast Welding. Welding Processes—PMC. 2018. Available online: https://www.fronius.com/en/welding-technology/our-expertise/welding-processes/pmc (accessed on 28 December 2022).
113. DAIHEN Corporation. Synchro-Feed. 2015 OTC Daihen Inc. Synchro-feed GMAW Robotic Welding. Available online: https://www.daihen-usa.com/product/synchrofeed-technology/ (accessed on 28 December 2022).
114. Zhang, C.; Gao, M.; Zeng, X. Workpiece vibration augmented wire arc additive manufacturing of high strength aluminum alloy. *J. Mater. Process. Technol.* **2019**, *271*, 85–92. [CrossRef]

materials

Article

The Effect of a Duplex Surface Treatment on the Corrosion and Tribocorrosion Characteristics of Additively Manufactured Ti-6Al-4V

Kelsey Ann Vella [1], Joseph Buhagiar [1], Glenn Cassar [1], Martina Marie Pizzuto [1], Luana Bonnici [1], Jian Chen [2], Xiyu Zhang [2], Zhiquan Huang [2] and Ann Zammit [1,*]

[1] Department of Metallurgy and Materials Engineering, University of Malta, MSD 2080 Msida, Malta
[2] School of Materials Science and Engineering, Southeast University, Nanjing 211189, China
* Correspondence: ann.zammit@um.edu.mt

Abstract: The use of additively manufactured components specifically utilizing titanium alloys has seen rapid growth particularly in aerospace applications; however, the propensity for retained porosity, high(er) roughness finish, and detrimental tensile surface residual stresses are still a limiting factor curbing its expansion to other sectors such as maritime. The main aim of this investigation is to determine the effect of a duplex treatment, consisting of shot peening (SP) and a coating deposited by physical vapor deposition (PVD), to mitigate these issues and improve the surface characteristics of this material. In this study, the additive manufactured Ti-6Al-4V material was observed to have a tensile and yield strength comparable to its wrought counterpart. It also exhibited good impact performance undergoing mixed mode fracture. It was also observed that the SP and duplex treatments resulted in a 13% and 210% increase in hardness, respectively. Whilst the untreated and SP treated samples exhibited a similar tribocorrosion behavior, the duplex-treated sample exhibited the greatest resistance to corrosion-wear observed by the lack of damage on the surface and the diminished material loss rates. On the other hand, the surface treatments did not improve the corrosion performance of the Ti-6Al-4V substrate.

Keywords: surface engineering; selective laser melting; tribocorrosion; corrosion; Ti6Al4V

Citation: Vella, K.A.; Buhagiar, J.; Cassar, G.; Pizzuto, M.M.; Bonnici, L.; Chen, J.; Zhang, X.; Huang, Z.; Zammit, A. The Effect of a Duplex Surface Treatment on the Corrosion and Tribocorrosion Characteristics of Additively Manufactured Ti-6Al-4V. *Materials* **2023**, *16*, 2098. https://doi.org/10.3390/ma16052098

Academic Editor: Costica Bejinariu

Received: 7 February 2023
Revised: 18 February 2023
Accepted: 25 February 2023
Published: 4 March 2023

1. Introduction

Ti-6Al-4V is a versatile titanium alloy with an excellent combination of material characteristics making it exceedingly useful for a large range of applications including in the marine sector. Such properties include its excellent corrosion resistance in saline environments, high specific strength, ability of providing a significant reduction in weight and its relatively maintenance free nature when compared to other materials [1,2]. Some current marine applications include its use for the production of propellers and propeller shafts and for power and transmission equipment [2,3].

Such components are traditionally manufactured through conventional subtractive methods; thus, the first main objective of this investigation was to determine whether additive manufacturing can be a suitable replacement allowing for potential reduction in lead times and associated costs. Additive manufacturing allows the rapid near net shape production of components irrespective of their complexity. Components produced through such means require no or limited post processing and result in the production of minimal material waste [4,5].

However, additive manufacturing also presents its own limitations. Improper control over printing parameters can result in highly porous components and an associated poor surface finish. Furthermore, the high temperatures required to melt and fuse the metallic powder result in tensile residual stresses [6]. All of which can be very detrimental to the mechanical, corrosion, anti-fouling, and wear characteristics of the material under typical marine conditions. Thus, to combat this limitation—as well as that of the material itself, which is its poor behavior under friction conditions—a duplex surface treatment is proposed.

This treatment comprises of the dual application of shot peening and a coating deposited via physical vapor deposition. In shot peening, a stream of shots impinges the surface at room temperature, making it a cold work process. This impingement causes work hardening at the surface and induces compressive residual stresses, which are beneficial as they mitigate any crack initiation and propagation [7]. The shot peening treatment is commonly applied to primarily enhance fatigue resistance. However, it can also improve the wear resistance due to the surface hardening it imparts and the induced dimples which may act as lubrication pockets [8].

One major characteristic of the SP process is its effect on the surface roughness of the material which tends to increase due to the impingement of the shots. This is not beneficial for both the material's corrosion and tribocorrosion performance. Present investigations provide varying results with respect to the effect of the SP treatment on the corrosion performance. Studies have shown that it tends to enhance the corrosion resistance as the treatment results in grain refinement, thus increasing the grain boundary density which promotes the formation of the passive film [9]. Zhang et al. [9] investigated the effect that SP and ultrasonic shot peening treatments have on the corrosion resistance of SLM Ti64 in 3.5 wt % NaCl. The authors observed that both treatments enhanced the corrosion resistance however, in this regard the SP treatment was not as effective as the ultrasonic variety. This was attributed to the rougher surface generated by the conventional-SP treatment. A decrease in corrosion resistance due to the induced roughness was also observed by Zhan et al. [10] when investigating the effect different SP processes have on the corrosion resistance of S30432 steel in 3.5 wt % NaCl. The authors concluded that a single SP treatment was detrimental to the corrosion resistance in comparison with dual and triple SP treatments which cause limited roughening. Such a finding also suggests that the treatment parameters play an important role [11].

Currently, a limited number of investigations into the use of SP to enhance the tribocorrosion resistance of Ti-6Al-4V are available. Tribocorrosion is a mechanism whereby degradation occurs through corrosion in conjunction with wear. With respect to wear resistance, mixed behaviors have been observed when peened Ti64 was subject to linear reciprocating ball-on-flat testing in air. Whilst Tsuji et al. [12] observed a decrease in material loss for the shot-peened sample due to the surface hardening induced, Bansal et al. [13] concluded that the roughness induced was detrimental as it produced high coefficient of friction values. When investigating the tribocorrosion behavior of SP AISI 4140 low-alloy steel, Bozkurt et al. [11] concluded that the SP treatment resulted in a decrease in material loss rates with the maximum resistance being provided by the sample shot peened at the highest intensity equivalent to 24A. This was attributed to the increase in sub-grains and surface hardness induced by the treatment. However, the authors also noted that if the optimal Almen intensity is surpassed, then the resistance decreases due to increasing both the surface energy and roughness.

Secondly, PVD coatings are commonly applied with the aim of enhancing both wear and corrosion resistance [14]. Coatings primarily increase the corrosion resistance of the substrate as they eliminate contact between the substrate and the corrosive medium [15]. Such coatings typically have high hardness values and are therefore not worn easily [15,16]. In turn, PVD-coated substrates typically exhibit an excellent corrosion and tribocorrosion behavior as observed by various investigations [17,18]. Furthermore, in this investigation, the coating applied has a multilayer construction. This is beneficial as compared to monolayer coatings, such coatings result in a dense structure with interfacial strengthening and have enhanced load carrying capabilities as the different layers aid to hinder the movement of dislocations [17]. Moreover, the various layers also prevent the propagation of the corrosive medium through any coating defects present in the substrate, leading to an overall good corrosion resistance [17,19,20]. This was confirmed by Çomaklı et al. [17] when investigating the corrosion-wear behavior of TiAlN/TiN multilayer and TiN and TiAlN monolayer coatings deposited onto Ti45Nb. The multilayer coated substrate obtained the lowest wear rate and coefficient of friction and exhibited the greatest resistance to corrosion.

The effect that the coating has on tribological characteristics is also dependent on the coating's adhesion to the substrate. The extent of adhesion determines the coating's performance as it determines the load transferring abilities of the coating–substrate system. Poor adhesion causes the coating to easily flake off impacting negatively its function [21]. In the case of duplex treatment, shot peening induces a certain amount of surface roughness which may affect the coating adhesion. The high surface roughness increases the real contact area and thus, may encourage mechanical locking between the surface and the coating. However, the amount of roughness requires careful control as above a critical amount, this effect is reversed. Another reason for enhanced adhesion is due to the work hardening which increases the substrate's resistance against deformation. Thus, the coating is better supported and its failure is delayed due to low shear stresses [22,23]. Zhang et al. [24] observed a decrease of nearly 50%, in the specific wear rate of duplex-treated, via high-energy shot-peening treatment and TiN coating, industrial pure titanium compared to the TiN coated-only sample. The duplex-treated sample also exhibited a lower coefficient of friction compared to the coated-only sample.

Whilst the corrosion and corrosion-wear behavior of additively manufactured Ti-6Al-4V have been actively studied, presently the application of a duplex treatment to improve the corrosion and tribocorrosion performance of the additively manufactured substrate has not. The proposed duplex treatment is quite novel and the interaction of the peened surface and the PVD coating is not yet extensively studied especially with respect to the effect that such a combination has on the properties and characteristics of printed metallic materials. Thus, this investigation aims to shed light on the effect of this combination of treatments on the corrosion and tribocorrosion characteristics of additively manufactured Ti64 via testing in an artificial sea water solution. The results obtained provide an indication regarding the suitability of the proposed treatment applied to AM metallic components, allowing the faster production and replacement of parts employed in marine applications.

2. Materials and Methods

2.1. Substrate Material and Sample Preparation

The cylindrical 20 mm diameter, 6 mm thick Ti-6Al-4V samples, having the composition listed in Table 1, were manufactured via selective laser melting (SLM) of a Ti-6Al-4V powder using an AmPro Innovations SP100 Metal 3D Printer (Suzhou, China) as per parameters listed in Table 2. The powder, supplied by Avimetal Powder Metallurgy Technology Co. Ltd. (Beijing, China), had a particle size ranging between 15 and 35 μm and composition listed in Table 1. The printed samples were subsequently heat treated using a TAV Dualjet TPH-200 (Lombardia, Italy) furnace, in a nitrogen atmosphere at 800 °C for 2 h followed by furnace cooling. The aim of this heat treatment is to transform the hard and brittle acicular α' martensite phase formed upon printing and relieve induced thermal stresses. The 20 mm × 20 mm × 3 mm wrought mill-annealed Ti-6Al-4V samples, having the composition listed in Table 1, were supplied by Daido Steel Co. (Nagoya, Japan). The wrought and printed samples were ground and polished to a mean surface roughness, R_a, of 0.006 μm and 0.049 μm, respectively.

Table 1. Chemical composition of 3D printed and wrought Ti-6Al-4V samples.

	Ti	Al	V	Fe	O	N
Ti-6Al-4V powder (wt %)	Bal.	6.13	4.00	0.16	0.09	0.01
3D Printed (wt %)	Bal.	7.24	2.89	/	/	/
Wrought (wt %)	Bal.	6.03	4.19	0.20	/	/

Table 2. 3D printing parameters.

Parameter	Value
Laser Power	100 W
Laser Speed	600 mms^{-1}
Beam Offset	10 μm
Layer Thickness	30 μm
Hatch Distance	70 μm
Internal Overlap	≥100 μm
Hatch Type	Stripe
Stripe Width	10 mm
Rotate Each Layer	Start Angle: 45°, Angle Change: 66.66°

2.2. Surface Treatments

The shot-peening treatment was carried out via an CBI Equipment Ltd. AB850 air blasting machine (Bournemouth, UK) at an Almen intensity of 0.20 mmA, 100% shot flow, 7 bar nozzle pressure using a nozzle of 80 mm length with 6 mm diameter and a nozzle-to-specimen distance of 100 mm. Zirshot Z300 ceramic shots were used having a diameter ranging from 300 to 435 μm.

Coating deposition on the printed and SP-treated substrates was carried out using a Teer UDP800 (Beijing, China) closed field unbalanced magnetron sputtering ion plating system. Before deposition, any oxide layers were removed by subjecting the surface to sputter cleaning via high energy bombardment at 600 V and 0.5 A for 10 min followed by 10 min of cooling. The coating deposited has a multilayer structure composed of the following layers: Ti, TiN, TiAlN, and TiAlCuN. The coating was deposited at a bias of −90 V, target currents of 1 A for Cu and 8 A for Ti and Al, 35% optical emission monitor voltage, deposition pressure of 0.23 Pa and a target to sample distance of 145 mm. The designations used for the various sample conditions studied are explained in Table 3.

Table 3. Designations used for the various sample conditions.

Designation	Sample Condition
Wrought (W)	Conventionally manufactured mill-annealed Ti-6Al-4V
Printed (P)	SLM + heat-treated + polished Ti-6Al-4V
Shot-peened (SP)	SLM + heat-treated + polished + shot-peened Ti-6Al-4V
Duplex-treated (DU)	SLM + heat-treated + polished + shot-peened + coated Ti-6Al-4V

2.3. Surface and Near-Surface Characterization

Micrographic analysis of the surface and cross-section of the different sample conditions was carried out using a Carl Zeiss Axioscope 5 optical microscope, for low magnification analysis and a Carl Zeiss Merlin Gemini (Oberkochen, Germany) scanning electron microscope (SEM), for higher magnification analysis. Prior to imaging work, the substrates were etched using Kroll's reagent composed of 5% HF, 13.5% HNO$_3$ and 81.5% H$_2$O. For chemical composition analysis, Ametek EDAX (Mahwah, NJ, USA) energy dispersive spectroscopy (EDS) analyzer was used in conjunction with the SEM.

Surface roughness measurements were carried out via an AEP Technology NanoMap-500LS (Santa Clara, USA) contact profilometer. Scans were performed over 2500 μm at 25 μms^{-1} and a lateral resolution of 1 μm. Five scans were performed for each sample condition.

Using a Mitutoyo MVK-H2 (Kawasaki, Japan) microhardness tester in conjunction with a pyramidal diamond indenter, Vickers microhardness measurements were obtained. A load of 100 gf was applied for 10 s. A series of five indentations were carried out.

2.4. Mechanical Testing

Tensile specimens having the geometry and dimensions as observed in Figure 1a, were tested using an Instron 5982 (Norwood, MA, USA) equipped with an Instron 2620-604 dynamic extensometer (USA). A strain rate of 3 mm min^{-1} was applied, and testing was carried out at a temperature of 25 °C and a humidity of 62%. The specimens were manufactured and tested as per ASTM E8/E8M-16—Standard Test Methods for Tension Testing of Metallic Materials. Charpy impact specimens with the geometry and dimensions shown Figure 1b were tested via an Instron 450MPX-J2 (USA) motorized pendulum impact testing system. The specimens were manufactured and tested as per ASTM E23-16—Standard Test Methods for Notched Bar Impact Testing of Metallic Materials.

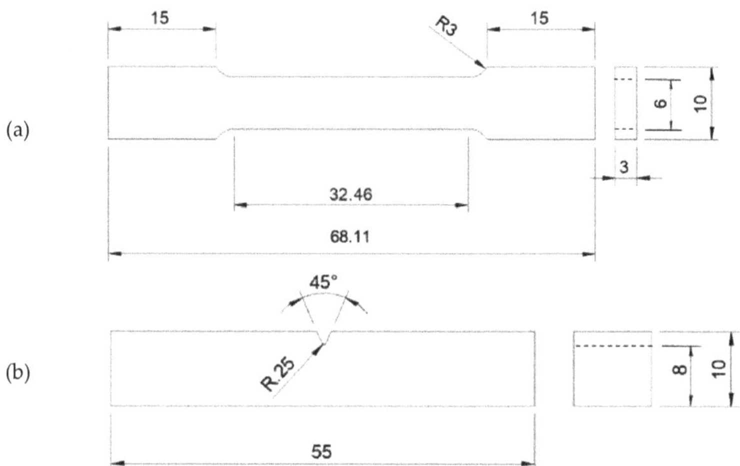

Figure 1. Dimensions of (**a**) tensile and (**b**) impact specimen. All units are in mm.

2.5. Corrosion Testing

Corrosion testing was carried out using a Gamry Interface 1000™ (Philadelphia, PA, USA) potentiostat connected to a 3-electrode setup. The working, counter and reference electrodes were the titanium-based test coupon, a platinum coated rod and a saturated calomel electrode (SCE) respectively. A surface area of approximately 0.785 cm^2 was exposed to 300 mL of artificial seawater, formulated in accordance with ASTM D1141-98 (2021)—Standard Practice for the Preparation of Substitute Ocean Water. The solution was kept at a temperature of 25.0 ± 0.2 °C to simulate the marine environment as much as possible. Table 4 provides the chemical composition of the substitute ocean water. The stock solution was diluted in 300 mL deionized water in addition to 1.23 g of Na$_2$SO$_4$ and 7.36 g of NaCl.

Initially, the open circuit potential (OCP) was monitored for 2 h to allow for its stabilization. This was followed by potentiodynamic polarization sweeps from a range of −0.2 mV versus OCP to +1.5 V versus reference at a sweep rate of 0.1667 mVs^{-1}. The test was repeated three times for each of the sample conditions to ensure repeatability. To account for the effect of the increased surface roughness, the actual area of the surface-treated samples was calculated using the developed interfacial area ratio, S$_{dr}$, obtained from height data obtained from profilometry. Following corrosion testing, the exposed areas were observed via optical and scanning electron microscopy.

Table 4. Chemical composition of stock solutions.

Chemical Compound	Concentration ($\mathrm{gL^{-1}}$)
Stock Solution No. 1	
$MgCl_2.6H_2O$	555.6
$CaCl_2$	57.9
$SrCl_2. 6H_2O$	2.1
Stock Solution No. 2	
KCl	69.5
$NaHCO_3$	20.1
KBr	10.0
H_3BO_3	2.7
NaF	0.3

2.6. Tribocorrosion Testing

For tribocorrosion testing, a Bruker UMT TriboLab (Billerica, MA, USA) set up with a reciprocating drive and a three-electrode tribocorrosion cell was used to analyze the corrosion-wear response of the samples. The cell was connected to a Gamry Interface 1000™ (Philadelphia, PA, USA) potentiostat to induce an anodic potential of 0.5 V with respect to the Ag/AgCl reference electrode. This value was determined from polarization curves obtained from potentiodynamic tests carried out. This potential ensures that all samples are in the passive regime when it is undergoing sliding wear.

The cell was filled with approximately 150 mL of artificial seawater formulated in accordance with ASTM D1141-98 (2021)—Standard Practice for the Preparation of Substitute Ocean Water. A 4.76 mm diameter Al_2O_3 counter-face was utilized. Alumina was chosen as it allows the sole characterization of the substrate due to its high hardness and inertness. A different ball for each test was utilized. The voltage was induced for 600 s without sliding followed by sliding for 2000 s at a load of 1 N (equivalent to a maximum calculated contact pressure of 678 MPa), frequency of 1 Hz and a stroke length of 3.5 mm. The load applied was selected after making calculations using Hertzian contact theory. This was determined by obtaining the load required to result in stress equivalent to the yield strength of untreated Ti64 (795 MPa). A value less than that acquired was chosen as the aim is to apply a contact pressure below the yield strength to avoid plastic deformation. The applied contact pressure in this investigation is approximately 15% less than the yield strength.

During sliding, the dynamic anodic current and coefficient of friction values were recorded. Once sliding stopped, the anodic potential was monitored for a further 600 s. All tests were carried out at room temperature and repeated three times.

Following testing, all wear tracks were analyzed via optical and scanning electron microscopy. Elemental analysis of any debris and artefacts present was also carried out. All wear track depths were measured via profilometry with three measurements taken for each wear track. From data collected during testing and measurements taken, the material loss rates were then quantified. The total wear and corrosion components can be quantified using the following Equations (1) and (2).

$$CW = C^* + W^* \tag{1}$$

where, CW is the total volumetric corrosion-wear rate ($\mathrm{mm^3 s^{-1}}$), W^* is the rate of loss due to the mechanical wear component ($\mathrm{mm^3 s^{-1}}$) and C^* is the rate of loss due to the corrosion component ($\mathrm{mm^3 s^{-1}}$). CW was obtained via measuring the total volume loss using the profilometer. C^* was obtained via Faraday's Law.

$$C^* = IM/nF\rho \tag{2}$$

where, I is the current and equivalent to the area under the graph divided by the sliding duration (A), M is the atomic mass ($\mathrm{gmol^{-1}}$), n is the charge no. for the oxidation reaction (obtained from the passive region of their respective Pourbaix diagrams), F is Faraday's constant (96,487 $Cmol^{-1}$), ρ is the density of the substrate (gcm^{-3}).

3. Results and Discussion

3.1. Micrographic Analysis

The wrought material has an equiaxed microstructure composed of the hexagonal α, the lighter phase, and cubic β; the darker phase as observed in Figure 2a. In comparison, the 3D printed sample following heat treatment, is composed from $\alpha + \beta$ phases forming a basket weave network in prior β columnar grains as observed in Figure 2b,c. This structure is typical for heat treatments carried out below the β transus temperature (980 °C, sub-transus), during which the initial acicular α' martensite structure coarsens and transforms into $\alpha + \beta$. The initial difference in microstructures is due to the rapid heating and cooling rates of the 3D printing process. The columnar grains are also a consequence of the process and form due to the epitaxial buildup of previously built layers which partially re-melt upon the formation of a new layer [25].

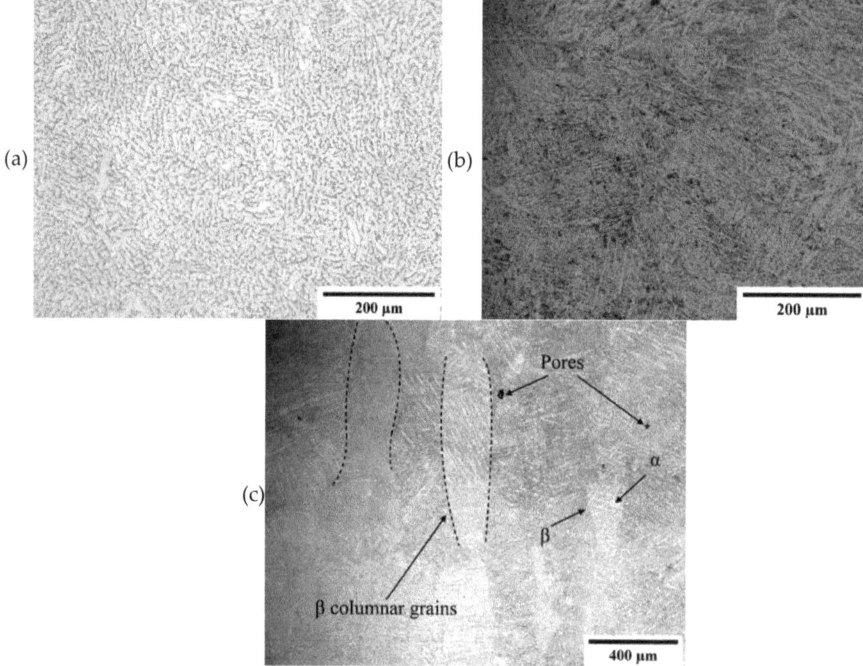

Figure 2. Low magnification micrographs of (**a**) surface of wrought Ti-6Al-4V, (**b**) surface of SLM Ti-6Al-4V, and (**c**) cross-section of SLM Ti-6Al-4V.

Following shot peening, the previously smooth surface of the printed sample loses its highly polished appearance due to the impact of the shots. On the microstructural level, the effect of grain refinement was not obvious. Moreover, the presence of twins was also not apparent which is a typical mechanism by which hcp crystals deform. Small initial grain size, peening intensity which does nott result in sufficient strain rates and high Al content which suppresses twin formation may all be contributing factors to the lack of twinning [26].

Upon the application of the coating, the underlying textured surface characteristic of SP could still be observed, no apparent levelling occurred. In addition, some artefacts typical of PVD coatings—including overgrowths, craters, and voids—were observed following coating deposition as observed in Figure 3. Overgrowths form due to the presence of surface features, such as asperities which act as nucleation sites on which a nodule composed of the coating grows. Craters form upon detachment of such overgrowths. Present voids are most likely shallow depressions or pits due to the shadowing effect, coating growth at the walls is slower compared to at the surface. Flaking observed could be attributed to the high compressive stresses induced upon deposition or due to the thermal stresses induced upon cooling [27]. Overall, the presence of defects was expected due to the rough nature of the underlying substrate preventing the formation of a perfectly uniform layer.

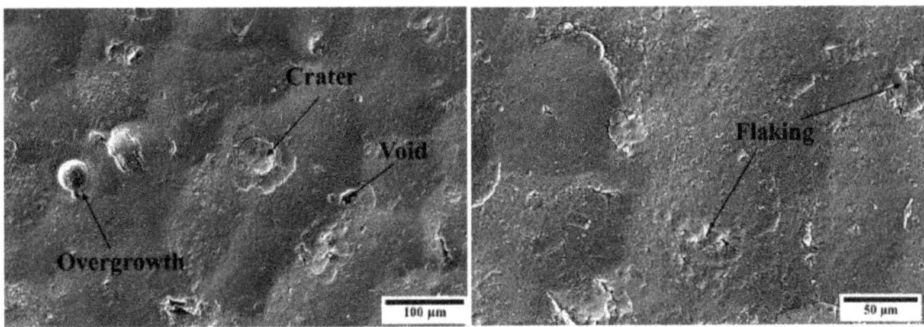

Figure 3. High-magnification micrographs of coating defects observed on the duplex-treated (DU) printed sample.

Further analysis of the coating's cross-section was carried out. The coating shown in Figure 4 was observed to have an average total thickness of 4.26 μm comprising the various layers visible. Due to the multilayer structure, the columnar growth is continuously interrupted enhancing the structural density of the coating and decreasing the grain size [17]. Such a dense coating is likely to protect the underlying substrate from corrosive media as the multiple layers hinder its propagation towards the bulk.

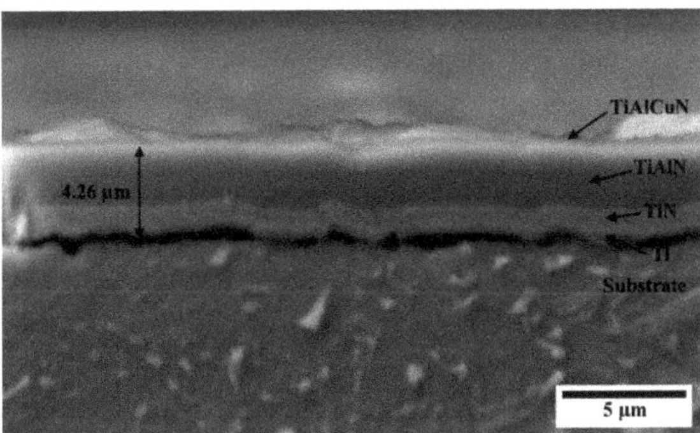

Figure 4. High-magnification micrograph of the cross-sectioned PVD coating deposited on the duplex-treated (DU) printed sample.

3.2. Surface Topography Analysis

Roughness measurements have been tabulated in Table 5. Compared to the wrought sample (W), the printed and polished (P) samples demonstrated a significantly higher R_a value due to retained porosity after polishing. However, compared to the initial as-printed condition, the polishing process resulted in a significant decrease in the surface roughness [28]. The R_a and R_z roughness values were the highest for the printed and shot-peened (SP) and duplex (DU)-treated samples. This was expected due to the roughening effect imparted by the SP treatment. The impact of shots results in sharp protrusions which in turn depend on the peening intensity. The maximum depth obtained is dependent on the shot diameter used [9]. Furthermore, the deposited coating did not enhance the surface finish. This is most likely due to the limited number of layers, thus the thicknesses are insufficient to cause smoothing or result in levelling. Jiang and Arnell [29] observed the opposite effect; however, it was observed that, due to levelling, the coating at asperities is thinner thus promoting crack initiation and increasing the susceptibility of the coating to failure via fragmentation since the layers are not evenly distributed.

Table 5. Roughness measurements obtained for all sample conditions.

Sample	Arithmetic Mean Roughness R_a (μm)	Mean Roughness Dept R_z (μm)
W	0.006 ± 0.001	0.072 ± 0.024
P	0.049 ± 0.014	1.325 ± 0.518
SP	1.470 ± 0.126	11.592 ± 0.950
DU	1.548 ± 0.294	10.781 ± 2.088

3.3. Hardness Analysis

The Vickers microhardness measurements obtained for the various samples are listed in Table 6. In literature, as-printed Ti-6Al-4V, which has not undergone any heat treatment, is reported to have a hardness as high as 409 HV due to the acicular α' martensite which is a hard and brittle phase [25,30]. This high hardness is then observed to decrease upon the application of a heat treatment depending on the heat treatment temperature [25]. In this investigation, the printed sample was heat treated and a hardness of around 330 $HV_{0.1}$ was recorded. Following SP, the hardness of the printed sample increased by 13% confirming that a degree of work hardening occurred. A similar increase in magnitude was observed by Zhang et al. [9] for SP SLM Ti64. Such hardening is due to grain refinement, which results in an increase in grain boundaries and thus a greater hindrance to dislocation movement and induced compressive residual stresses.

Table 6. Surface hardness measurements.

Sample	Hardness ($HV_{0.1}$)
W	348 ± 18
P	330 ± 3
SP	375 ± 5
DU	1024 ± 58

The effect of the treatment was further confirmed by the profile hardness measurements. The hardness values were highest at the surface and decreased gradually to the bulk hardness resulting in an affected depth of around 180 μm as observed in Figure 5. Depths of similar magnitudes were observed by other authors [9,31]. The duplex treatment provided an increase in surface hardness of 210% over the hardness of the 3D printed sample. High hardness values for such coatings are expected; moreso in this case, since the multilayer structure hinders dislocation movement, thus enhancing the load-carrying capabilities of the coating [17]. Compared to a multilayer Ti/TiN/TiAlN coating deposited onto a carbide substrate studied by Vereschaka et al. [32], where values ranging between 1764 and 2471 $HV_{0.05}$ were recorded, lower values were recorded in the current investigation. This is most likely due to the softer Ti64 substrate which contributes to this diminished value as the indentation depth is most likely greater than 10% of the coating thickness when obtaining microhardness measurements.

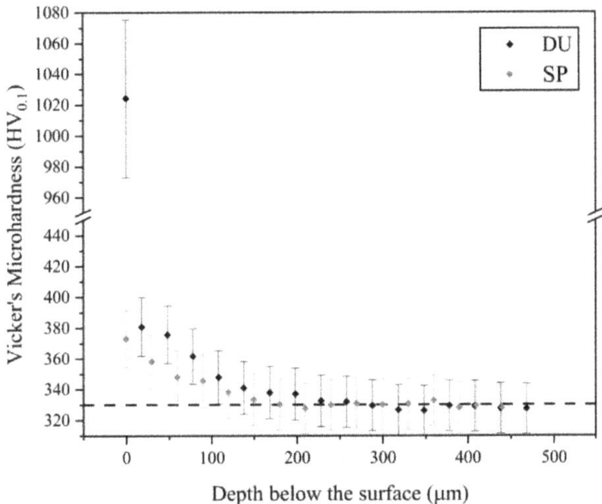

Figure 5. Plot of profile microhardness measurement for the printed and shot-peened sample (SP) and the printed and duplex-treated sample (DU).

3.4. Mechanical Testing

3.4.1. Tensile Testing

Table 7 presents the results obtained following tensile testing of the printed and heat-treated specimens, in comparison with values observed in literature for wrought and printed and heat-treated Ti64. For the latter, values listed were from investigations where a similar heat treatment as this investigation was applied.

Table 7. Results obtained from tensile testing and values quoted in literature for the mechanical properties of wrought Ti-6Al-4V.

Sample	UTS (MPa)	YS (MPa)	Elongation (%)
Printed + HT	873 ± 25	804 ± 20	6 ± 1
Printed + HT (Literature) [33]	837–1176	768–1104	5–13
Wrought Ti-6Al-4V [34]	900	830	14

As observed, values obtained are within the range of values observed in literature. However, the elongation obtained for the printed and heat-treated samples is on the lower end of the range. Typically, larger elongation values are observed following heat-treatment as the brittle martensitic phase is transformed into alpha and beta phases [35–37]. Furthermore, the 3D printed material is observed to have properties comparable to the wrought material, except for the elongation. This is a commonly observed difference, and it is likely attributed to the different microstructures obtained by the additive manufacturing and conventional manufacturing routes, where the latter includes processes such as forging, casting, and rolling [33].

In this investigation, it appears that the chosen HT temperature did not impact the ductility significantly. This may be due to defects such as micro-cracks, lack of fusion and pores which the 3D printing process is prone to, and which also affect the mechanical properties as they act as stress concentrators [38,39]. Plessis et al. [39] noted that 1% porosity had a significant effect on the mechanical properties in cases where printing parameters were not optimal resulting in lack of fusion defects which are large and irregularly shaped.

Furthermore, the build orientation also plays a significant role. In this investigation, the specimens were horizontally built, thus a relatively low elongation was expected since—during tensile testing—tension is applied perpendicular to the columnar grains. Therefore, the number of effective grain boundaries which can hinder crack growth is decreased when samples are horizontally built [35].

3.4.2. Impact Testing

Compared to typical values observed in literature, the specimens exhibited a very good impact performance since the typical impact energy values vary between 4 J and 10 J for the as-printed condition. Such low values are due to the high brittle martensitic phase fraction [40–44]. In this investigation, for printed and heat-treated Ti64, a total energy value of approximately 14 J was observed.

Compared to wrought samples, a reduced impact energy was obtained since for the former values up to 49 J are typically observed [45]. However, this value is heavily dependent on the microstructure, and for additively manufactured specimens it also depends on the residual stresses present. Lee et al. [42] noted that for Ti64 a wide variation of energy values exists, depending on the volume fraction of the phases.

Following testing, the fractured surfaces were analyzed using a scanning electron microscope to determine the type of fracture present. Figure 6 depicts high magnification micrographs obtained of the fractured surface of the printed and heat-treated specimen. Visually, the fracture appears to be brittle as there was no considerable plastic deformation. Cleavage facets can be observed which are a characteristic of brittle fracture. However, some dimples can also be observed which are a typical characteristic of ductile fracture. Thus, the fracture type is most likely mixed-fracture, as observed by Kazachenok et al. [45] for SLM Ti-6Al-4V.

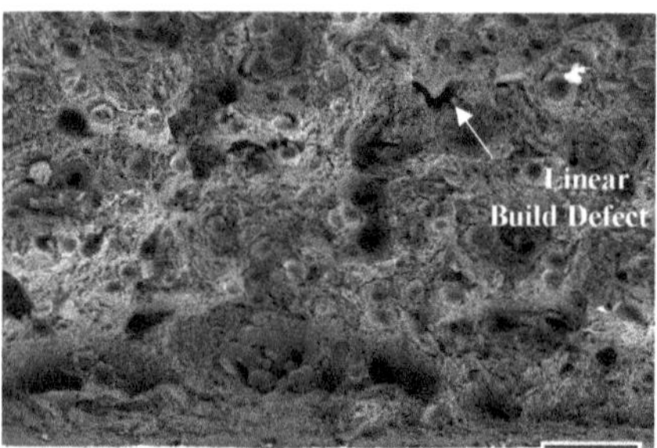

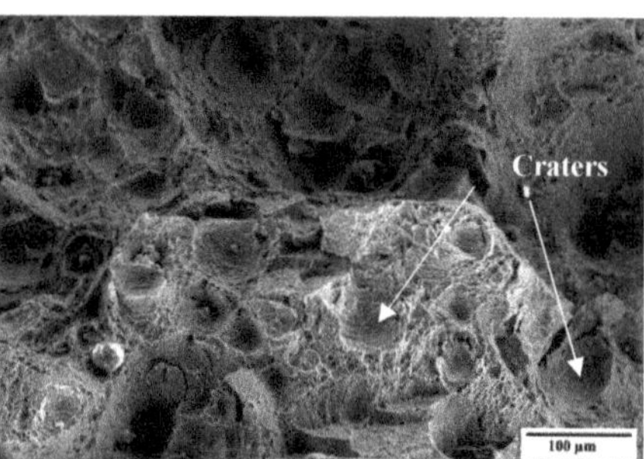

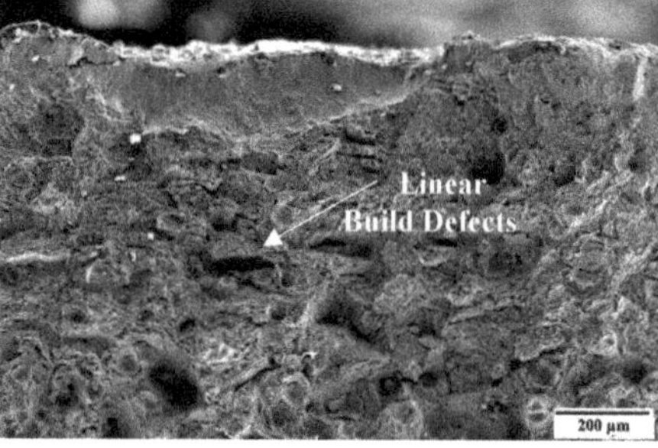

Figure 6. High magnification micrographs of the fractured heat-treated specimen.

This decrease in toughness for the printed samples compared to the wrought, may also be related to defects such as pores and presence of impurities which the printing process is prone to [40,41]. In Figure 6, the presence of pores, solid un-melted powder particles and other build defects are highlighted. Several linear building defects can be observed. Similar characteristics were observed by Wu et al. [44]. Such defects indicate that the build quality is not consistent from one layer to another. These were observed to negatively impact the load bearing capacity resulting in a decreased toughness. Furthermore, a significant number of crater-like defects or larger dimples can be observed having features of cleavage fracture inside. These appear to represent the occurrence of particle detachment or pull-out.

3.5. Corrosion Testing

Representative potentiodynamic (PD) curves obtained from the three repeats carried out for all sample conditions tested in simulated ocean water can be observed in Figure 7. For comparison bar graphs of the E_{corr} and E_{break}, i_{corr} and passive current density at 400 mV vs SCE values were plotted as observed in Figures 8–10. The plots depict the range of the three repeats.

From the PD curves and as observed from the E_{corr}, i_{corr} and passive current density at 400 mV vs SCE values (Figures 8–10), the printed and polished Ti-6Al-4V samples exhibit a slightly better corrosion performance compared to the wrought samples exhibited through the more positive E_{corr} and smaller current density values obtained. On the other hand, compared to the printed and polished sample, the shot-peening and duplex treatment—which were applied to printed Ti64 only—resulted in similar E_{corr} values; however, a greater i_{corr} than the polished sample was recorded for both conditions.

Previous investigations observed no change or a decrease in corrosion resistance following shot peening when testing in 3.5 wt % NaCl, 0.9 wt % NaCl and Ringer's solutions [9,46,47]. This lack of improvement in corrosion resistance can be correlated to the high surface roughness as it increases the interaction area [9,48]. Furthermore, as observed in Figure 7, for the shot-peened and duplex-treated samples a steep increase in current values was exhibited at around 1000 mV and 1300 mV, respectively. This value is regarded as the pitting potential, E_{break} or E_{pit} and was observed for all three repeats (Figure 8). It indicates that, at these potentials, pits are starting to form. On the other hand, the wrought and the printed sample do not exhibit this phenomenon, most likely due to having a more stable passive film which was less likely to breakdown [49]. The high surface roughness induced by the peening treatment of the shot-peened (SP), compared to that of the wrought (W) and printed and polished sample (P), is likely contributing to the formation of an nonuniform and unstable passive film with more active sites that encourage its breakdown, as evidenced by the pitting observed [9]. The breakdown of the film occurs as chloride anions adsorb at defective sites reacting with the TiO_2 passive film causing it to weaken and forms soluble high-energy coordination complexes [50].

Furthermore, shot peening induces dislocations as well as stress. Whilst some authors observed that induced dislocations promoted passive film formation [9], others observed an opposite effect. Buhagiar and Dong [51], in their investigation into the corrosion performance of low temperature carburized austenitic stainless steel in Ringer's solution, observed that the treatment applied generated a significant number of dislocations, similar to shot peening. The authors observed that such dislocations caused the passive film to rupture exposing the underlying substrate to the solution. This led to the anodic dissolution of the exposed site prior to the re-passivation of the film, resulting in crevice and pitting corrosion conditions.

To better understand why poor corrosion resistance was observed for the shot-peened (SP) and the duplex-treated (DU) samples, the samples were observed using a scanning electron microscope. As observed in Figure 11, for the wrought (W) and printed and polished (P) samples negligible corrosion damage was observed. However, as seen in Figures 11 and 12, for both the shot-peened (SP) and duplex-treated (DU) samples, what appear to be pits were observed. Craters and cracks observed on the shot-peened sample's surface are attributed to damage sustained during the application of the peening treatment. Such defects allow the solution to penetrate and become trapped, resulting in a more corrosive environment thus increasing the probability of localized corrosion to take place. For the duplex-treated sample, several defects—such as cracks and delamination—were observed. Additionally, in the test area several pits formed in the coating deposited could be seen. Thus, its poor performance similar to that of the shot-peened sample is attributed to such defects which allow the propagation of the corrosive medium towards the substrate enhancing the corrosion under the coating rendering the coating unprotective.

The formation of such defects is also linked to the high roughness of the underlying substrate. In literature, as the underlying substrate's surface roughness is increased, the corrosion protection provided by the coating is observed to decrease [52,53]. As PVD coatings are deposited, their nucleation and growth are dependent on the surface topography. A smooth surface promotes a lower defect density and improves adhesion. In comparison, a rougher surface encourages the formation of porosity and growth defects. This results in a local loss of adhesion and increases the coating's permeability, decreasing its corrosion-barrier performance [53].

The corrosion current density, i_{corr}, provides a good indication regarding the corrosion resistance of the samples. A smaller corrosion current density value indicates that the surface has a higher affinity to passivation and a lower corrosion rate [49]. From Figure 10, it was observed that at around 400 mV, where the samples are all in the passive regime, all conditions exhibit a similar corrosion behaviour, thus confirming that at this point the application of the surface treatments neither increased nor decreased the corrosion resistance of the material.

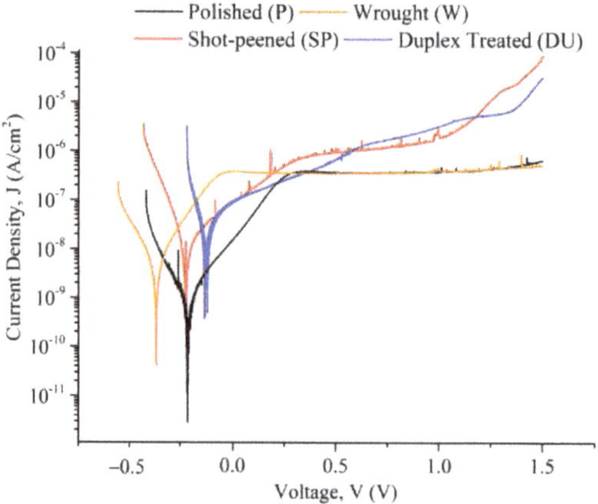

Figure 7. Potentiodynamic polarization curves obtained for each of the sample conditions tested in a simulated ocean water solution.

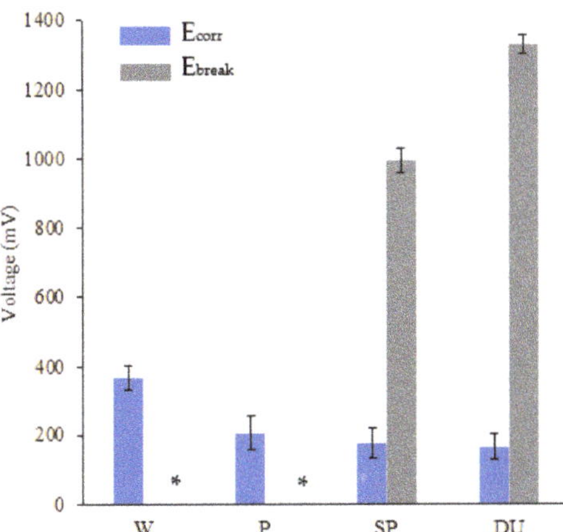

Figure 8. Plot of E_{corr} and E_{break} values for the different sample conditions. * For the wrought (W) and printed and polished (P) sample, the E_{break} value is equal to 0 as for both conditions, this phenomenon was not observed.

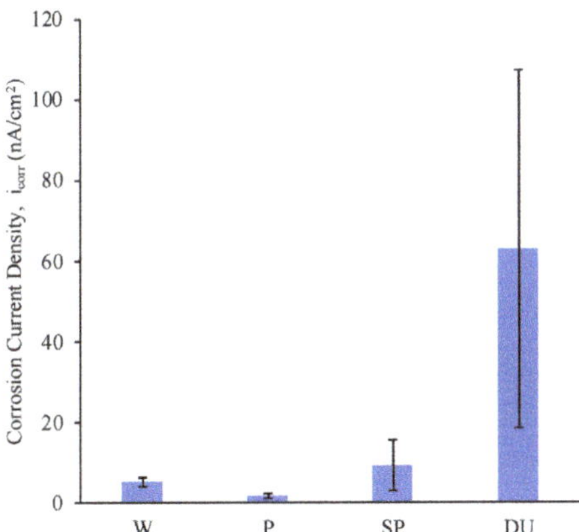

Figure 9. Plot of i_{corr} values for the different sample conditions.

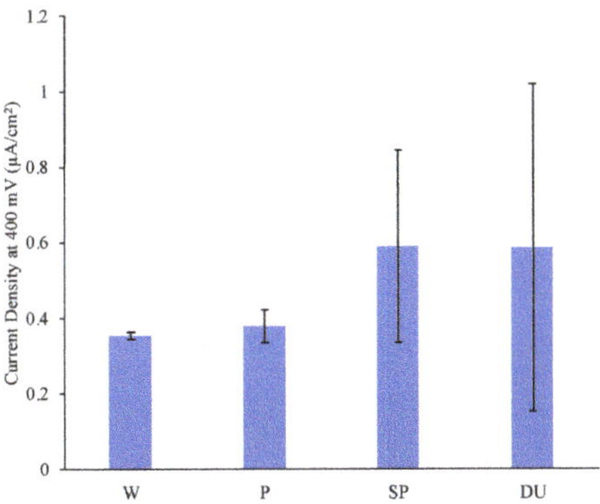

Figure 10. Plot of current density values at 400 mV for the different sample conditions.

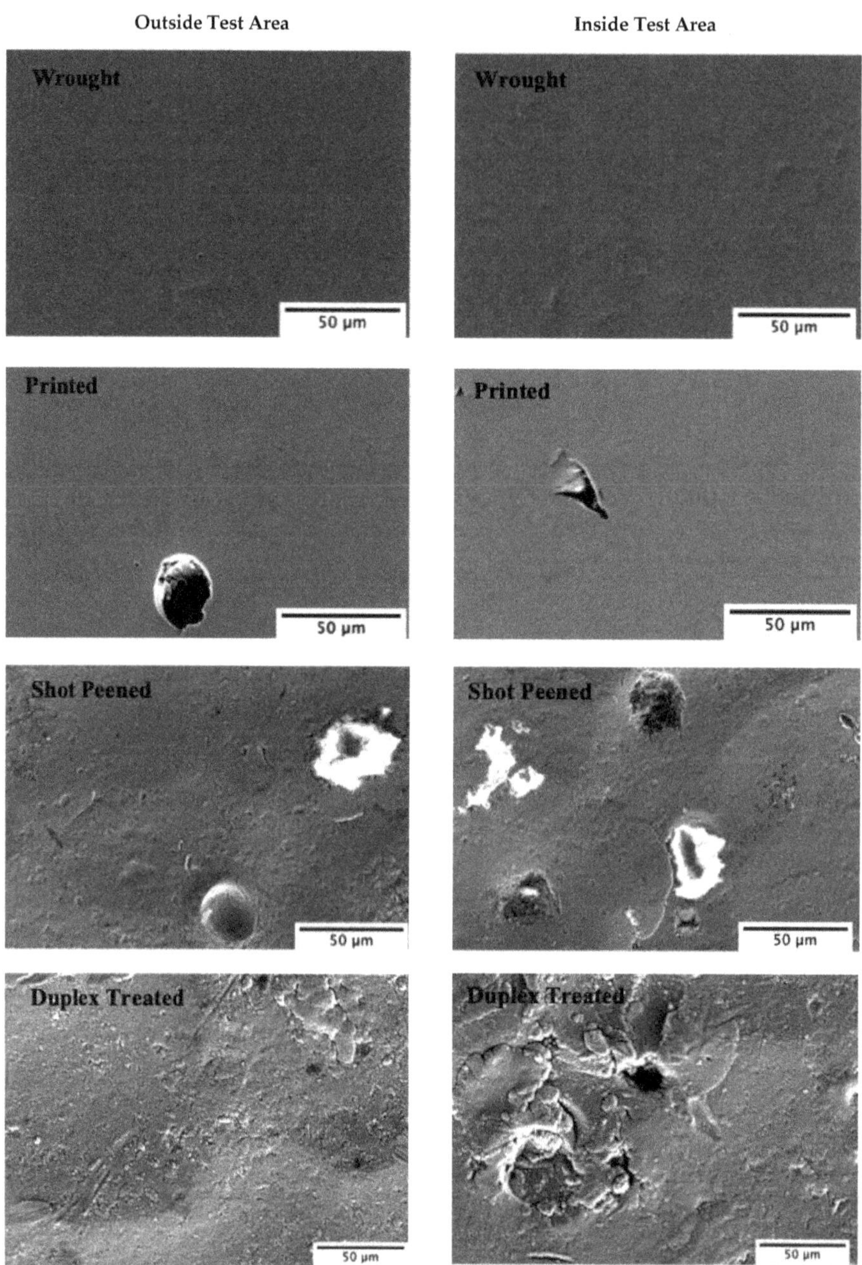

Figure 11. High-magnification images of outside and inside the corrosion test area for all sample conditions.

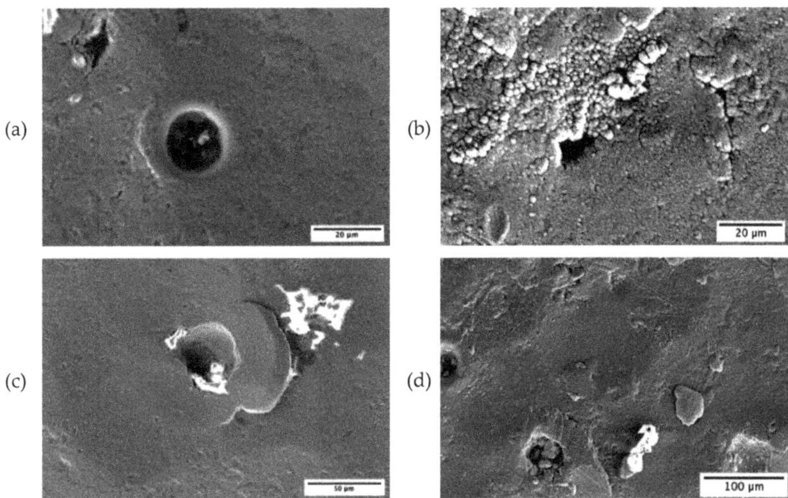

Figure 12. High magnification micrographs obtained post corrosion testing in simulated ocean water, of (**a**) a pit and (**c**) surface damage in the form of cracking and craters, observed on the shot-peened (SP) sample, (**b**) a pit and (**d**) coating defects including cracks and delamination, observed on the duplex-treated (DU) sample.

3.6. Tribocorrosion Testing

3.6.1. Dynamic Current and COF Measurements

As observed in Figure 13, for all sample conditions upon sliding, an increase in anodic current was observed which dropped back to initial values once sliding stopped. This increase is due to the removal of the passive film exposing the underlying substrate to oxidation [54,55]. The anodic current and coefficient of friction (COF) measurements recorded against time can be observed in Figure 10. The current values recorded for the polished and printed (P) sample dropped from currents ranging from 40 to 70 µA to currents ranging from 10 to 25 µA (Figure 13a). Moreover, the printed and shot-peened (SP) sample also exhibited a similar behavior to the P sample, whereby the anodic current and COF exhibited an inversely proportional relationship. The decrease in current implies a decrease in the amount of new substrate exposed to the electrolyte and it is attributed to the formation of oxidized patches which provide this twofold effect [55].

For the P sample, the COF increased from 0.30 to 0.47, similar to values observed in literature [55]. Initial low values could be due to smoothing of the sliding surfaces, values then increase due to micro-fragmentation and adhesive wear resulting in a wear track which is no longer smooth [56]. Furthermore, for the SP sample several spikes, once current values stabilized at lower values, were also observed (Figure 13b). Such spikes are related to the development of wear particles where their formation results in de-passivation and subsequent re-passivation of the substrate. Once the particle is ejected or crushed, the COF increases whilst the current decreases again [57]. This increase in COF may also be related to the formation of asperities which decrease the contact area; thus, the contact pressure at such areas increases, resulting in roughening.

For the SP samples, currents as high as 80 µA were recorded suggesting that the substrate was immediately exposed hence confirming further that the SP treatment was ineffective. The high COF values observed are typical for SP-treated Ti64 since the treatment increases the surface roughness. This increased roughness was most likely detrimental to the tribocorrosion behavior [58].

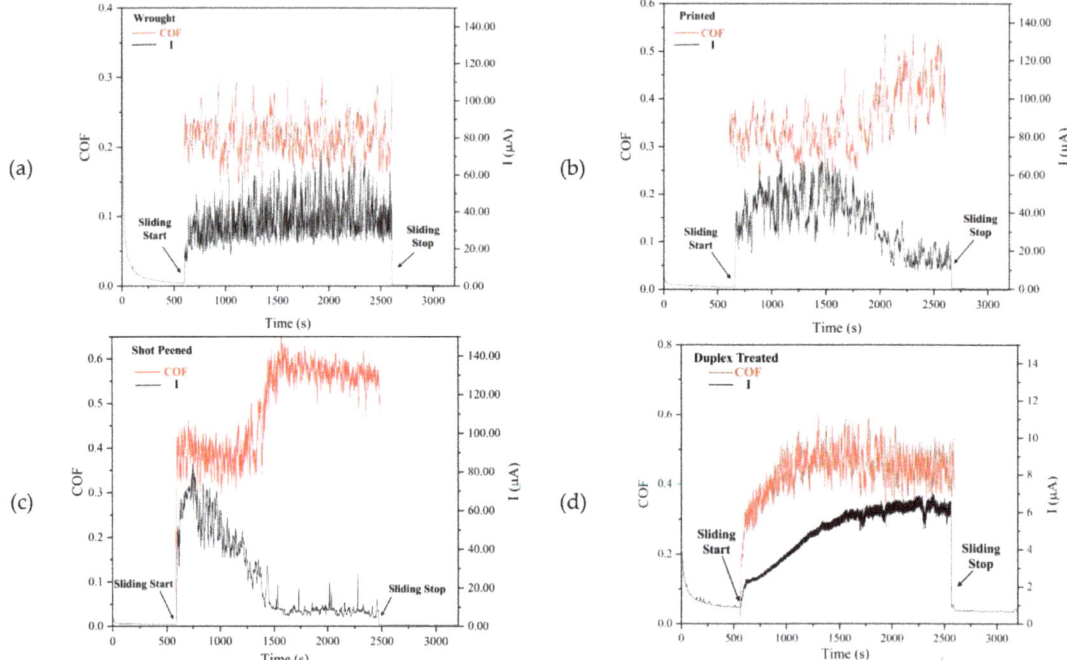

Figure 13. Dynamic anodic current and COF plots (**a**) wrought, (**b**) printed and polished, (**c**) printed + shot-peened, and (**d**) printed + duplex-treated Ti-6Al-4V. The plot in (**d**) has a current y-axis scale that is 10 times smaller in magnitude compared to the y-axis scale in (**a–c**).

For the printed and duplex-treated sample (DU) sample, low current values around 6 µA were recorded which are typical for PVD coatings subjected to such testing mostly due to their high hardness [17]. Both the COF and current are observed to increase with time (Figure 13c) as wear occurs gradually at a diminished rate until the coating undergoes complete failure. The coatings can undergo three different wear-corrosion mechanisms [59]:

- Type I: the removal or damage of the passive layer present at the coating surface and its subsequent re-passivation.
- Type II: the substrate undergoes galvanic attack causing the coating to blister or become completely removed during sliding.
- Type III: the counter-body undergoes galvanic attack causing it to roughen resulting in abrasive wear.

The DU sample underwent Type I corrosion-wear. Type I is also applicable to the uncoated substrates (W, P, and SP) and is reflected in the constant fluctuations of corrosion-current, corresponding to the continuous re-passivation and de-passivation, as observed in Figure 13a–c. Moreover, from the low currents obtained it can be noted that the coated was effective in reducing the Type I corrosion-wear significantly. Furthermore, no blistering of the coating was observed thus Type II corrosion-wear could not have occurred and Type III corrosion-wear was not possible since an alumina ball was used. Thus, no galvanic coupling could ensue. A COF value of 0.42 was obtained, a high COF value was expected due to its deposition on a previously shot-peened surface. In previous investigations, this increased roughness has been observed to be detrimental to the corrosion-wear resistance of the treated sample, negatively impacting the coating's adhesion and wear resistance [60,61]. However, despite the high COF values, very low anodic current values were recorded reflecting the effectiveness of the duplex treatment which is attributed to the multilayer structure of the coating. Such coatings result in a small increase in corrosion current over time due to: (i) their lower tendency to dissolution accredited to their high chemical stability; or (ii) the small ratio of the active to passive regions in the track [62]. A diminished active area is obtained due to the high hardness of the coating which results in a reduced wear track width.

3.6.2. Material Losses

Figure 14 provides the corrosion-wear data calculated from the anodic current and wear track volume measurements. The data confirm that the DU sample has superior corrosion-wear resistance, showing a reduction in rate of material loss by almost a factor of 6 compared to the P and SP samples. Thus, the treatment was effective mainly due to the increased hardness and corrosion resistance provided by the multilayer coating [55,60]. No significant difference between the corrosion-wear resistance of the W and P sample is present however the slight improvement, observed by the lower current values recorded, of the P sample is attributed to the oxidized patches formed, as observed via optical imaging (Figure 15), and analyzed via EDS, which decreasee the contact between the sample surface and the solution. The SP treatment also appears to be ineffective as the resistance did not improve relative to the P sample. The increased roughness rather enhances the degradation mechanisms as on average greater material loss due to mechanical wear (W*) is observed compared to the P sample.

From the material losses, the ratio of loss due to the corrosion component and loss due to the wear component (C*/W*) provides an insight in the dominant degradation mechanisms present [55,63]. The ratios obtained can be observed in Table 8. Since all values are between 0.1 and 1, this suggests that synergism is present, and the most dominant mechanism is wear-corrosion. Therefore, the greatest losses are caused by mechanical wear with limited contribution from corrosion which is also reflected in the results obtained. The former contributed to 69%, 62%, 63%, and 66% of the corrosion-wear for the W, P, SP, and DU samples respectively.

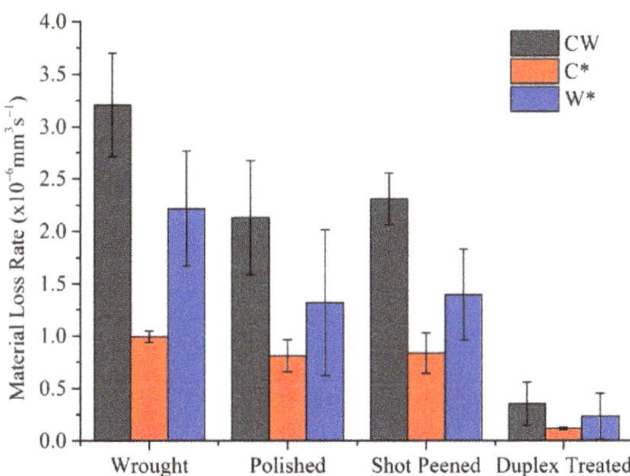

Figure 14. Plot of material losses obtained from testing the sample conditions over a sliding duration of 2000 s at a load of 1 N and frequency of 1 Hz against an alumina counter face in simulated ocean water.

The wear tracks obtained for the printed (P), printed and shot-peened (SP), and printed and duplex-treated (DU) samples are observed in Figure 15. Grooves parallel to the direction of motion are observed for all sample conditions and are evidence of abrasive wear. For the DU sample, these grooves are distinctively shallow confirming its increased wear resistance. Both the P and SP samples show signs of oxidation in the form of patches all over the wear track. These were confirmed to be a build-up of TiO_2 by EDS analysis and a result of adhesive wear. Strong adhesion is experienced between the opposing surfaces at asperities resulting in high frictional forces. As these asperities are removed, debris is generated which is then compacted resulting in patches which protect the underlying substrate. The patches may result in an increase in COF values however they also aid with load carrying due to their higher hardness. If not compacted, the debris freely moves, forming grooves and scratches by abrasion [55,60]. The occurrence of adhesive wear was also evidenced by the unstable COF and high wear rates of the P and SP samples [55].

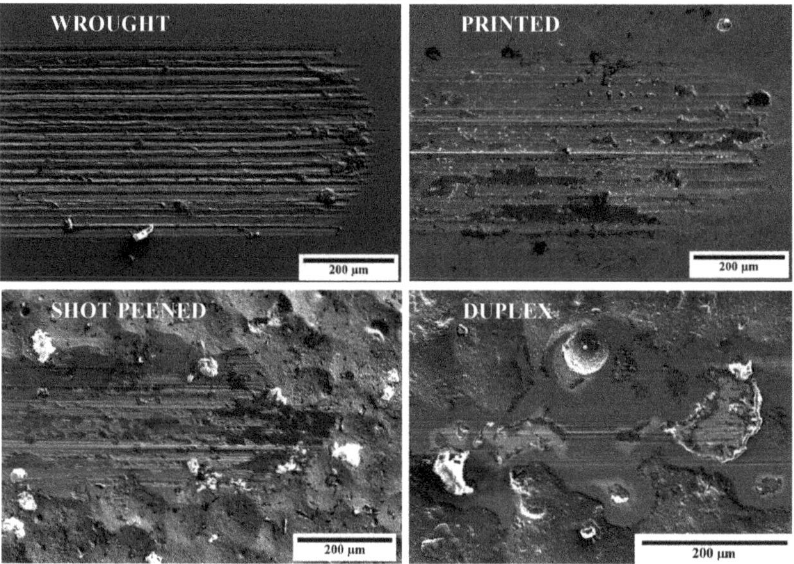

Figure 15. High-magnification micrographs of the wear tracks obtained for the various sample conditions.

When comparing the P and SP sample wear tracks, no significant difference can be observed which is opposed to that observed by Ganesh et al. [7] when subjecting wrought SP Ti64 to a pin-on-disc wear test against a hardened steel counter-body. The authors observed thicker and coarser grooves in the wear track of the treated material, which suggests that it exhibited a greater resistance to abrasive wear due to the higher hardness of the material. It appears that the limited increased hardness due to the peening treatment was not sufficient to hinder material removal as opposed to the duplex-treated sample.

For the DU-treated substrate, minimal wear was observed, and this mainly occurred at asperities as a non-uniform track was obtained. At certain areas, significant material removal occurred likely due to the high contact forces experienced at asperities. However, this was not accompanied with a significant increase in current, as the ball did not wear the coating enough such that contact was made with the exposed underlying substrate. The exposed substrate was then protected by the passive layer. Furthermore, no blistering was observed, thus confirming that the coating underwent Type I corrosion-wear. Despite the improvement, compared to other coated substrates tested under similar conditions degradation endured in the current investigation was more significant [24,55,64]. A contributing factor may be due to the high surface roughness which appears to be aggravating the wear mechanisms present locally. However, the overall reduced wear endured by the DU samples is also evidenced by the decreased wear track depths obtained as observed in Table 8.

Table 8. Wear track depths and C*/W* ratios for the different sample conditions.

Sample	Wear Track Depth (μm)	Ratio of Loss due to the Corrosion Component and Wear Component (C*/W*)
W	7.24 ± 0.54	0.45
P	5.38 ± 0.87	0.61
SP	6.42 ± 0.92	0.60
DU	2.84 ± 0.49	0.50

4. Conclusions

Mechanical, corrosion, and tribocorrosion testing were carried out on additive manufactured Ti-6Al-4V alloy which was surface treated by shot peening and a duplex treatment, consisting of shot peening and a TiAlCuN coating deposited by physical vapor deposition. The following conclusions were drawn from the results obtained:

- The shot-peened and duplex-treated samples showed an increase in the R_a and R_z roughness by a magnitude of 30 and 8 respectively, while the surface hardness increased by 13% and 210%, respectively. The hardened depth was equivalent to around 180 μm.
- The printed and heat-treated material was observed to have comparable mechanical properties to the wrought material, except for the elongation. It also exhibited a decreased impact energy absorbed compared to wrought. In both cases, this was attributed to the different microstructures resulting from the different manufacturing processes and printing process defects.
- The printed and polished sample was observed to have a better corrosion resistance than the wrought, when testing in simulated ocean water at a temperature of 25.0 ± 0.2 °C.
- Both surface treatments were ineffective in improving the corrosion performance of the printed sample. This was attributed to the high surface roughness induced by the peening process. For the shot-peened sample, it resulted in a non-uniform passive film prone to corrosion attack. For the duplex-treated sample, it increased the coating's defect density, making the substrate more susceptible to pitting and crevice corrosion.
- The corrosion-wear resistance increased in the following sequence, wrought < polished ≈ shot peened < duplex. The excellent performance of the duplex-treated sample is credited to the increased surface hardness and to the barrier protection provided by the coating which diminished Type-I corrosion-wear significantly.
- The similar behavior of the printed and shot-peened sample indicates that the peening treatment was ineffective in enhancing the corrosion-wear resistance, as it does not protect the substrate from Type-I corrosion-wear.

From the results obtained, it can be concluded that printed Ti64 can be used in place of wrought Ti64 in applications where high strength is required. With further printing and heat treatment parameter optimization, a perfect combination of mechanical and impact properties can be achieved. The proposed duplex treatment proved to be beneficial in applications where the component is under friction conditions in a corrosive environment as experienced by components used in the marine transportation industry.

Author Contributions: Conceptualization, A.Z., J.C. and J.B.; Methodology, K.A.V., M.M.P., X.Z. and Z.H.; Formal analysis, K.A.V.; Investigation, K.A.V., A.Z., J.B. and G.C.; Writing—original draft preparation, K.A.V.; Writing—review and editing, K.A.V., A.Z., J.B., G.C., J.C. and L.B.; Supervision, A.Z., J.B., G.C. and J.C.; Project administration, A.Z. and J.C.; Funding acquisition, A.Z. and J.C. All authors have read and agreed to the published version of the manuscript.

Funding: This study was part of the SEAM—Surface Engineering for Additive Manufactured parts used in marine transportation project, funded by the Malta Council for Science and Technology (MCST) as part of the Sino-Malta fund 2019-02 and the Ministry of Science and Technology (MOST) of the People's Republic of China (2019YFE0191500).

Institutional Review Board Statement: Not applicable.

Informed Consent Statement: Not applicable.

Data Availability Statement: Not applicable.

Conflicts of Interest: The authors declare no conflict of interest.

References

1. Carlton, J.S. *Marine Propellers and Propulsion*, 2nd ed.; Butterworth-Heinemann: Oxford, UK, 2007.
2. Zhu, J.; Qu, H.; Yang, J.; Zhang, Z.; Chen, T.; Lin, C.; Qi, L.; Shu, Y.; O, W.; Yang, J.; et al. The Application and Prospect of Titanium Materials in Marine Engineering Equipments. *Mater. Sci.* **2011**, *3*, 2268–2271.
3. Leyens, C.; Peters, M. (Eds.) *Titanium and Titanium Alloys: Fundamentals and Applications*, 1st ed.; Wiley: Hoboken, NJ, USA, 2003. [CrossRef]
4. Shunmugavel, M.; Polishetty, A.; Littlefair, G. Microstructure and Mechanical Properties of Wrought and Additive Manufactured Ti-6Al-4V Cylindrical Bars. *Procedia Technol.* **2015**, *20*, 231–236. [CrossRef]
5. Losertová, M.; Kubeš, V. Microstructure and mechanical properties of selective laser melted Ti6Al4V alloy. *IOP Conf. Series Mater. Sci. Eng.* **2017**, *266*, 012009. [CrossRef]
6. Vrancken, B.; Thijs, L.; Kruth, J.-P.; Van Humbeeck, J. Heat treatment of Ti6Al4V produced by Selective Laser Melting: Microstructure and mechanical properties. *J. Alloy Compd.* **2012**, *541*, 177–185. [CrossRef]
7. Ganesh, B.; Sha, W.; Ramanaiah, N.; Krishnaiah, A. Effect of shotpeening on sliding wear and tensile behavior of titanium implant alloys. *Mater. Des.* **2013**, *56*, 480–486. [CrossRef]
8. Zammit, A. Shot Peening of Austempered Ductile Iron. In *Advanced Surface Engineering Research*; Chowdhury, M.A., Ed.; IntechOpen: Rijeka, Croatia, 2018. [CrossRef]
9. Zhang, Q.; Xu, S.; Wang, J.; Zhang, X.; Wang, J.; Si, C. Microstructure change and corrosion resistance of selective laser melted Ti-6Al-4V alloy subjected to pneumatic shot peening and ultrasonic shot peening. *Surf. Topogr. Metrol. Prop.* **2022**, *10*, 015010. [CrossRef]
10. Zhan, K.; Zhang, Y.; Zhao, S.; Yang, Z.; Zhao, B.; Ji, V. Tribological Behavior and Corrosion Resistance of S30432 Steel after Different Shot Peening Processes. *J. Mater. Eng. Perform.* **2021**, *31*, 1250–1258. [CrossRef]
11. Bozkurt, Y.; Kovacı, H.; Yetim, A.; Çelik, A. Tribocorrosion properties and mechanism of a shot peened AISI 4140 low-alloy steel. *Surf. Coatings Technol.* **2022**, *440*, 128444. [CrossRef]
12. Tsuji, N.; Tanaka, S.; Takasugi, T. Effects of combined plasma-carburizing and shot-peening on fatigue and wear properties of Ti–6Al–4V alloy. *Surf. Coatings Technol.* **2009**, *203*, 1400–1405. [CrossRef]
13. Bansal, D.; Eryilmaz, O.; Blau, P. Surface engineering to improve the durability and lubricity of Ti–6Al–4V alloy. *Wear* **2011**, *271*, 2006–2015. [CrossRef]
14. Davis, J.R. *Surface Engineering for Corrosion and Wear Resistance*; ASM International: Almere, The Netherlands, 2001.
15. Oliveira, V.; Aguiar, C.; Vazquez, A.; Robin, A.; Barboza, M. Improving corrosion resistance of Ti–6Al–4V alloy through plasma-assisted PVD deposited nitride coatings. *Corros. Sci.* **2014**, *88*, 317–327. [CrossRef]
16. Ceschini, L.; Lanzoni, E.; Martini, C.; Prandstraller, D.; Sambogna, G. Comparison of dry sliding friction and wear of Ti6Al4V alloy treated by plasma electrolytic oxidation and PVD coating. *Wear* **2008**, *264*, 86–95. [CrossRef]
17. Çomaklı, O. Improved structural, mechanical, corrosion and tribocorrosion properties of Ti45Nb alloys by TiN, TiAlN monolayers, and TiAlN/TiN multilayer ceramic films. *Ceram. Int.* **2020**, *47*, 4149–4156. [CrossRef]
18. Li, H.; Chen, Z.W.; Ramezani, M. Wear behaviours of PVD-TiN coating onTi-6Al-4V alloy processed by laser powder bed fusion or conventionally processed. *Int. J. Adv. Manuf. Technol.* **2021**, *113*, 1389–1399. [CrossRef]
19. Ma, F.; Li, J.; Zeng, Z.; Gao, Y. Structural, mechanical and tribocorrosion behaviour in artificial seawater of CrN/AlN nano-multilayer coatings on F690 steel substrates. *Appl. Surf. Sci.* **2018**, *428*, 404–414. [CrossRef]
20. Wang, Y.; Zhang, J.; Zhou, S.; Wang, Y.; Wang, C.; Wang, Y.; Sui, Y.; Lan, J.; Xue, Q. Improvement in the tribocorrosion performance of CrCN coating by multilayered design for marine protective application. *Appl. Surf. Sci.* **2020**, *528*, 147061. [CrossRef]
21. Lupicka, O.; Warcholinski, B. The Adhesion of CrN Thin Films Deposited on Modified 42CrMo4 Steel. *Adv. Mater. Sci. Eng.* **2017**, *2017*, 4064208. [CrossRef]
22. Zammit, A.; Attard, M.; Subramaniyan, P.; Levin, S.; Wagner, L.; Cooper, J.; Espitalier, L.; Cassar, G. Investigations on the adhesion and fatigue characteristics of hybrid surface-treated titanium alloy. *Surf. Coatings Technol.* **2021**, *431*, 128002. [CrossRef]
23. Ravi, N.; Markandeya, R.; Joshi, S.V. Effect of substrate roughness on adhesion and tribological properties of nc-TiAlN/a-Si3N4nanocomposite coatings deposited by cathodic arc PVD process. *Surf. Eng.* **2016**, *33*, 7–19. [CrossRef]
24. Zhang, C.; Zheng, M.; Wang, Y.; Gao, P.; Gan, B. Effect of high energy shot peening on the wear resistance of TiN films on a TA2 surface. *Surf. Coat. Technol.* **2019**, *378*, 124821. [CrossRef]
25. Lekoadi, P.; Tlotleng, M.; Annan, K.; Maledi, N.; Masina, B. Evaluation of Heat Treatment Parameters on Microstructure and Hardness Properties of High-Speed Selective Laser Melted Ti6Al4V. *Metals* **2021**, *11*, 255. [CrossRef]
26. Xie, L.; Wen, Y.; Zhan, K.; Wang, L.; Jiang, C.; Ji, V. Characterization on surface mechanical properties of Ti–6Al–4V after shot peening. *J. Alloy Compd.* **2016**, *666*, 65–70. [CrossRef]

27. Panjan, P.; Čekada, M.; Panjan, M.; Kek-Merl, D. Growth defects in PVD hard coatings. *Vacuum* **2009**, *84*, 209–214. [CrossRef]
28. Wang, M.; Wu, Y.; Lu, S.; Chen, T.; Zhao, Y.; Chen, H.; Tang, Z. Fabrication and characterization of selective laser melting printed Ti–6Al–4V alloys subjected to heat treatment for customized implants design. *Prog. Nat. Sci.* **2016**, *26*, 671–677. [CrossRef]
29. Jiang, J.; Arnell, R. The effect of substrate surface roughness on the wear of DLC coatings. *Wear* **2000**, *239*, 1–9. [CrossRef]
30. Zhang, L.-C.; Attar, H. Selective Laser Melting of Titanium Alloys and Titanium Matrix Composites for Biomedical Applications: A Review. *Adv. Eng. Mater.* **2015**, *18*, 463–475. [CrossRef]
31. Yang, Q.; Zhou, W.; Niu, Z.; Zheng, X.; Wang, Q.; Fu, X.; Chen, G.; Li, Z. Effect of different surface asperities and surface hardness induced by shot-peening on the fretting wear behavior of Ti-6Al-4V. *Surf. Coatings Technol.* **2018**, *349*, 1098–1106. [CrossRef]
32. Vereschaka, A.S.; Vereschaka, A.A.; Sladkov, D.; Aksenenko, A.; Sitnikov, N. Control of Structure and Properties of Nanostructured Multilayer Composite Coatings Applied to Cutting Tools as a Way to Improve Efficiency of Technological Cutting Operations. *J. Nano Res.* **2015**, *37*, 51–57. [CrossRef]
33. Bartolomeu, F.; Gasik, M.; Silva, F.S.; Miranda, G. Mechanical Properties of Ti6Al4V Fabricated by Laser Powder Bed Fusion: A Review Focused on the Processing and Microstructural Parameters Influence on the Final Properties. *Metals* **2022**, *12*, 986. [CrossRef]
34. Callister, W.D. *Materials Science and Engineering: An Introduction*, 7th ed.; John Wiley and Sons, Inc.: New York, NY, USA, 2007.
35. Tao, P.; Li, H.-X.; Huang, B.-Y.; Hu, Q.-D.; Gong, S.-L.; Xu, Q.-Y. Tensile behavior of Ti-6Al-4V alloy fabricated by selective laser melting: Effects of microstructures and as-built surface quality. *China Foundry* **2018**, *15*, 243–252. [CrossRef]
36. Rafi, H.K.; Starr, T.L.; Stucker, B.E. A comparison of the tensile, fatigue, and fracture behavior of Ti–6Al–4V and 15-5 PH stainless steel parts made by selective laser melting. *Int. J. Adv. Manuf. Technol.* **2013**, *69*, 1299–1309. [CrossRef]
37. He, J.; Li, D.; Jiang, W.; Ke, L.; Qin, G.; Ye, Y.; Qin, Q.; Qiu, D. The Martensitic Transformation and Mechanical Properties of Ti6Al4V Prepared via Selective Laser Melting. *Materials* **2019**, *12*, 321. [CrossRef] [PubMed]
38. Li, F.; Qi, B.; Zhang, Y.; Guo, W.; Peng, P.; Zhang, H.; He, G.; Zhu, D.; Yan, J. Effects of Heat Treatments on Microstructures and Mechanical Properties of Ti6Al4V Alloy Produced by Laser Solid Forming. *Metals* **2021**, *11*, 346. [CrossRef]
39. du Plessis, A.; Yadroitsava, I.; Yadroitsev, I. Effects of defects on mechanical properties in metal additive manufacturing: A review focusing on X-ray tomography insights. *Mater. Des.* **2020**, *187*, 108385. [CrossRef]
40. Yasa, E.; Deckers, J.; Kruth, J.-P.; Rombouts, M.; Luyten, J. Charpy impact testing of metallic selective laser melting parts. *Virtual Phys. Prototyp.* **2010**, *5*, 89–98. [CrossRef]
41. Hernández, D.G. Mechanical Behaviour Assessment of the Ti6Al4V Alloy Obtained by Additive Manufacturing towards Aeronautical Industry. Master's Thesis, Instituto Superior Técnico (IST), Lisboa, Portugal, 2014; p. 10.
42. Lee, K.-A.; Kim, Y.-K.; Yu, J.-H.; Park, S.-H.; Kim, M.-C. Effect of Heat Treatment on Microstructure and Impact Toughness of Ti-6Al-4V Manufactured by Selective Laser Melting Process. *Arch. Met. Mater.* **2017**, *62*, 1341–1346. [CrossRef]
43. Muiruri, A.M.; Maringa, M.; Du Preez, W.; Masu, L. Variation of Impact Toughness of As-Built Dmls Ti6al4v (Eli) Specimens with Temperature. *S. Afr. J. Ind. Eng.* **2018**, *29*, 284–298. [CrossRef]
44. Wu, M.-W.; Lai, P.-H.; Chen, J.-K. Anisotropy in the impact toughness of selective laser melted Ti–6Al–4V alloy. *Mater. Sci. Eng. A* **2016**, *650*, 295–299. [CrossRef]
45. Kazachenok, M.; Panin, A.; Panin, S.; Vlasov, I. Impact toughness of Ti–6Al–4V parts fabricated by additive manufacturing. In Proceedings of the International Conference on Advanced Materials with Hierarchical Structure for New Technologies and Reliable Structures 2019, Tomsk, Russia, 1–5 October 2019; p. 020153. [CrossRef]
46. Bagherifard, S.; Hickey, D.J.; Fintová, S.; Pastorek, F.; Fernandez-Pariente, I.; Bandini, M.; Webster, T.J.; Guagliano, M. Effects of nanofeatures induced by severe shot peening (SSP) on mechanical, corrosion and cytocompatibility properties of magnesium alloy AZ31. *Acta Biomater.* **2017**, *66*, 93–108. [CrossRef]
47. Chen, G.; Fu, Y.; Cui, Y.; Gao, J.; Guo, X.; Gao, H.; Wu, S.; Lu, J.; Lin, Q.; Shi, S. Effect of surface mechanical attrition treatment on corrosion fatigue behavior of AZ31B magnesium alloy. *Int. J. Fatigue* **2019**, *127*, 461–469. [CrossRef]
48. Ahmed, A.A.; Mhaede, M.; Wollmann, M.; Wagner, L. Effect of micro shot peening on the mechanical properties and corrosion behavior of two microstructure Ti–6Al–4V alloy. *Appl. Surf. Sci.* **2016**, *363*, 50–58. [CrossRef]
49. Yu, Z.; Chen, Z.; Qu, D.; Qu, S.; Wang, H.; Zhao, F.; Zhang, C.; Feng, A.; Chen, D. Microstructure and Electrochemical Behavior of a 3D-Printed Ti-6Al-4V Alloy. *Materials* **2022**, *15*, 4473. [CrossRef] [PubMed]
50. Szklarska-Smialowska, Z. *Pitting and Crevice Corrosion*; NACE International: Houston, TX, USA, 2005.
51. Buhagiar, J.; Dong, H. Corrosion properties of S-phase layers formed on medical grade austenitic stainless steel. *J. Mater. Sci. Mater. Med.* **2011**, *23*, 271–281. [CrossRef] [PubMed]
52. Durst, O.; Ellermeier, J.; Berger, C. Influence of plasma-nitriding and surface roughness on the wear and corrosion resistance of thin films (PVD/PECVD). *Surf. Coatings Technol.* **2008**, *203*, 848–854. [CrossRef]
53. Daure, J.; Voisey, K.; Shipway, P.; Stewart, D. The effect of coating architecture and defects on the corrosion behaviour of a PVD multilayer Inconel 625/Cr coating. *Surf. Coatings Technol.* **2017**, *324*, 403–412. [CrossRef]
54. Vilhena, L.M.; Shumayal, A.; Ramalho, A.; Ferreira, J.A.M. Tribocorrosion Behaviour of Ti6Al4V Produced by Selective Laser Melting for Dental Implants. *Lubricants* **2020**, *8*, 22. [CrossRef]
55. Toptan, F.; Alves, A.C.; Carvalho, Ó.; Bartolomeu, F.; Pinto, A.M.; Silva, F.; Miranda, G. Corrosion and tribocorrosion behaviour of Ti6Al4V produced by selective laser melting and hot pressing in comparison with the commercial alloy. *J. Mater. Process. Technol.* **2018**, *266*, 239–245. [CrossRef]

56. Molinari, A.; Straffelini, G.; Tesi, B.; Bacci, T.; Pradelli, G. Effects of load and sliding speed on the tribological behaviour of Ti6Al4V plasma nitrided different temperatures. *Wear* **1997**, *203–204*, 447–454. [CrossRef]
57. Landolt, D. Electrochemical and materials aspects of tribocorrosion systems. *J. Phys. D Appl. Phys.* **2006**, *39*, 3121–3127. [CrossRef]
58. Hammood, A.S.; Thair, L.; Altawaly, H.D.; Parvin, N. Tribocorrosion Behaviour of Ti–6Al–4V Alloy in Biomedical Implants: Effects of Applied Load and Surface Roughness on Material Degradation. *J. Bio-Tribo-Corrosion* **2019**, *5*, 1–12. [CrossRef]
59. Mallia, B.; Dearnley, P.A. The corrosion–wear response of Cr–Ti coatings. *Wear* **2007**, *263*, 679–690. [CrossRef]
60. Pougoum, F.; Jedrzejczak, A.; Azzi, M.; Martinu, L.; Klemberg-Sapieha, J.E. Effect of interface roughness on the tribo-corrosion behavior of diamond like carbon coatings on titanium alloy. *J. Vac. Sci. Technol. A* **2022**, *40*, 033405. [CrossRef]
61. Tan, A.W.-Y.; Sun, W.; Bhowmik, A.; Lek, J.Y.; Song, X.; Zhai, W.; Zheng, H.; Li, F.; Marinescu, I.; Dong, Z.; et al. Effect of Substrate Surface Roughness on Microstructure and Mechanical Properties of Cold-Sprayed Ti6Al4V Coatings on Ti6Al4V Substrates. *J. Therm. Spray Technol.* **2019**, *28*, 1959–1973. [CrossRef]
62. Lane, I.R.; Cavallaro, J.L. Metallurgical and Mechanical Aspects of the Sea-Water Stress Corrosion of Titanium. *Appl. Relat. Phenom. Titan. Alloys* **1968**, *432*, 147–169.
63. Ruan, H.; Wang, Z.; Wang, L.; Sun, L.; Peng, H.; Ke, P.; Wang, A. Designed Ti/TiN sub-layers suppressing the crack and erosion of TiAlN coatings. *Surf. Coatings Technol.* **2022**, *438*, 128419. [CrossRef]
64. Naghibi, S.; Raeissi, K.; Fathi, M. Corrosion and tribocorrosion behavior of Ti/TiN PVD coating on 316L stainless steel substrate in Ringer's solution. *Mater. Chem. Phys.* **2014**, *148*, 614–623. [CrossRef]

Article

Nanofunctionalization of Additively Manufactured Titanium Substrates for Surface-Enhanced Raman Spectroscopy Measurements

Marcin Pisarek [1,*], Robert Ambroziak [1], Marcin Hołdyński [1], Agata Roguska [1], Anna Majchrowicz [2], Bartłomiej Wysocki [3] and Andrzej Kudelski [4]

[1] Institute of Physical Chemistry, Polish Academy of Sciences, Kasprzaka 44/52, 01-224 Warsaw, Poland; rambroziak@ichf.edu.pl (R.A.); mholdynski@ichf.edu.pl (M.H.); aroguska@ichf.edu.pl (A.R.)

[2] Faculty of Materials Science and Engineering, Warsaw University of Technology, Wołoska 141, 02-507 Warsaw, Poland; aankam@gmail.com

[3] Center of Digital Science and Technology, Cardinal Stefan Wyszynski University in Warsaw, Woycickiego 1/3, 01-938 Warsaw, Poland; b.wysocki@uksw.edu.pl

[4] Faculty of Chemistry, University of Warsaw, Pasteura 1, 02-093 Warsaw, Poland; akudel@chem.uw.edu.pl

* Correspondence: mpisarek@ichf.edu.pl

Abstract: Powder bed fusion using a laser beam (PBF-LB) is a commonly used additive manufacturing (3D printing) process for the fabrication of various parts from pure metals and their alloys. This work shows for the first time the possibility of using PBF-LB technology for the production of 3D titanium substrates (Ti 3D) for surface-enhanced Raman scattering (SERS) measurements. Thanks to the specific development of the 3D titanium surface and its nanoscale modification by the formation of TiO_2 nanotubes with a diameter of ~80 nm by the anodic oxidation process, very efficient SERS substrates were obtained after deposition of silver nanoparticles (0.02 mg/cm^2, magnetron sputtering). The average SERS enhancement factor equal to 1.26×10^6 was determined for pyridine (0.05 M + 0.1 M KCl), as a model adsorbate. The estimated enhancement factor is comparable with the data in the literature, and the substrate produced in this way is characterized by the high stability and repeatability of SERS measurements. The combination of the use of a printed metal substrate with nanofunctionalization opens a new path in the design of SERS substrates for applications in analytical chemistry. Methods such as SEM scanning microscopy, photoelectron spectroscopy (XPS) and X-ray diffraction analysis (XRD) were used to determine the morphology, structure and chemical composition of the fabricated materials.

Keywords: additive manufacturing; powder bed fusion (PBF); CP titanium; anodic oxidation; TiO_2 nanotubes; Ag nanoparticles; SERS platforms; plasmonic substrates

Citation: Pisarek, M.; Ambroziak, R.; Hołdyński, M.; Roguska, A.; Majchrowicz, A.; Wysocki, B.; Kudelski, A. Nanofunctionalization of Additively Manufactured Titanium Substrates for Surface-Enhanced Raman Spectroscopy Measurements. *Materials* **2022**, *15*, 3108. https://doi.org/10.3390/ma15093108

Academic Editor: Ana Paula Piedade

Received: 17 March 2022
Accepted: 20 April 2022
Published: 25 April 2022

1. Introduction

Surface-enhanced Raman spectroscopy (SERS) offers a wide range of possibilities in the study of various types of analytes on surfaces appropriately prepared for this purpose. Usually, these are nanostructured substrates with a high degree of surface development based on plasmonic metals [1]. From a historical point of view, the first substrates used in SERS spectroscopy were based on nanostructured silver obtained by electrochemical pretreatment [2]. On such surfaces, a very strong Raman signal from adsorbed pyridine was observed [1–4]. Since then, there has been a continuous search for durable, stable and repeatable substrates that generate high amplification of the recorded Raman spectra [5]. Such investigations are mainly focused on designing nanoscale platforms in terms of topography and chemical composition, where plasmonic nanoparticles such as silver, gold, copper and their alloys play a key role [1,6,7]. SERS-active substrates utilize the effect of localized surface plasmon resonance (LSPR), which occurs when an electromagnetic wave with a frequency tuned to the frequency of the localized surface plasmons interacts with plasmonic nanoparticles [8,9]. For molecules in close proximity to such substrates, this effect leads to an increase in the cross-section of scattering and absorption, as well as causing the appearance of strong electromagnetic fields around the plasmonic nanoparticles (local field enhancement (LFE)) [9–12]. The magnitude of the induced electromagnetic field depends on the size, shape and distribution of the plasmonic nanoparticles (possible plasmonic couplings) [5,9,13,14]. Therefore, when plasmonic nanostructures are deposited on nonplamonic substrates, the achievable SERS enhancement factors strongly depend on the surface topography and physical properties of semiconductors [10,15]. These make it possible to control the resonance at the nanoscale and adapt it to the wavelengths to be used in the planned applications [9,10,14]. All these factors are extremely important when designing new platforms for SERS investigations. Therefore, the aim of this study was to use for the first time the powder bed fusion process to fabricate a macrorough titanium surface, which could be later functionalized by anodic oxidation to create a nanoporous surface in the form of nanotubes [16–20]. The nanostructure prepared in this way was covered with a deposit of silver nanoparticles by the magnetron sputtering technique (0.02 mg/cm^2). The use of the powder bed fusion process made it possible to generate an original geometry of the SERS substrate in the submicron scale, which was further functionalized at the nanoscale. This solution represents a new proposed strategy for the design of active SERS substrates. According to our knowledge, this approach to designing active SERS platforms is an innovative solution. Until now, printed substrates based on polymers with metals were used for this purpose [21], which were then functionalized by applying, for example, a layer of silver [21]. The electroplating technique, which produced metallic layers with controlled thickness and surface morphology, turned out to be a particularly effective method in this respect [22]. The advantage of printed substrates is that the geometry of the sample surface can be freely shaped on the submicron scale [23]. Such surface geometry, however, does not meet the criterion of applicability in SERS spectroscopy, because the phenomenon of Raman signal amplification occurs most effectively at the nanoscale [11]. Therefore, further functionalization of the printed substrates is required. Typically, for this purpose, various types of coatings containing nanoparticles of plasmonic metals are used [24]. An alternative way to obtain this type of substrate can be via lithographic methods [6,25], but they are much more difficult to apply and more expensive to use. Nevertheless, their advantage is the fact that the substrates are designed and manufactured on the nanoscale from start to finish (the horizontal resolution of the method is much better than that of standard 3D printing) [6,25]. The idea of using the PBF process seems in this case an interesting solution for designing active SERS platforms.

2. Materials and Methods

2.1. Fabrication of 3D Titanium Substrates by Powder Bed Fusion Process Using a Laser Beam

Samples with a disc shape (diameter of about 14 mm and a thickness of about 0.25 mm) were fabricated by powder bed fusion technology using the Realizer SLM50 (Realizer GmbH, Borchen, Germany) selective laser melting machine. Spherical, gas-atomized and contamination-free CP titanium Grade 1 metallic powder (ECKART TLS, Bitterfeld-Wolfen, Germany) was used for the background of SERS substrates. According to the manufacturer data, the CP Ti Gr 1 powder had a diameter below 45 μm and met the American Society for Testing and Materials (ASTM, West Conshohocken, PN, USA) titanium Grade 1 requirements, and its purity was minimally 99.5 wt.% (max. 0.20 Fe, max. 0.08% C, max. 0.03% N, max. 0.015% H, max. 0.18 O, balance Ti). The substrates were fabricated using laser power of 43 W, summarized scanning speed of 325 mm/s (40 μs exposure time and a 20 μm point distance between each next point) and layer thickness set at 25 μm. The total energy density was 130 J/mm^3. The laser scanning strategy for the individual disc alternated with a 45° rotation on each layer, while the distance between each laser scanning vector (hatch distance) was 40 μm. The support structure for the disc was fabricated using a lower laser power (30 W) but using the same laser exposure time (40 µs) in each random point of the support structure's cross-section [10], as shown in Figure 1.

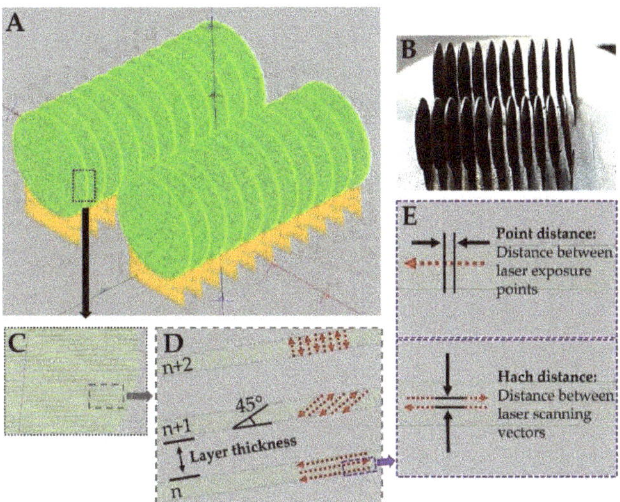

Figure 1. CAD models of sample discs (green) with support structures (yellow) after PBF-LB process parameters (**A**); fabricated 3D titanium samples (**B**); representative disc sample with visible layers in CAD model (**C**); alternating scanning strategy with visible distance between layers (**D**); laser exposure points (point distance) and distance between laser scanning vectors (hatch distance) (**E**).

2.2. Formation of TiO₂ Nanotubes on Ti Substrates Produced by PBF-LB Process

TiO$_2$ NTs were fabricated in one-step anodic oxidation of 3D Ti substrates (Ti 3D). The anodization of the titanium substrates was performed in an optimized electrolyte: a glycerol/water mixture (volume ratio 50:50) with 0.27 M NH$_4$F at a constant voltage of 20 V and a time of 2 h using a two-electrode system (anode: -Ti, cathode: Pt) [26,27]. After anodization, the samples were cleaned with deionized water through long-term rinsing (24 h) and subsequently dried in air. Such a procedure leads to the cleaning of the sample surface from some organic contaminants coming from the electrolyte. Thermal annealing in air was performed at 450 °C for 3 h in order to transform the TiO$_2$ NTs structure from amorphous to crystalline structure [26,28–30] and remove the rest of the contaminants from the surfaces of the samples.

2.3. Deposition of Metal Nanoparticles

The structures obtained were covered with Ag (0.02 mg/cm^2) by the DC magnetron sputtering technique using a Leica EM MED020 apparatus (Leica Microsystems GmbH, Wetzlar, Germany). The average amount of metal deposited per cm^2 was strictly controlled by a quartz microbalance in situ (Leica EM QSG100, Leica Microsystems GmbH, Wetzlar, Germany). The configuration of the setup was perpendicular to the surface of the sample. This configuration was the most suitable for the silver to penetrate into the TiO$_2$ nanopores.

2.4. Characterization

The surface morphology and chemical composition of the samples were examined using a scanning electron microscope (SEM, an FEI Nova NanoSEM 450, Brno, Czech Republic). For typical imaging, low-energy electron detectors, an Everhart–Thornley detector (ETD) and a through-the-lens (TLD) detector, were used, in the low- and high-resolution modes, respectively. All modes were performed in the same configuration, at a primary beam energy of 10 kV.

The chemical states of individual elements were verified by X-ray photoelectron spectroscopy (XPS) using a Microlab 350 (Thermo Electron, East Grinstead, UK) spectrometer. For this purpose, the X-ray excitation source (AlKα anode: power 300 W, voltage 15 kV, beam current 20 mA) was used. The lateral resolution of XPS analysis was about 0.2 cm^2. The high-resolution XPS spectra were recorded using the following parameters: pass energy 40 eV, energy step size 0.1 eV. A smart function of background subtraction was used to obtain the XPS signal intensity. All the collected XPS peaks were fitted using an asymmetric Gaussian/Lorentzian mixed function. The measured binding energies were corrected in reference to the energy of C 1s at 285.0 eV. Avantage-based data system software (Version 5.9911, Thermo Fisher Scientific, Waltham, MA, USA) was used to process the data.

X-ray powder diffraction data were collected on a PANalytical Empyrean (Malvern Panalytical Ltd., Malvern, UK) diffractometer fitted with a X'Celerator detector using Ni-filtered Cu Kα radiation (λ_1 = 1.54056 Å and λ_2 = 1.54439 Å). Data were collected on a flat plate with θ/θ geometry on a spinning sample holder. All presented data were collected in the 2θ range 10–90°, in intervals of 0.0167°, with a scan time of 30 s per interval.

The SERS Raman spectra were measured with a Horiba Jobin-Yvon Labram HR800 spectrometer (Longjumeau, France) equipped with a Peltier-cooled CCD detector (1024 $\times$ 256 pixels), a 600 groove/mm holographic grating and an Olympus BX40 microscope (Tokyo, Japan) with a long-distance 50$\times$ objective. All Raman spectra were recorded using a diode-pumped, frequency-doubled Nd:YAG laser (532 nm). A 200 µL volume of 0.05 M pyridine in 0.1 M KCl was applied to the substrate, and the measurement itself was performed before being drop-dried. For each substrate, 400 spectra were collected from a square surface with an edge length of 50 µm.

3. Results and Discussion

Figure 2 shows the topography of the 3D titanium substrates (Ti 3D) fabricated by the powder bed fusion process using a laser beam. Microscopic observations revealed a large development of the surface in the form of spherical growths, which are unmelted fully titanium powders occurring typically after the PBF process [31,32]. This phenomenon is explained by the nature of the melt pool during fabrication. The high-energy laser beam produces sufficient heat to create the melt pool at the surface of the powder bed [33]. The temperature surrounding the melt pool is immediately increased by heat conduction. Some of the powder particles, regardless of the CAD design, placed in the heat conduction zone do not melt but may become lightly sintered to the surface of the part [34,35]. In our study, the unmelted powder particles on the disc's surfaces are also caused by heat transfer, between the melt pool and unmelted powder in the bed, affecting their sintering with the surface [36]. The size of unmelted fully powder particles is within the range of the used raw powders and is from a few to several dozen micrometers. Slightly sintered titanium particles remain on the surface even after washing in an ultrasonic cleaner. As a result, an interesting topography of the sample surface was created for the SERS applications. Nevertheless, the obtained microscale morphology may not meet the key requirements for the design of active SES substrates after deposition of a plasmonic metal such as Ag. Thus, such a surface generates some irregularities that may have an influence on the variation of the intensity of the measured SERS spectra, knowing that the phenomenon of localized resonance of surface plasmons occurs most effectively in smaller nanostructures [37,38].

Therefore, considering the above criterion, the as-made laser powder bed fused 3D titanium substrates underwent further surface functionalization. Figure 3 shows the SEM images of the 3D titanium substrate surface morphology after anodic oxidation in an electrolyte based on glycerin and water with the addition of ammonium fluoride at a voltage of 20 V. This type of treatment led to the formation of titanium oxide nanotubes, which are well visible. The nanotubes accurately reflected the state of the initial surface, thanks to which development of the 3D titanium surface at the nanoscale was obtained, as shown in Figure 3. The nanotubes evenly covered both the spaces between the spherical accretions and the accretions themselves. The size of the nanotubes formed at 20 V was around 80 nm, which is in line with our previous research in this field [39]. The size of nanotubes can be freely changed depending on the value of the applied voltage of the anodic oxidation process. As the voltage increases, the size of the nanotubes grows larger and larger depending on the electrolyte used [17,26,29,40].

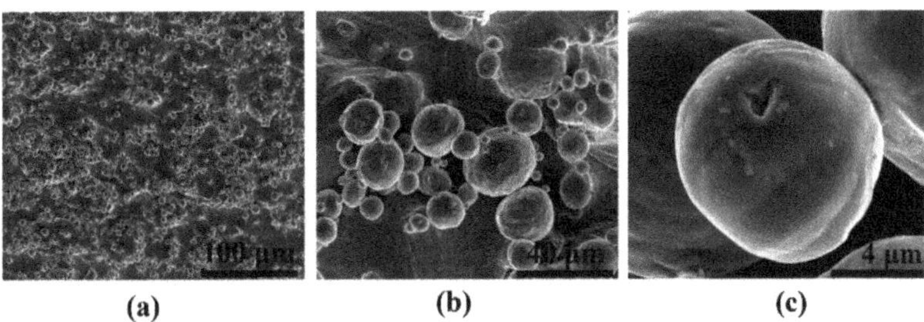

Figure 2. SEM images of the surface topography of the Ti 3D after PBF-LB process. Magnification of images is: (**a**) 100×, (**b**) 1000× and (**c**) 10,000×.

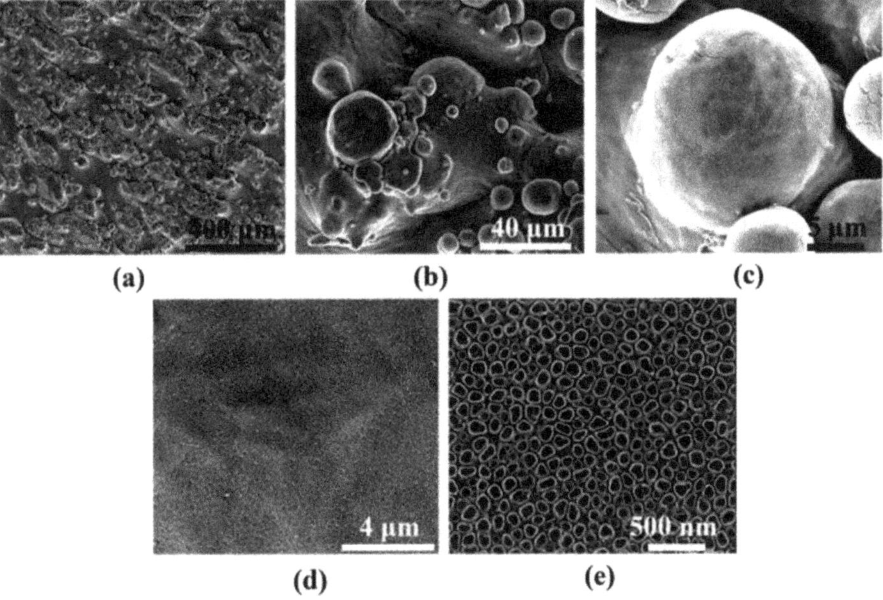

Figure 3. SEM images of Ti 3D surface after anodic oxidation in an electrolyte based on glycerin and water with ammonium fluoride and heating at 450 °C (low and high magnification). Magnification of images is: (**a**) 100×, (**b**) 1000×, (**c**) 5000×, (**d**) 10,000× and (**e**) 50,000×.

The samples with nanotubes were then heated to 450 °C in order to transform the structure of titanium oxide from an amorphous form directly after anodization to a crystalline form, anatase, as shown in Figure 4 [26,28–30,37,39,41]. The XRD spectra show characteristic peaks from metallic Ti (3D-printed substrate) and from the titanium oxide phase in the form of anatase (3D-printed substrate after annealing at 450 °C). Typically, anatase occurs above 300 °C, which is consistent with the observations of other researchers for this type of system [26,40]. In addition, the heat treatment procedure mechanically stabilized the obtained nanostructure on the 3D titanium surface, because a barrier layer is formed at the oxide/metal interface, as we have shown in our previous work [41,42]. Annealing also leads to almost complete loss of the fluorides at around 300 °C and reduced other surface contamination such as hydrocarbons coming from anodic oxidation process [29]. Moreover, based on our earlier research, we know that annealing at 450 °C does not lead to any significant changes in the size of the nanotubes [39].

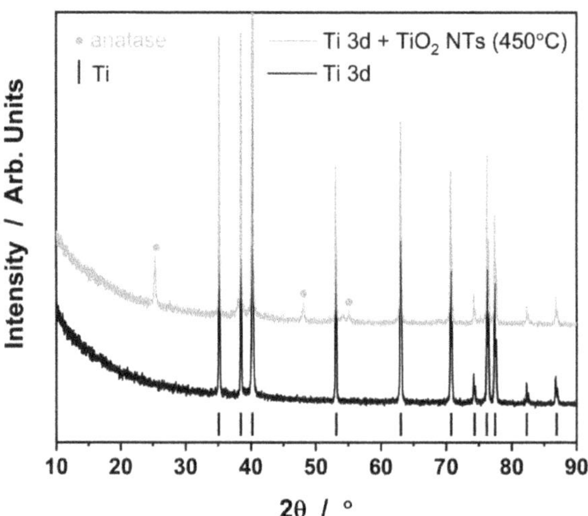

Figure 4. XRD patterns for the Ti 3D substrate after PBF-LB process and annealed at 450 °C.

Our studies and also other research groups showed that heated TiO_2 nanotubes coated with Ag, Au and Cu plasmonic nanoparticles turned out to be extremely active substrates for basic SERS research [17,18,43]. The nanotubes with this type of deposit acted as nanoresonators, amplifying the electromagnetic field at nanovolumes under the influence of the selected laser light with a wavelength close to the energy of the plasmonic resonance of the nanomaterials [17,18,20,43]. The resulting plasmonic nanostructures were characterized by a regular, highly ordered surface development, where the distances between the nanotubes of metal deposit and the resulting gaps favored the formation of "hot spots", which were responsible for the enhancement of the SERS signal of the adsorbed probe molecule [9]. This was in accordance with electromagnetic theory, which states that the vibrations of the molecules normal to the surface of the plasmonic nanostructure are the most strongly enhanced, and so the best morphology for the occurrence of surface plasma resonance is small particles (<100 nm) with an atomically uneven surface [7,9,44,45]. In addition to the electromagnetic mechanism (EM) of the enhancement of the intensity of the Raman spectra, for this type of substrate, a chemical (so-called charge transfer (CT)) effect of the enhancement of the SERS spectra may also appear. This mechanism is associated with molecular orbital interaction between the analyte and the metal nanoparticles (NPs) deposited on the sensing platform [46]. The contribution of the charge-transfer (CT) mode to SERS is mainly based on the change of the molecule polarizability attached to the nanostructured surface, which can lead to new metal–analyte complex formation. Chemical enhancement is much weaker than EM enhancement [9]. Such a structure is shown in Figure 5. Titanium oxide nanotubes coated with silver nanoparticles (0.02 mg/cm^2) smaller than 100 nm can be observed on the 3D titanium surface. Spherical nanoparticles forming agglomerates are located at the edges of the nanotubes, but also decorate the inner and outer walls because the nanotubes are separated from each [17,28]. The side view shows exactly how the silver deposit is arranged. The topography of nanotubes ensures a homogeneous distribution of silver on their surface, and at the same time generates the formation of privileged slits, cavities and gaps between Ag NPs, which affect the SERS activity.

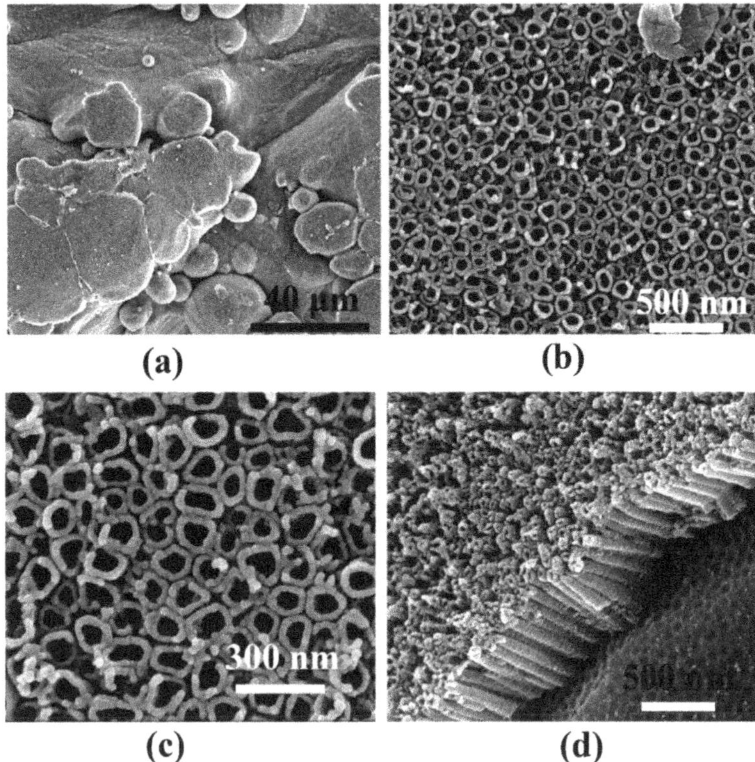

Figure 5. SEM images of Ti 3D surface after anodic oxidation and silver deposition by magnetron sputtering (0.02 mg/cm^2). Magnification of images is: (**a**) 1000×, (**b**) 50,000× and (**c**) 100,000×. Side view of TiO$_2$ NTs with silver deposit (**d**).

The literature shows that types of SERS-active surfaces can be distinguished that effectively generate the SERS signal:

- "With first-generation hot spots that appear as a result of interaction between a single plasmonic nano-object and incident radiation;
- With second-generation hot spots that are produced by coupled plasmonic nano-objects with controllable interparticle distances;
- With third-generation hot spots that are a product of superposition of electromagnetic field originating from metal NPs and electromagnetic field scattered from the backing platform [9]".

Taking into account the above data and looking at Figure 5, it can be seen that the dominant mechanism of creating hot spots in our case may be second-generation interaction, because the proper distance between nano-objects is created by the surface topography of the nanotubes. Nanotubes form a densely packed structure and are separated from each other (see Figure 3).

A different surface topography was obtained for the 3D titanium substrate without titanium oxide nanotubes but with a silver deposit of the same amount (0.02 mg/cm^2) (see Figure 6). In this case, we did not observe such a strong development of the nanoscale surface. Silver, just like nanotubes, precisely covers the surface of the 3D titanium substrate, and the resulting silver structure appears to be morphologically slightly roughened, without significant topographic changes. The only changes are due to microscale spherical accretions.

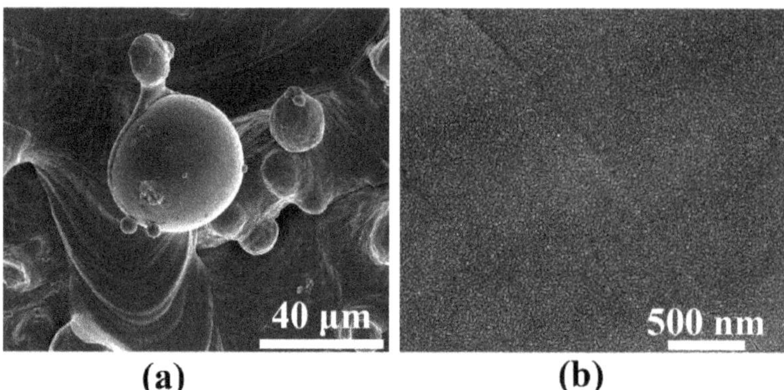

Figure 6. SEM images of Ti 3D surface after PBF-LB process with silver deposit (0.02 mg/cm^2) in (a) low and (b) high magnification.

The obtained materials, which were microscopically characterized, were also analyzed for their chemical composition using XPS spectroscopy. Figure 7 shows XPS Ti2p high-resolution spectra for 3D titanium substrate in the as-made state (PBF-LP process), as well as after anodic oxidation and annealing at a temperature of 450 °C and silver deposition on both substrates. For all the tested samples, it can be seen that the peak position of the Ti2p$_{3/2}$ peak corresponds to an energy of ~458.8 eV, which suggests the presence of stoichiometric titanium oxide TiO$_2$ [47]. Moreover, it can be observed in the Ti2p spectrum of the Ti 3D substrate that two additional components can be distinguished at lower binding energies. The position of these additional peaks is attributed to the presence of nitrides (454.5 eV) [48,49] and metallic titanium and nonstoichiometric oxides or oxynitrides of Ti (456.6 eV) [49,50]. This is confirmed by the signal recorded from the nitrogen N1s, where, after peak deconvolution, the characteristic positions of the peak maxima for nitrides at energies 397.2 (TiN) [48,51] and 396.1 eV (TiO$_x$N$_y$) [49] were distinguished. The presence of nitrides in the 3D titanium initial state is related to the manufacturing process itself, where it is impossible to remove from the preparation chamber all nitrogen and oxygen and keep the process in pure argon atmosphere [52]. The construction of the Realizer SLM50 machine where the rotary pump is unable to remove all gas residues from the building chamber may favor the formation of nitrides/oxynitrides during the powder bed fusion process using a laser beam. However, for the Ti 3D sample with TiO$_2$ nanotubes, after annealing at 450 °C, no signals from nitrides were observed, only signals from titanium oxides (~458.8 eV (TiO$_2$), ~460.9 eV (TiO$_x$)). The absence of nitrides in this case is related to the fact that the thickness of the nanoporous layer after anodizing and annealing at 450 °C is about 800 nm, which we observed in our previous work [41]. It should be noted here that the depth resolution of the XPS method is several—tens of nm depending on the type of material analyzed [53–56]. Another reason is that the Ti-N bonds decompose under the influence of temperature, which was observed during the introduction of nitrogen into the TiO$_2$ structure under the influence of thermochemical treatment [56]. Similar results were obtained for the sample with the silver deposit (see Figure 8).

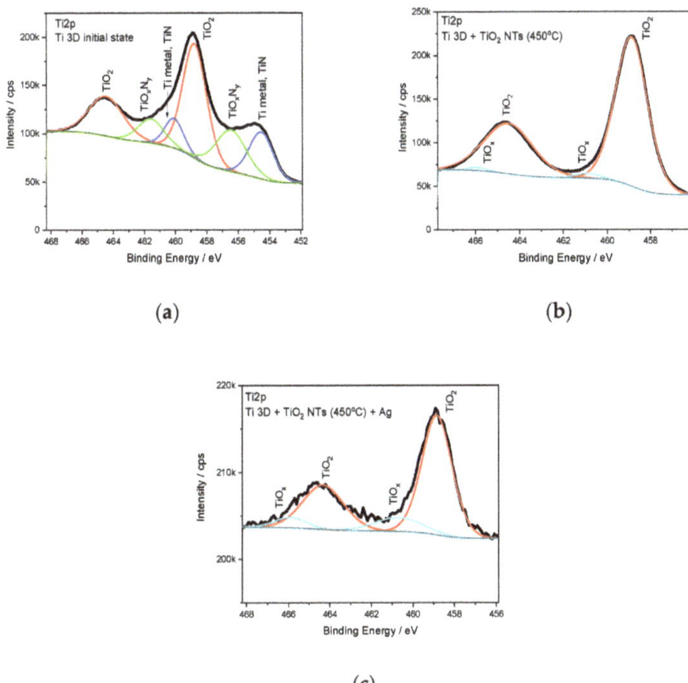

Figure 7. High-resolution XPS spectra of titanium Ti2p (**a**) for the Ti 3D substrate, (**b**) after anodic oxidation and annealing at 450 °C and (**c**) after silver deposition (0.02 mg/cm²) on both substrates.

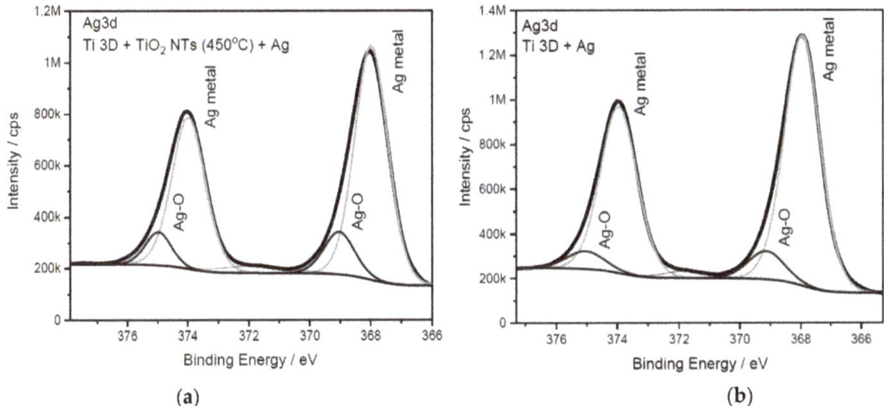

Figure 8. (**a**) XPS high-resolution spectra of silver Ag3d for Ti 3D substrate (**b**) after anodic oxidation process and annealing at 450 °C.

After the silver deposition (see Figure 8), the XPS analysis showed the presence of this element. A strong Ag3d silver signal was recorded. The deconvolution of this signal into two components contributed to assigning the main maximum to the metallic silver (~368.0 eV) and the signal of lower intensity to silver oxide (~369.0 eV) [37,57]. It should be noted here that the Ag3d silver peak for both substrates was over 96% of the sum of all other signals from the analyzed surfaces.

In the next step, the prepared substrates with silver deposit were used for SERS measurements. Figure 8 shows the spectra of adsorbed pyridine on the surface of 3D titanium subjected to surface functionalization and the sample in its initial state. The spectra show the characteristic positions of the bands of pyridine resulting from the vibrations of the aromatic ring (at about 1004 cm^{-1} due to the ν_1 vibration, and at 1032 cm^{-1} due to the ν_{12} vibration), which are typical bands of pyridine adsorbed on the nanostructured silver [58]. Comparing the two spectra, it can be seen that the SERS intensity is several hundred times higher for the sample with TiO$_2$ nanotubes than for the sample not subjected to nanofunctionalization. The characteristic distribution of nanoparticles on the surface of the nanotubes (nanorings) favors the generation of strong electromagnetic fields that amplify the signal from the adsorbed probe molecule under the influence of laser light. According to the theory of plasmon resonance, at each measurement step, the radiation is proportional to the square of the electric field gain, which yields a total gain of the SERS signal proportional to the fourth power of the field enhancement [5,9,45]. This is closely related to the formation of "hot spots", of which there are definitely more on the laser powder bed fused 3D titanium surface covered with nanotubes and silver. Therefore, they can form second-generation hot spots produced by coupled plasmonic nano-objects with controllable interparticle distances. A lack of an appropriate nanotopography is not conducive to the generation of strong electromagnetic fields around the nanoparticles; therefore, for the sample without nanotubes, a SERS spectrum of much lower intensity was recorded (see Figure 9). Apart from the characteristic bands from pyridine, the signals that can be seen in the spectra are related to carbon species at 1100 cm^{-1} and ~1600 cm^{-1} [59].

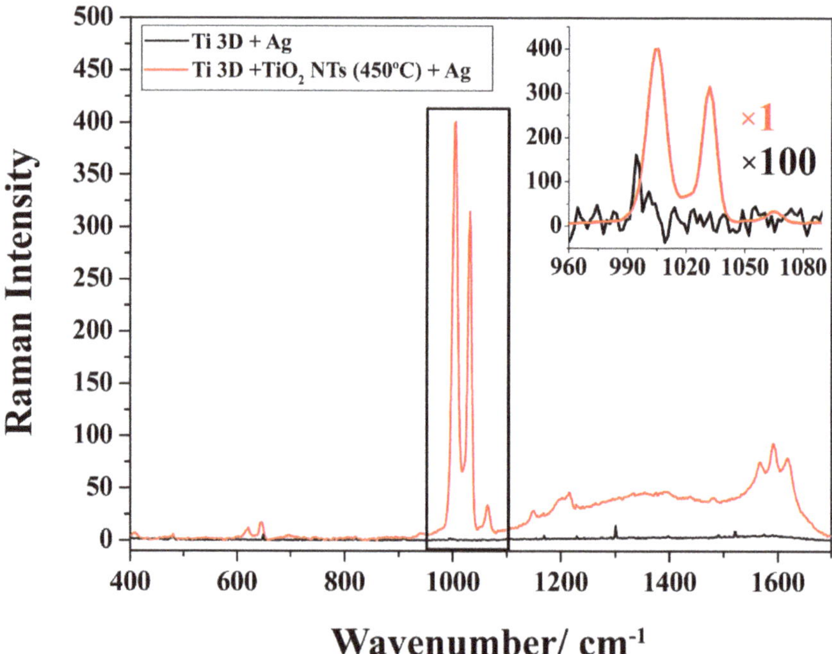

Figure 9. SERS spectra of adsorbed pyridine (0.05 M + 0.1 M KCl) on the surface of samples Ti 3D + TiO$_2$ NTs (450 °C) + 0.02 mg/cm^2 and Ti 3D + 0.02 mg/cm^2.

As mentioned earlier, the distribution of silver on the nanotubes and the specific nanotopography are of key importance in enhancing the SERS signal. Figure 10 shows maps of the SERS signal distribution on the surface of the silver-decorated samples. These results clearly show the influence of the substrate on the change in the intensity of the recorded spectra. The concentration of hot spots is much greater on the powder bed fused 3D titanium surface with TiO_2 nanotubes and silver deposit than it is on the 3D titanium without nanotubes, as can be seen on the scale showing the change in SERS signal intensity (see legend). In the case of the sample without nanotubes, it can be seen that the signal is uniform over the entire analyzed surface. On the other hand, in the case of the 3D titanium substrate with nanotubes, the signal intensity changes very clearly with the appearance of local gain maxima. This effect can be attributed to the influence of the sample topography, which is attributed to the 3D titanium substrate after the PBF-LB process (see Figure 2) and the nanotopography after surface functionalization by anodic oxidation (see Figures 4 and 5).

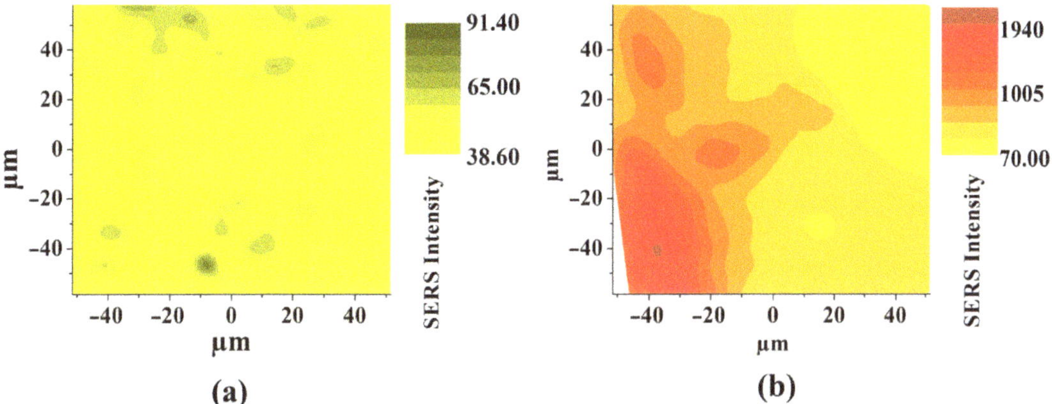

Figure 10. SERS spectrum distribution maps recorded on the (**a**) Ti 3D with 0.02 mg/cm^2 silver deposit, (**b**) Ti 3D with TiO_2 nanotubes and Ag deposit (0.02 mg/cm^2).

Based on the SERS spectra collected for the materials produced with silver deposit, the enhancement factor (E_F) for adsorbed pyridine was estimated (for the band v_1 at 1004 cm^{-1}), based on the following formula: $E_F = I_{SERS}/I_{REF} \times hC_{REF}/N_{SURF}$, where I_{SERS} and I_{REF} are the Raman intensities obtained from the SERS and normal Raman (NR) investigations, respectively, C_{REF} is the concentration of pure pyridine in the NR measurements and h is the depth-of-focus of the laser beam. The average number of adsorbed molecules of pyridine per geometrical surface area unit participating in the SERS measurements (N_{SURF}) was calculated assuming that the adsorbed molecules are spheres closely packed on a plane to form a hexagonal lattice [10,18,28]. Therefore, the calculated E_F for the powder bed fused 3D Ti substrate without nanotubes was 2.24×10^3 and for the substrate with nanotubes 1.26×10^6. This huge change in the E_F for both substrates is apparently due to the influence of the surface morphology. On the substrate with nanotubes, there are definitely more privileged places generating very strong electromagnetic field amplification around the silver nanoparticles. These are places such as the gaps between the nanotubes, where the appropriate distance between the silver nanoparticles is maintained, as well as the nanopores themselves, which are covered with silver [18,20,41,43]. In addition, the original geometry of the titanium sample after the PBF-LP process with characteristic spherical accretions also affects the enhancement effect, as can be seen on the map shown in Figure 10.

The results obtained were additionally compared with our previous works, in which we used titanium oxide nanotubes to design and manufacture SERS platforms. For comparative purposes, a substrate was selected where nanotubes of the same size (~80 nm) on Ti foil were decorated with silver (0.02 mg/cm^2). Silver nanoparticles were deposited on a Ti foil nanotube substrate using the same method as in this study, using the sputtering method in a vacuum [60]. The collected results are presented in Figure 11 and compared with the reference sample of the electrochemically roughened silver electrode [28,57]. It can be seen that the TiO_2 nanotubes formed on the 3D Ti surface have the highest enhancement factor 1.26×10^6. Lower value of E_F for the sample with TiO_2 nanotubes on Ti foil and with a silver deposit of 0.02 mg/cm^2 ($E_F = 6.83 \times 10^5$) was obtained and for the electrochemically roughened Ag electrode ($EF = 4.55 \times 10^5$, reference platform) [28,57]. This means that the E_F, in addition to the surface nanotopography, is influenced by the effect associated with the substrate in its original state. For flat surfaces such as Ti foil, the E_F is lower than for printed titanium. This comparison clearly shows that the surface topography of the printed titanium in its initial state (see Figure 2) has a significant impact on the final result related to the estimate of the E_F of the produced platforms, despite the similar surface nanotopography after surface functionalization. Other researchers who used printed platforms for SERS applications came to similar conclusions. They showed that printed Cu−PLA composite disks with silver film deposited by galvanic displacement proved extremely attractive for detection of environmental contaminants in water. They also noted that the combination of 3D printing with SERS measurements opens up new possibilities in the development of this method [21].

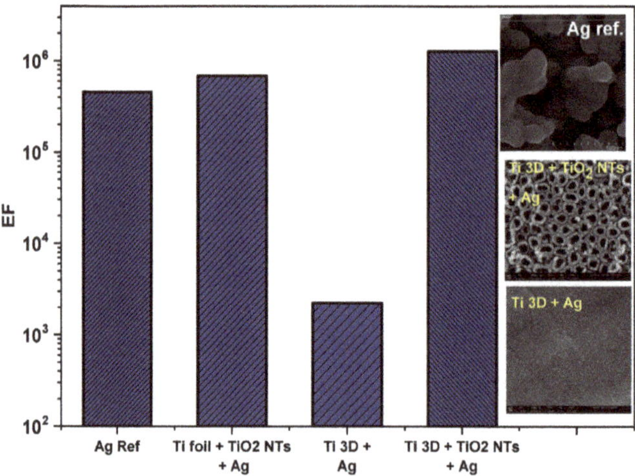

Figure 11. Comparison of the estimated E_F from adsorbed pyridine (0.05 M + 0.1 M KCl) on the fabricated platforms in this work with TiO_2 nanotube-based substrates formed on a Ti foil with a silver deposit. An electrochemically roughened silver electrode was used as a reference. Illustrative SEM images for the discussed samples are attached to the Figure.

4. Conclusions

This work shows the possibility of the application of powder bed fusion using laser beam (PBF) technology to design and manufacture 3D titanium SERS platforms. The 3D titanium-based platforms obtained by the PBF-LB process, after proper functionalization of the nanoscale surface by anodic oxidation, had an enhancement factor of 1.26×10^6 estimated for pyridine, which was used as a probe molecule in the SERS studies. The main factors influencing the final enhancement of the produced platforms were topographic effects, both in the initial state and after the nanofunctionalization of the surface. Therefore, this kind of substrate can form second-generation SERS hot spots produced by coupled plasmonic nano-objects (Ag NPs) with controllable distances. The distance is controlled by titanium oxide nanotubes produced by anodic oxidation, which form a dense structure on the surface of the printed titanium. The gaps, slits and cavities play a crucial role in enhancing the SERS signal, where the electromagnetic mechanism is the most important factor responsible for this phenomenon. Preliminary studies on the use of titanium powders in PBF-LB technology for the preparation of SERS platforms have shown that this process opens up new design possibilities, in particular in analytical chemistry and biochemistry. The only limitation of this process is a lateral resolution of the formation of the structures (microscale printing). Therefore, further surface functionalization of the materials produced in this way is required at the nanoscale, where the plasmonic effects are much more effective.

Author Contributions: Conceptualization, M.P.; methodology, M.P.; formal analysis, M.H., R.A.; investigation, M.H., A.R., A.M., R.A. and B.W.; resources, A.M.; data curation, R.A.; writing—original draft preparation, M.P., B.W. and A.K.; writing—review and editing, M.P. and A.K.; visualization, R.A. and B.W.; supervision, M.P.; project administration, M.P.; funding acquisition, M.P. All authors have read and agreed to the published version of the manuscript.

Funding: This study was supported by NCN (National Science Center) which provided financial support to project Synthesis and characterization of novel biomaterials based on three–dimensional (3D) multifunctional titanium substrates (Grant No. 2017/25/B/ST8/01599). Moreover, this study was supported by the project MCB - Multidisciplinary Research Center of The Cardinal Stefan Wyszynski University in Warsaw (RPMA.01.01.00-14-8496/17).

Institutional Review Board Statement: Not applicable.

Informed Consent Statement: Not applicable.

Data Availability Statement: Not applicable.

Acknowledgments: The authors are grateful to W. Święszkowski for technical support and to M. Lewandowska for helpful discussions.

Conflicts of Interest: The authors declare no conflict of interest.

References

1. Pettinger, B.; Wetzel, H. Surface Enhanced Raman Scattering from Pyridine, Water, and Halide Ions on Au, Ag, and Cu Electrodes. *Berichte der Bunsengesellschaft für Phys. Chemie* **1981**, *85*, 473–481. [CrossRef]
2. Fleischmann, M.; Hendra, P.J.; McQuillan, A.J. Raman Spectra of Pyridine Adsorbed at a Silver Electrode. *Chem. Phys. Lett.* **1974**, *26*, 163–166. [CrossRef]
3. Albrecht, M.G.; Creighton, J.A. Anomalously Intense Raman Spectra of Pyridine at a Silver Electrode. *J. Am. Chem. Soc.* **1977**, *99*, 5215–5217. [CrossRef]
4. Jeanmaire, D.L.; Van Duyne, R.P. Surface raman spectroelectrochemistry: Part I. Heterocyclic, aromatic, and aliphatic amines adsorbed on the anodized silver electrode. *J. Electroanal. Chem. Interfacial Electrochem.* **1977**, *84*, 1–20. [CrossRef]
5. Jing, Y.; Wang, R.; Wang, Q.; Xiang, Z.; Li, Z.; Gu, H.; Wang, X. An Overview of Surface-Enhanced Raman Scattering Substrates by Pulsed Laser Deposition Technique: Fundamentals and Applications. *Adv. Compos. Hybrid Mater.* **2021**, *4*, 885–905. [CrossRef]
6. Mosier-Boss, P. Review of SERS Substrates for Chemical Sensing. *Nanomaterials* **2017**, *7*, 142. [CrossRef]
7. Sharma, B.; Frontiera, R.R.; Henry, A.-I.; Ringe, E.; Van Duyne, R.P. SERS: Materials, Applications, and the Future. *Mater. Today* **2012**, *15*, 16–25. [CrossRef]
8. Moskovits, M. Surface-Enhanced Raman Spectroscopy: A Brief Perspective. In *Surface-Enhanced Raman Scattering*; Springer: Berlin/Heidelberg, Germany, 1977; pp. 1–17.
9. Shvalya, V.; Filipič, G.; Zavašnik, J.; Abdulhalim, I.; Cvelbar, U. Surface-Enhanced Raman Spectroscopy for Chemical and Biological Sensing Using Nanoplasmonics: The Relevance of Interparticle Spacing and Surface Morphology. *Appl. Phys. Rev.* **2020**, *7*, 031307. [CrossRef]
10. Pilot, R.; Signorini, R.; Durante, C.; Orian, L.; Bhamidipati, M.; Fabris, L. A Review on Surface-Enhanced Raman Scattering. *Biosensors* **2019**, *9*, 57. [CrossRef]
11. Schlücker, S. Surface-Enhanced Raman Spectroscopy: Concepts and Chemical Applications. *Angew. Chem. Int. Ed.* **2014**, *53*, 4756–4795. [CrossRef]
12. Anger, P.; Bharadwaj, P.; Novotny, L. Enhancement and Quenching of Single-Molecule Fluorescence. *Phys. Rev. Lett.* **2006**, *96*, 113002. [CrossRef] [PubMed]
13. Temperini, M.L.A.; Sala, D.; Lacconi, G.I.; Gioda, A.S.; Macagno, V.A.; Arvia, A.J. Correlation between SERS of Pyridine and Electrochemical Response of Silver Electrodes in Halide-Free Alkaline Solutions. *Langmuir* **1988**, *4*, 1032–1039. [CrossRef]
14. Aroca, R. *Surface-Enhanced Vibrational Spectroscopy*; John Wiley & Sons Ltd.: Hoboken, NJ, USA, 2006; ISBN 9780471607311.
15. Dong, Y.; Ji, X.; Laaksonen, A.; Cao, W.; An, R.; Lu, L.; Lu, X. Determination of the Small Amount of Proteins Interacting with TiO_2 Nanotubes by AFM-Measurement. *Biomaterials* **2019**, *192*, 368–376. [CrossRef] [PubMed]
16. Michalska-Domańska, M. An Overview of Anodic Oxides Derived Advanced Nanocomposites Substrate for Surface Enhance Raman Spectroscopy. In *Assorted Dimensional Reconfigurable Materials*; IntechOpen: London, UK, 2020.

17. Ling, Y.; Zhuo, Y.; Huang, L.; Mao, D. Using Ag-Embedded TiO_2 Nanotubes Array as Recyclable SERS Substrate. *Appl. Surf. Sci.* **2016**, *388*, 169–173. [CrossRef]

18. Jimenez-Cisneros, J.; Galindo-Lazo, J.P.; Mendez-Rojas, M.A.; Campos-Delgado, J.R.; Cerro-Lopez, M. Plasmonic Spherical Nanoparticles Coupled with Titania Nanotube Arrays Prepared by Anodization as Substrates for Surface-Enhanced Raman Spectroscopy Applications: A Review. *Molecules* **2021**, *26*, 7443. [CrossRef]

19. Lamberti, A.; Virga, A.; Chiadò, A.; Chiodoni, A.; Bejtka, K.; Rivolo, P.; Giorgis, F. Ultrasensitive Ag-Coated TiO_2 Nanotube Arrays for Flexible SERS-Based Optofluidic Devices. *J. Mater. Chem. C* **2015**, *3*, 6868–6875. [CrossRef]

20. Roguska, A.; Kudelski, A.; Pisarek, M.; Lewandowska, M.; Dolata, M.; Janik-Czachor, M. Raman Investigations of TiO_2 Nanotube Substrates Covered with Thin Ag or Cu Deposits. *J. Raman Spectrosc.* **2009**, *40*, 1652–1656. [CrossRef]

21. Jaitpal, S.; Chavva, S.R.; Mabbott, S. 3D Printed SERS-Active Thin-Film Substrates Used to Quantify Levels of the Genotoxic Isothiazolinone. *ACS Omega* **2022**, *7*, 2850–2860. [CrossRef]

22. Mersagh Dezfuli, S.; Sabzi, M. Deposition of Ceramic Nanocomposite Coatings by Electroplating Process: A Review of Layer-Deposition Mechanisms and Effective Parameters on the Formation of the Coating. *Ceram. Int.* **2019**, *45*, 21835–21842. [CrossRef]

23. Gross, B.; Lockwood, S.Y.; Spence, D.M. Recent Advances in Analytical Chemistry by 3D Printing. *Anal. Chem.* **2017**, *89*, 57–70. [CrossRef]

24. Lee, S.; Ongko, A.; Kim, H.Y.; Yim, S.-G.; Jeon, G.; Jeong, H.J.; Lee, S.; Kwak, M.; Yang, S.Y. Sub-100 Nm Gold Nanohole-Enhanced Raman Scattering on Flexible PDMS Sheets. *Nanotechnology* **2016**, *27*, 315301. [CrossRef] [PubMed]

25. Ding, T.; Sigle, D.O.; Herrmann, L.O.; Wolverson, D.; Baumberg, J.J. Nanoimprint Lithography of Al Nanovoids for Deep-UV SERS. *ACS Appl. Mater. Interfaces* **2014**, *6*, 17358–17363. [CrossRef] [PubMed]

26. Roy, P.; Berger, S.; Schmuki, P. TiO_2 Nanotubes: Synthesis and Applications. *Angew. Chemie Int. Ed.* **2011**, *50*, 2904–2939. [CrossRef] [PubMed]

27. Park, J.; Cimpean, A.; Tesler, A.B.; Mazare, A. Anodic TiO_2 Nanotubes: Tailoring Osteoinduction via Drug Delivery. *Nanomaterials* **2021**, *11*, 2359. [CrossRef]

28. Pisarek, M.; Holdynski, M.; Roguska, A.; Kudelski, A.; Janik-Czachor, M. TiO_2 and Al_2O_3 Nanoporous Oxide Layers Decorated with Silver Nanoparticles—Active Substrates for SERS Measurements. *J. Solid State Electrochem.* **2014**, *18*, 3099–3109. [CrossRef]

29. Macak, J.M.; Tsuchiya, H.; Ghicov, A.; Yasuda, K.; Hahn, R.; Bauer, S.; Schmuki, P. TiO_2 Nanotubes: Self-Organized Electrochemical Formation, Properties and Applications. *Curr. Opin. Solid State Mater. Sci.* **2007**, *11*, 3–18. [CrossRef]

30. Scaramuzzo, F.A.; Dell'Era, A.; Tarquini, G.; Caminiti, R.; Ballirano, P.; Pasquali, M. Phase Transition of TiO_2 Nanotubes: An X-ray Study as a Function of Temperature. *J. Phys. Chem. C* **2017**, *121*, 24871–24876. [CrossRef]

31. Chmielewska, A.; Wysocki, B.; Żrodowski, Ł.; Święszkowski, W. Hybrid Solid-Porous Titanium Scaffolds. *Trans. Addit. Manuf. Meets Med.* **2019**, *1*, 2–3. [CrossRef]

32. Wysocki, B.; Idaszek, J.; Zdunek, J.; Rożniatowski, K.; Pisarek, M.; Yamamoto, A.; Święszkowski, W. The Influence of Selective Laser Melting (SLM) Process Parameters on In-Vitro Cell Response. *Int. J. Mol. Sci.* **2018**, *19*, 1619. [CrossRef]

33. Lavery, N.P.; Brown, S.G.R.; Sienz, J.; Cherry, J. A Review of Computational Modelling of Additive Layer Manufacturing—Multi-Scale and Multi-Physics. *Sustain. Des. Manuf.* **2014**, *1*, 651–673. [CrossRef]

34. Li, Y.; Gu, D. Parametric Analysis of Thermal Behavior during Selective Laser Melting Additive Manufacturing of Aluminum Alloy Powder. *Mater. Des.* **2014**, *63*, 856–867. [CrossRef]

35. Chmielewska, A.; Jahadakbar, A.; Wysocki, B.; Elahinia, M.; Święszkowski, W.; Dean, D. Chemical Polishing of Additively Manufactured, Porous, Nickel–Titanium Skeletal Fixation Plates. *3D Print. Addit. Manuf.* **2021**. [CrossRef]

36. Nasab, M.H.; Gastaldi, D.; Lecis, N.F.; Vedani, M. On Morphological Surface Features of the Parts Printed by Selective Laser Melting (SLM). *Addit. Manuf.* **2018**, *24*, 373–377. [CrossRef]

37. Wang, Y.; Li, M.; Wang, D.; Han, C.; Li, J.; Wu, C.; Xu, K. Fabrication of Highly Uniform Ag Nanoparticle-TiO_2 Nanosheets Array Hybrid as Reusable SERS Substrates. *Colloid Interface Sci. Commun.* **2020**, *39*, 100324. [CrossRef]

38. Krajczewski, J.; Ambroziak, R.; Kudelski, A. Substrates for Surface-Enhanced Raman Scattering Formed on Nanostructured Non-Metallic Materials: Preparation and Characterization. *Nanomaterials* **2020**, *11*, 75. [CrossRef]

39. Roguska, A.; Pisarek, M.; Belcarz, A.; Marcon, L.; Holdynski, M.; Andrzejczuk, M.; Janik-Czachor, M. Improvement of the Bio-Functional Properties of TiO_2 Nanotubes. *Appl. Surf. Sci.* **2016**, *388*, 775–785. [CrossRef]

40. Lee, K.; Mazare, A.; Schmuki, P. One-Dimensional Titanium Dioxide Nanomaterials: Nanotubes. *Chem. Rev.* **2014**, *114*, 9385–9454. [CrossRef]

41. Pisarek, M.; Roguska, A.; Kudelski, A.; Andrzejczuk, M.; Janik-Czachor, M.; Kurzydłowski, K.J. The Role of Ag Particles Deposited on TiO_2 or Al_2O_3 Self-Organized Nanoporous Layers in Their Behavior as SERS-Active and Biomedical Substrates. *Mater. Chem. Phys.* **2013**, *139*, 55–65. [CrossRef]

42. Roguska, A.; Pisarek, M.; Andrzejczuk, M.; Dolata, M.; Lewandowska, M.; Janik-Czachor, M. Characterization of a Calcium Phosphate–TiO_2 Nanotube Composite Layer for Biomedical Applications. *Mater. Sci. Eng. C* **2011**, *31*, 906–914. [CrossRef]

43. Roguska, A.; Kudelski, A.; Pisarek, M.; Opara, M.; Janik-Czachor, M. Surface-Enhanced Raman Scattering (SERS) Activity of Ag, Au and Cu Nanoclusters on TiO_2-Nanotubes/Ti Substrate. *Appl. Surf. Sci.* **2011**, *257*, 8182–8189. [CrossRef]

44. Kudelski, A. Raman Spectroscopy of Surfaces. *Surf. Sci.* **2009**, *603*, 1328–1334. [CrossRef]

45. Samriti; Rajput, V.; Gupta, R.K.; Prakash, J. Engineering Metal Oxide Semiconductor Nanostructures for Enhanced Charge Transfer: Fundamentals and Emerging SERS Applications. *J. Mater. Chem. C* **2022**, *10*, 73–95. [CrossRef]

46. Hajipour, P.; Bahrami, A.; Mehr, M.Y.; van Driel, W.D.; Zhang, K. Facile Synthesis of Ag Nanowire/TiO$_2$ and Ag Nanowire/TiO$_2$/GO Nanocomposites for Photocatalytic Degradation of Rhodamine B. *Materials* **2021**, *14*, 763. [CrossRef]

47. Moulder, J.F.; Chastain, J. *Handbook of X-ray Photoelectron Spectroscopy: A Reference Book of Standard Spectra for Identification and Interpretation of XPS Data*; Physical Electronics Division, Perkin-Elmer Corporation: Eden Praire, MN, USA, 1992.

48. Jiang, X.; Wang, Y.; Pan, C. High Concentration Substitutional N-Doped TiO$_2$ Film: Preparation, Characterization, and Photocatalytic Property. *J. Am. Ceram. Soc.* **2011**, *94*, 4078–4083. [CrossRef]

49. Mohan, L.; Anandan, C.; Rajendran, N. Effect of Plasma Nitriding on Structure and Biocompatibility of Self-Organised TiO$_2$ Nanotubes on Ti–6Al–7Nb. *RSC Adv.* **2015**, *5*, 41763–41771. [CrossRef]

50. Biesinger, M.C.; Lau, L.W.M.; Gerson, A.R.; Smart, R.S.C. Resolving Surface Chemical States in XPS Analysis of First Row Transition Metals, Oxides and Hydroxides: Sc, Ti, V, Cu and Zn. *Appl. Surf. Sci.* **2010**, *257*, 887–898. [CrossRef]

51. Sait, R.; Govindarajan, S.; Cross, R. Nitridation of Optimised TiO$_2$ Nanorods through PECVD towards Neural Electrode Application. *Materialia* **2018**, *4*, 127–138. [CrossRef]

52. Wysocki, B.; Maj, P.; Krawczyńska, A.; Rożniatowski, K.; Zdunek, J.; Kurzydłowski, K.J.; Święszkowski, W. Microstructure and Mechanical Properties Investigation of CP Titanium Processed by Selective Laser Melting (SLM). *J. Mater. Process. Technol.* **2017**, *241*, 13–23. [CrossRef]

53. Stevie, F.A.; Donley, C.L. Introduction to X-ray Photoelectron Spectroscopy. *J. Vac. Sci. Technol. A* **2020**, *38*, 063204. [CrossRef]

54. Greczynski, G.; Hultman, L. X-ray Photoelectron Spectroscopy: Towards Reliable Binding Energy Referencing. *Prog. Mater. Sci.* **2020**, *107*, 100591. [CrossRef]

55. McCafferty, E.; Wightman, J. An X-ray Photoelectron Spectroscopy Sputter Profile Study of the Native Air-Formed Oxide Film on Titanium. *Appl. Surf. Sci.* **1999**, *143*, 92–100. [CrossRef]

56. Pisarek, M.; Krawczyk, M.; Hołdyński, M.; Lisowski, W. Plasma Nitriding of TiO$_2$ Nanotubes: N-Doping in Situ Investigations Using XPS. *ACS Omega* **2020**, *5*, 8647–8658. [CrossRef]

57. Ambroziak, R.; Hołdyński, M.; Płociński, T.; Pisarek, M.; Kudelski, A. Cubic Silver Nanoparticles Fixed on TiO$_2$ Nanotubes as Simple and Efficient Substrates for Surface Enhanced Raman Scattering. *Materials* **2019**, *12*, 3373. [CrossRef]

58. Zuo, C.; Jagodzinski, P.W. Surface-Enhanced Raman Scattering of Pyridine Using Different Metals: Differences and Explanation Based on the Selective Formation of α-Pyridyl on Metal Surfaces. *J. Phys. Chem. B* **2005**, *109*, 1788–1793. [CrossRef] [PubMed]

59. Kudelski, A.; Pettinger, B. SERS on Carbon Chain Segments: Monitoring Locally Surface Chemistry. *Chem. Phys. Lett.* **2000**, *321*, 356–362. [CrossRef]

60. Pisarek, M.; Krajczewski, J.; Hołdyński, M.; Płociński, T.; Krawczyk, M.; Kudelski, A.; Janik-Czachor, M. Titanium (IV) Oxide Nanotubes in Design of Active SERS Substrates for High Sensitivity Analytical Applications: Effect of Geometrical Factors in Nanotubes and in Ag-n Deposits. In *Raman Spectroscopy*; InTech: London, UK, 2018.

Article

Al–Al₃Ni In Situ Composite Formation by Wire-Feed Electron-Beam Additive Manufacturing

Artem Dobrovolskii, Andrey Chumaevskii *, Anna Zykova, Nikolay Savchenko, Denis Gurianov, Aleksandra Nikolaeva, Natalia Semenchuk, Sergey Nikonov, Pavel Sokolov, Valery Rubtsov and Evgeny Kolubaev

Institute of Strength Physics and Materials Science, Siberian Branch of Russian Academy of Sciences, 634055 Tomsk, Russia; artdobrov@ispms.ru (A.D.); zykovaap@mail.ru (A.Z.); savnick@ispms.tsc.ru (N.S.); desa-93@mail.ru (D.G.); philip371g@gmail.com (A.N.); natali.t.v@ispms.ru (N.S.); sergrff@ngs.ru (S.N.); sps11@sibmail.com (P.S.); rvy@ispms.ru (V.R.); eak@ispms.ru (E.K.)
* Correspondence: tch7av@gmail.com; Tel.: +7-(3822)-28-68-63

Abstract: The regularities of microstructure formation in samples of multiphase composites obtained by additive electron beam manufacturing on the basis of aluminum alloy ER4043 and nickel superalloy Udimet-500 have been studied. The results of the structure study show that a multicomponent structure is formed in the samples with the presence of $Cr_{23}C_6$ carbides, solid solutions based on aluminum -Al or silicon -Si, eutectics along the boundaries of dendrites, intermetallic phases Al_3Ni, $AlNi_3$, $Al_{75}Co_{22}Ni_3$, and Al_5Co, as well as carbides of complex composition $AlCCr$, Al_8SiC_7, of a different morphology. The formation of a number of intermetallic phases present in local areas of the samples was also distinguished. A large amount of solid phases leads to the formation of a material with high hardness and low ductility. The fracture of composite specimens under tension and compression is brittle, without revealing the stage of plastic flow. Tensile strength values are significantly reduced from the initial 142–164 MPa to 55–123 MPa. In compression, the tensile strength values increase to 490–570 MPa and 905–1200 MPa with the introduction of 5% and 10% nickel superalloy, respectively. An increase in the hardness and compressive strength of the surface layers results in an increase in the wear resistance of the specimens and a decrease in the coefficient of friction.

Keywords: wire feed electron beam additive manufacturing; electron beam freeform fabrication; aluminum alloys; nickel superalloys; multiphase materials; in situ composites

Citation: Dobrovolskii, A.; Chumaevskii, A.; Zykova, A.; Savchenko, N.; Gurianov, D.; Nikolaeva, A.; Semenchuk, N.; Nikonov, S.; Sokolov, P.; Rubtsov, V.; et al. Al–Al₃Ni In Situ Composite Formation by Wire-Feed Electron-Beam Additive Manufacturing. *Materials* **2023**, *16*, 4157. https://doi.org/10.3390/ma16114157

Academic Editors: Bartłomiej Wysocki, Joseph Buhagiar, Tomasz Durejko and Gábor Harsányi

Received: 8 March 2023
Revised: 11 May 2023
Accepted: 22 May 2023
Published: 2 June 2023

1. Introduction

Cast Al-Si alloys are commonly used in automotive engines due to their good castability, high strength-to-weight ratio, excellent corrosion resistance, and low expansion ratio coefficient [1,2]. Aluminum-silicon alloys (silumins) are used to produce body parts, shaped castings, and drawn and welding wires [1,2]. The negative characteristics of silumins are low tensile strength (135–235 MPa) and the tendency to brittle fracture (relative elongation less than 4%) [3,4]. Heat treatment, work hardening, and alloying can be used to improve the properties of aluminum alloys [5,6]. Ni, Ti, Mo, Cu, Mg, and other elements are used as alloying components [7–12]. To improve low mechanical properties, the fabrication of composite materials can be considered.

The production of composite materials with a metal matrix on the basis of aluminum alloys has wide application prospects, since it allows us to obtain products with a hardened structure of individual components of the product, while retaining the properties of the base metal with low cost and density, as well as high plasticity [13–17]. The use of modern production methods makes it possible to produce lightweight nickel-modified aluminum products that are resistant to frictional wear due to hardening with Al₃Ni intermetallic particles [18–33]. The bulk intermetallic phase of Al₃Ni exhibits an orthorhombic crystal structure and has a density of 4000 kg/m³ and an elastic modulus of 140 MPa [34].

The intermetallic phase Al_3Ni is formed in alloys with a high content of Al alloyed with Ni. This solid phase is a promising reinforcing material for improving the properties of aluminum alloys at high temperatures. The brittle nature of the Al_3Ni intermetallic phase limits its use unless it is uniformly dispersed in softer metal matrices. The primary Al_3Ni intermetallic phase is observed in Al–Ni alloys containing more than 5 wt. % Ni [35], while, conversely, the AlNi intermetallic phase becomes the main phase at an Ni content of 45 wt. % and higher [36].

Several processes for manufacturing Al_3Ni–Al composites exist, such as mechanical alloying [28], equal-channel angular pressing [29], directional solidification [30], the electromagnetic separation method [31], and friction stir processing [32,33].

One of the modern methods of manufacturing composite materials is two-wire electron beam additive manufacturing using wire filaments. This method is a type of electron beam additive manufacturing. The essence of the method is the sequential deposition of wire material layer by layer on a cooled substrate. The filament deposition is performed by electron beam melting in a vacuum chamber, while the filament is fed into the melting zone by wire feeders [37–41]. A feature of the process is the simultaneous feeding of two wires into the melt zone, resulting in the local alloying of a model product with the formation of a more complex multiphase structure [42,43]. This method is of great interest to the scientific community because it allows the production of a wide variety of materials based on standard metal wires by using two wire feeders with controlled filament feed into the melt bath. As a result, both the formation of composite materials with a non-standard combination of structure and properties and the production of parts with a directional structure in various areas are possible [44,45].

At present, although there are a number of publications in the literature on the preparation of composite and functionally graded materials by wire electron beam technology, there is only a small amount of information on the preparation of multicomponent composite materials, especially those based on light aluminum or titanium alloys [46–48]. In this paper, we considered the formation of the structure of multiphase composites based on the aluminum-silicon alloy ER4043 ($AlSi_5$) and the nickel superalloy Udimet-500 and its influence on the mechanical properties of the obtained materials. Such a combination of the base metal and the hardening additive will allow the study of the patterns of material formation during the interaction of the base metals of the alloys and their alloying elements. From the point of view of practical application, the preparation of these materials has prospects in the manufacture of parts based on aluminum alloys with surface layers of high hardness and wear resistance, as well as a strong and ductile base.

2. Materials and Methods

The samples were fabricated on a three-axis multi-beam electron beam additive manufacturing tool in a vacuum environment on a cooled $AlMg_5$ aluminum alloy substrate. Model products were fabricated in the form of thin walls 8 mm thick, 80 mm high, and 120 mm long by linearly moving the coordinate table along one of the axes, layer by layer. The electron gun and wire feeders were static. ER4043 ("MetPromStar", Moscow, Russia) and Udimet-500 (ESAB, North Bethesda, MD, USA) welding wires with a diameter of 1.2 mm were used as filaments. A Niton XL3t 980 GOLDD (Thermo Fisher Scientific, Waltham, MA, USA) X-ray fluorescence analyzer was used to control the chemical composition of the wires and the resulting model products (Table 1). In addition to these components, the Udimet-500 alloy contains up to 0.1% carbon.

Printing was performed according to the scheme shown in Figure 1. Pattern 1 was formed on the surface of the AA5056 alloy substrate 2 by feeding filaments 4 through nozzles 3 into the printing zone. The filaments 4 were melted by an electron beam 5 supplied from an electron gun 6 through a magnetic focusing system 7. As a result, a melt bath 8 was formed in the printing zone.

Table 1. Results of X-ray fluorescence analysis, filaments used, and samples obtained.

Material	Elements, wt. %											
	Al	Mn	Fe	Zn	Cr	Co	Mo	Ti	Si	Zr	Ni	V
ER4043 (AlSi5)	bal.	0.032	0.09	0.07	-	-	-	0.01	5.1	-	-	0.021
Udimet-500	1.6	0.1	0.4	0.04	17.3	13.2	4.3	2.7	0.3	0.04	bal.	-
Al–5% Ni	82.18	-	0.127	-	1.88	1.33	-	0.324	7.36	-	6.31	0.035
Al-10%Ni	66.53	-	1.08	-	5.16	3.44	-	0.654	5.35±	-	16.29	0.06

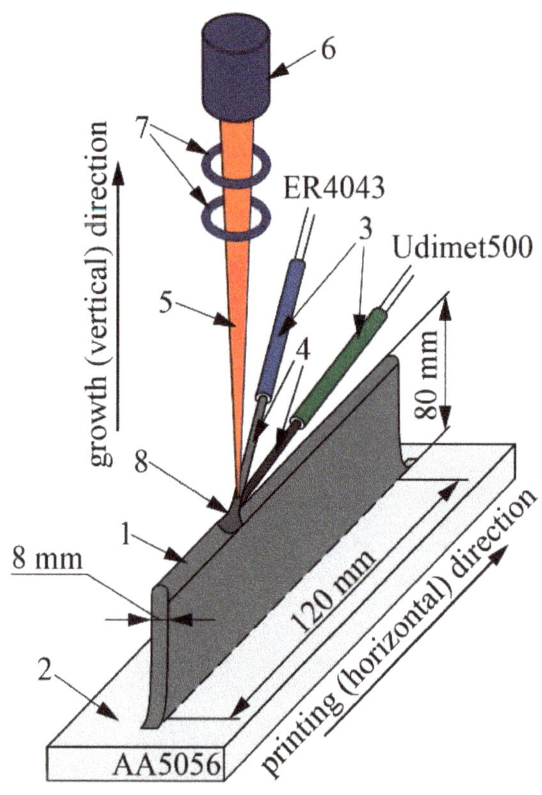

Figure 1. Schematic diagram of the printing process of multimetal composites based on AlSi5 aluminum alloy and Udimet-500 nickel superalloy using two-wire electron beam additive technology.

The pattern was formed layer by layer. The wire feed was not controlled during the printing process. Initially, the feed was adjusted based on achieving a volume concentration of the injected nickel alloy of 5 and 10 vol% of the composite volume. The comparison of the structure and properties of the composite materials was carried out with a pure AlSi5 alloy printed using the method of electron beam additive technology. This scheme was previously used for the preparation of multicomponent metal-matrix composites based on aluminum-manganese bronze with the introduction of nickel superalloys [48,49].

To reveal the structural elements, the samples were etched with Keller's reagent (2.5 mL HNO_3, 1.5 mL HCl, 1 mL HF, 95 mL H_2O) after pregrinding and polishing. An Altami-MET 1C (Altami Ltd., Saint-Petersburg, Russia) optical microscope and an Olympus LEXT 4100 (Olympus NDT, Inc., Waltham, MA, USA) confocal microscope were used to examine the microstructure of the samples. Microscopic studies were carried out on an Apreo

2 S (Thermo Fisher Scientific, Waltham, MA, USA) scanning electron microscope. X-ray diffraction analysis was performed with an X-ray diffractometer XRD-7000S (Yekaterinburg, Russia), CoKα. Transmission microscopy was performed on a JEOL-2100 (JEOL Ltd., Akishima, Japan) universal transmission microscope. Mechanical tests were performed on a UTS-110M (Testsystems, Ivanovo, Russia) machine. The size of the compression specimens was $3 \times 3 \times 6$ mm. The size of the working part of the tensile specimens was $12 \times 2.5 \times 2.5$ mm. The strain rate during compression and tension was 1 mm/min. Tribological tests for dry friction according to the "pin on disk" scheme were carried out on a TRIBOTECHNIC (Tokyo Boeki Group, Japan, Tokyo) tribometer paired with a counter body in the form of a 5 mm diameter ball at sliding speeds of 0.06, 0.15, and 0.24 m/s, and a normal compressive force of 10 N. The size of samples for tribological tests was $3 \times 3 \times 10$ mm. The rotational speed was 200 rpm and the test duration was 300 min. The change in the friction rate occurred due to the displacement of the mating body at a distance of 3, 7, and 12 mm from the center of the disk.

3. Results

In the presented optical images of the microstructure of the hypoeutectic aluminum-silicon alloy AlSi5 obtained using the EBAM method, aluminum-based α-solid solution dendrites, and fine-grained eutectics between the dendrites are observed (Figure 2). Such a structure is typical for aluminum-silicon alloys obtained by casting or electron beam printing [50]. The structure of the obtained material differs significantly in the upper (Figure 2a), middle (Figure 2b), and lower (Figure 2c) parts of the sample. In the upper part of the sample, the volume fraction of the eutectic between the α-Al dendrites increases and the size of the dendritic branches decreases.

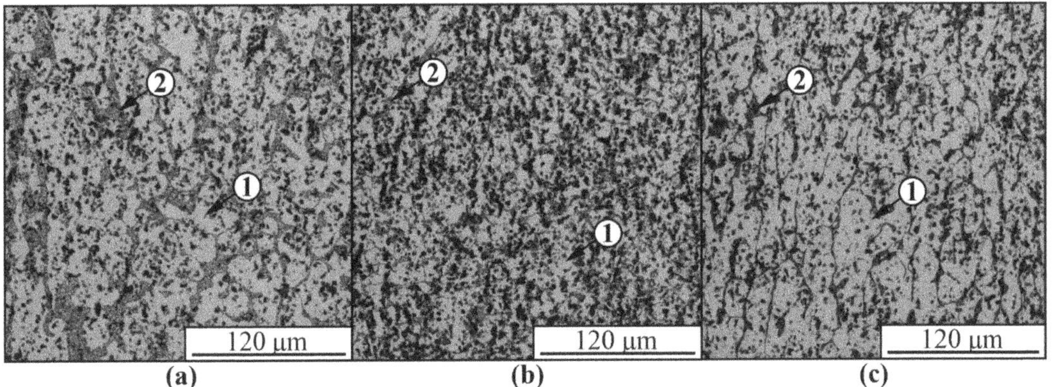

Figure 2. Optical images of the microstructure of ER4043 alloy obtained in the upper (**a**), middle (**b**), and lower (**c**) parts: 1—aluminum-based α-solid solution; 2—Al-Si eutectic.

The formation of the Al-Ni5% composite material leads, on the one hand, to the formation of various intermetallic and other phases in the structure of the samples and, on the other hand, to the formation of sufficiently large fragments of the Udimet-500 alloy, which only partially interacted with the aluminum matrix (Figure 3). The appearance of relatively large fragments of the nickel alloy, which only partially reacted with the matrix, is due to its high melting point compared to the AlSi5 alloy, as a result of which the material introduced into the melt bath crystallizes much earlier than the aluminum alloy and settles in the lower part of the layer, incompletely mixing with aluminum.

The size of the aluminum-based solid solution dendrites 1 in the composite is significantly smaller than in the unmodified material due to the presence of intermetallic compounds 3 and partially reacted nickel alloy 4 in the structure, which limit the growth of aluminum dendrites (Figure 2). The volume fraction of eutectic 2 is significantly reduced compared to the pure AlSi5 alloy (Figures 2 and 3). As a result of the interaction of nickel alloy components with an aluminum matrix, many structural components of different chemical compositions with different particle morphologies from globular to acicular inclusions are formed. In the structure of composite materials, the presence of microcracks is observed as a result of thermal stresses during cooling of the material, differences in the coefficients of thermal expansion of the components and low plasticity of intermetallic phases.

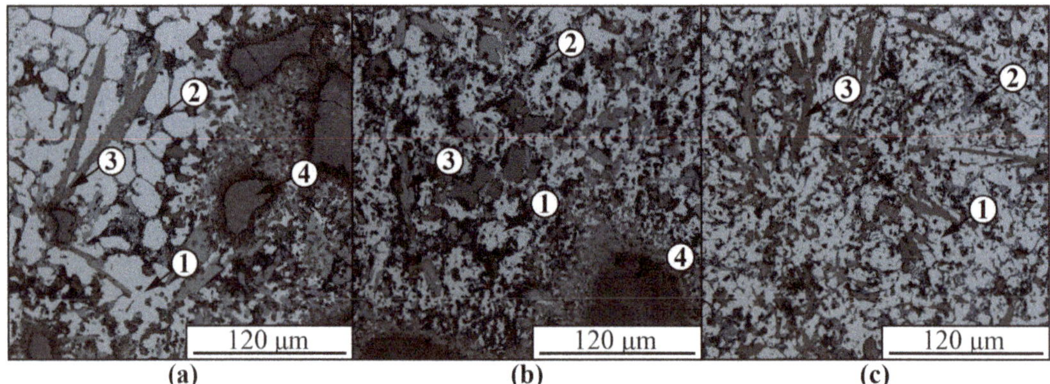
(a)   (b)   (c)

Figure 3. Optical images of the Al-Ni5% composite microstructure obtained in the upper (**a**), middle (**b**), and lower (**c**) parts: 1—aluminum-based α-solid solution; 2—Al-Si eutectic; 3—intermetallic phases; 4—a fragment of Udimet-500 nickel alloy partially reacted with the aluminum matrix.

As the nickel alloy concentration increases, the proportion of intermetallic inclusions increases and the size and volume fraction of α-solid solution dendrites in aluminum decreases (Figure 4). In the composite structure in Figure 4, practically no eutectics (α-Al+β-Si) are released. Intermetallic phases are formed from lamellar to almost equiaxed. The structure in the upper part of the samples is significantly enlarged compared to the lower part. Large fragments of a partially reacted nickel alloy are also present, consisting partly of intermetallic phases of different compositions and partly of nickel superalloy fragments.

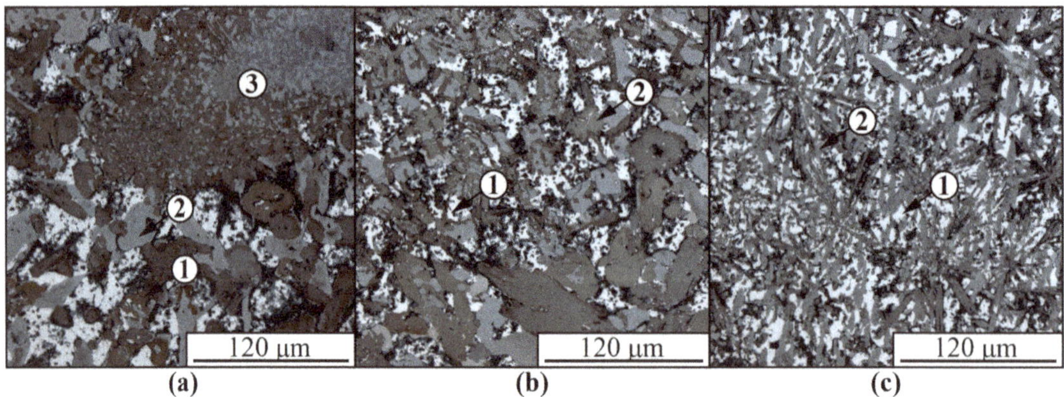
(a)   (b)   (c)

Figure 4. Optical images of the microstructure of the Al-Ni10% composite obtained in the upper (**a**), middle (**b**), and lower (**c**) parts. 1—Aluminum-based α-solid solution; 2—intermetallic phases; 3—a fragment of the nickel alloy Udimet-500 partially reacted with the aluminum matrix.

The phase composition of the samples, determined by X-ray diffraction analysis, is identical for different concentrations of the components contained in the composites

(Figure 5). In addition to the initial components of the aluminum alloy α-Al and β-Si, the structure of the composites contains intermetallic compounds Al$_3$Ni, Al$_{75}$Co$_{22}$Ni$_3$ and Al$_5$Co, as well as carbides of complex composition AlCCr, Al$_8$SiC$_7$. The structure in the lower, upper, and middle parts of the specimens also does not differ practically in the phase composition.

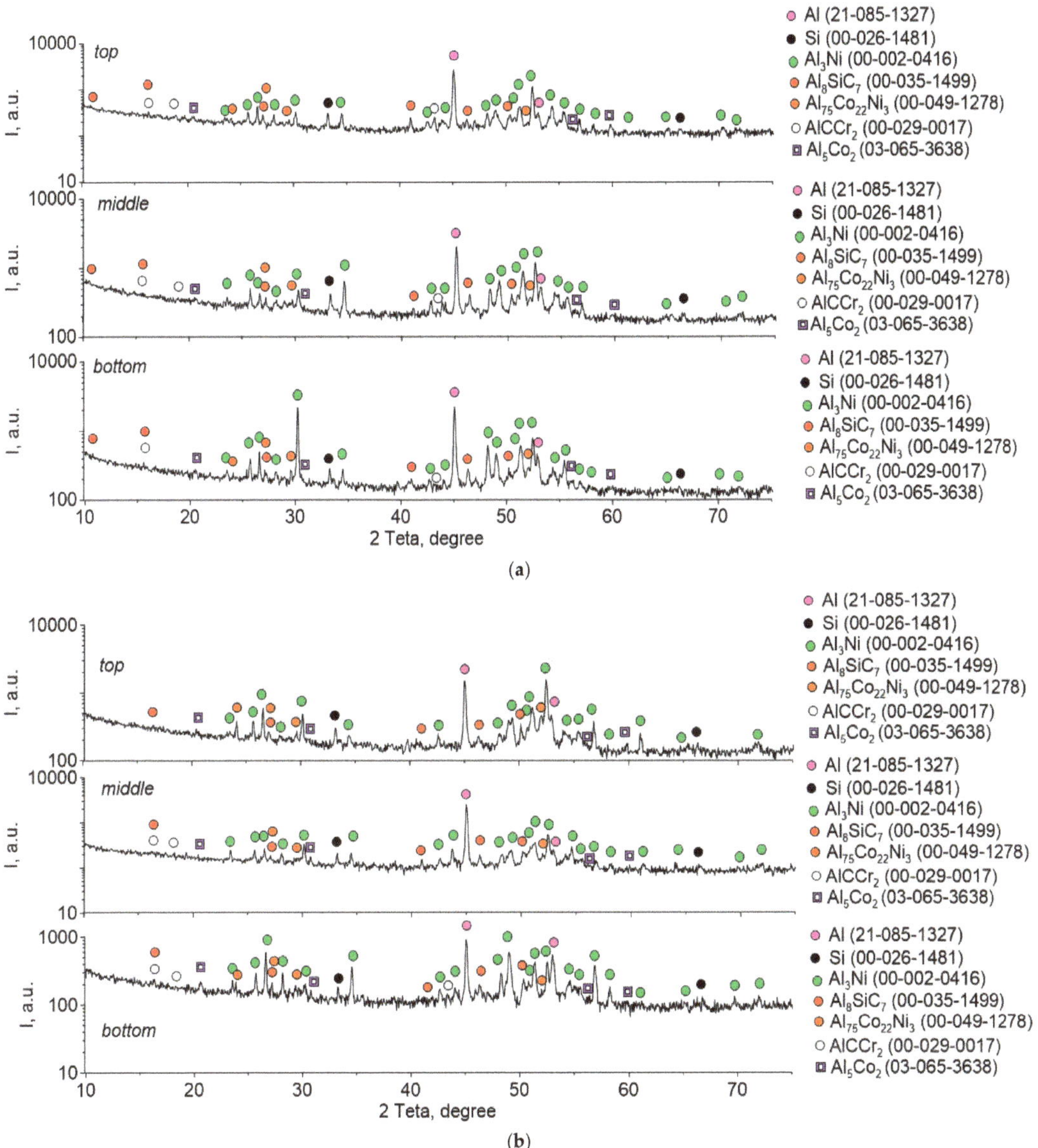

Figure 5. X-ray patterns of samples containing 5% (**a**) and 10% (**b**) nickel alloy Udimet-500 in an aluminum matrix.

The microstructure of AlSi5 alloy is characterized by dendritic cells of solid solution, along the boundaries of which the eutectic (α-Al+β-Si) is located (Figures 2 and 6). This is confirmed by TEM data, according to which the reflections (1$\bar{1}$1) and (00$\bar{2}$) contain -Si and -Al phases, respectively (spectrum 1–3 in Figure 7). According to EDS analysis, up to 1.5 at.% Fe also dissolves in the silicon particles (spectrum 2 and 3 in Figure 7a). The microstructure of the AlSi5 alloy is similar to that of a cast aluminum-silicon alloy. However, unlike the AlSi5 cast alloy obtained by the wire additive electron beam technique, the eutectic Si is not in the form of plates, but rather spherical particles of various sizes (Figure 6). In addition to fine silicon particles, the eutectic also contains rather large silicon particles, reaching a size of 3–5 μm (Figure 6).

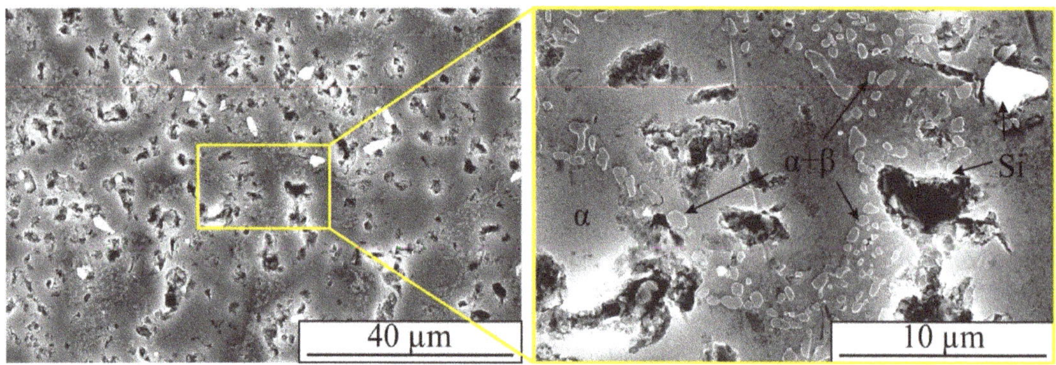

Figure 6. SE mode SEM images of the microstructure of the Al-8Si alloy obtained by EBAM.

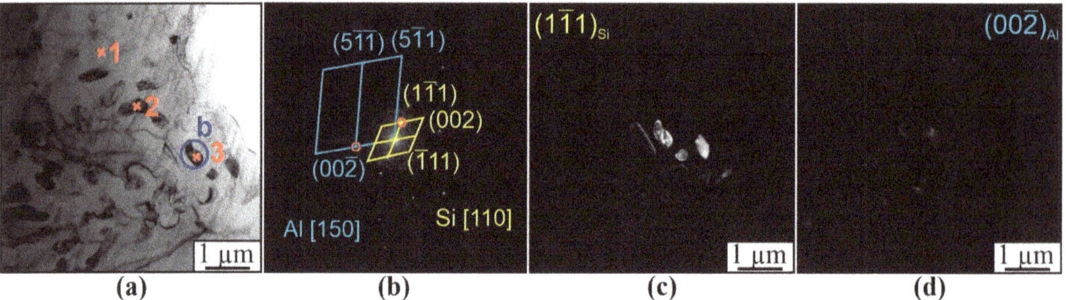

Figure 7. Bright-field TEM image of the microstructure of the ER4043 alloy (**a**), microdiffraction pattern (**b**) taken from a fragment of the area (**a**), dark-field images in the reflection (1$\bar{1}$1)$_{Si}$ (**c**) and in the reflection (00$\bar{2}$)$_{Al}$ (**d**).

In the area of uneven mixing of the components of 2 wires of the composite material, a chemical composition gradient is formed (Figure 8a,b). This leads to the formation of a complex multiphase structure. The regions of partially mixed components are a mixture of AlNi and Al$_3$Ni intermetallic grains enriched in Cr (up to ~5 at.%), Co (~6 at.%), and Si (~5 at.%) (Figure 8a, spectra 1–3, Table 2). Moving away from such regions, the solid solution contains Al$_3$Ni particles, in which the total concentration of Cr, Co, and Si does not exceed 7 at.% (Figure 8a, spectrum 6, Table 2). An insignificant number of impurities are observed in the α-Al solid solution (Figure 8a,d, spectra 5 and 7, Table 2). In general, in the composite where the components were completely mixed, the structure is characterized by a solid solution with eutectic (α+β), grains of supersaturated solid solution of Co, Ni, Si in -Al, and Al$_3$Ni particles (Figure 8d–i).

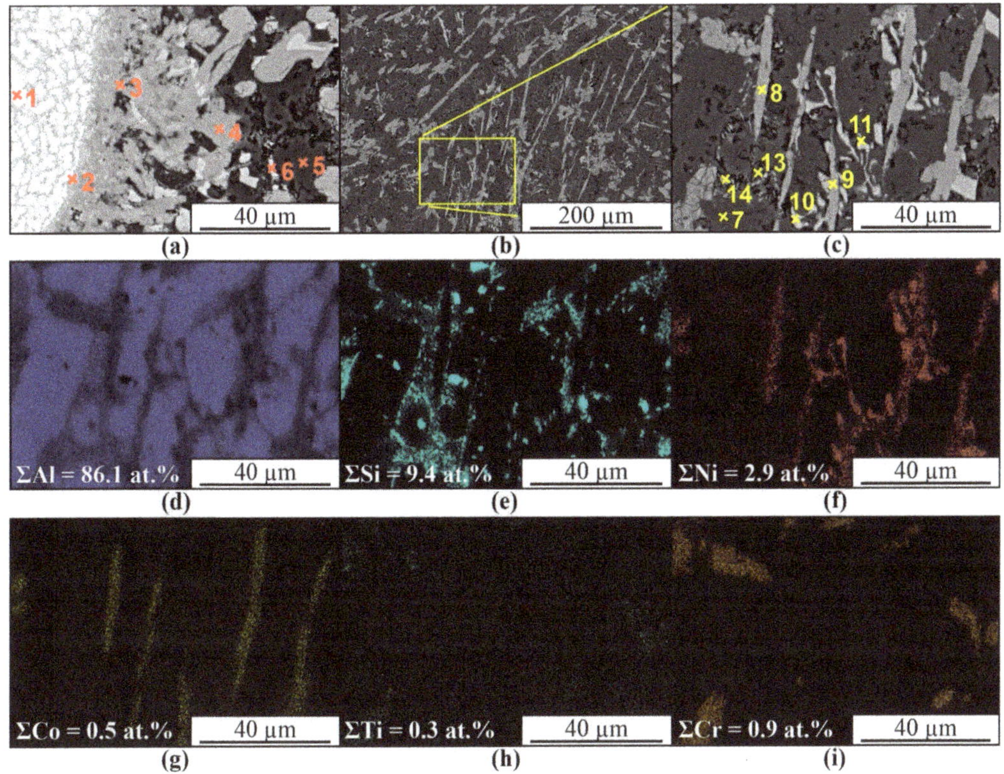

Figure 8. SEM images in BSE mode (**a**) and SE (**c**,**d**), EDS maps of element distribution (**e**–**i**) obtained from plot (**d**); (**b**) EDS analysis obtained along the line from plot (**a**).

Table 2. The chemical composition of the Al-5%Ni composite in various areas shown in Figure 8.

Spectrum	Chemical Composition, at.%						Possible Phase
	Al	Si	Ti	Cr	Co	Ni	
1	54.5	1.3	1.0	4.6	6.7	31.9	AlNi
2	63.5	5.4	1.0	4.6	5.0	20.5	Al$_3$Ni
3	75.8	5.1	1.0	6.6	3.2	8.2	Al$_3$(Ni,Cr,Si)
4	82.8	2.4	0.3	1.4	6.4	6.8	Al(Co,Ni)
5	97.9	1.2	0.4	0.3	0.1	0.2	α-Al
6	74.8	6.1	0.1	0.4	0.5	18.0	Al$_3$Ni
7	98.8	0.7	0.2	0.1	-	0.1	α-Al
8	81.3	4.7	0.1	0.1	4.3	9.6	Al(Co,Ni, Si)
9	80.5	1.7	0.1	0.2	5.2	12.2	Al(Co,Ni, Si)
10	78.4	1.2	0.1	-	0.6	19.8	Al$_3$Ni
11	80.6	3.6	0.1	-	0.5	15.3	Al$_3$Ni
12	84.9	1.2	0.1	0.1	0.3	13.4	Al$_3$Ni
13	72.0	18.3	0.1	-	0.3	9.4	α-Al, β-Si
14	73.0	18.7	0.1	0.1	0.2	7.8	α-Al, β-Si

The SEM-EDS data compare well with the TEM-EDS data (Figure 9 and Table 3). Particles of lamellar (Figure 9a–f) or spherical (Figure 9g–i) shape are also formed. The composition of particles of similar shape is identical. The major part of the material is occupied by solid solutions of α-Al and β-Si (3, 4, 5, 6, 9 at. Figure 9 and Table 3). Intermetallics of the complex composition Al₃(Ni,Cr,Si) and Al(Co,Ni) (1, 2, 7, 8 at. Figure 9 and Table 3) are also distinguished in the structure. This is due to the presence of chromium and cobalt in the Udimet-500 alloy and silicon in the aluminum alloy, resulting in the formation of intermetallic compounds of a more complex composition than Al₃Ni or AlNi. The intermetallic phases are predominantly lamellar or irregular, while the silicon particles are spherical (Figure 9).

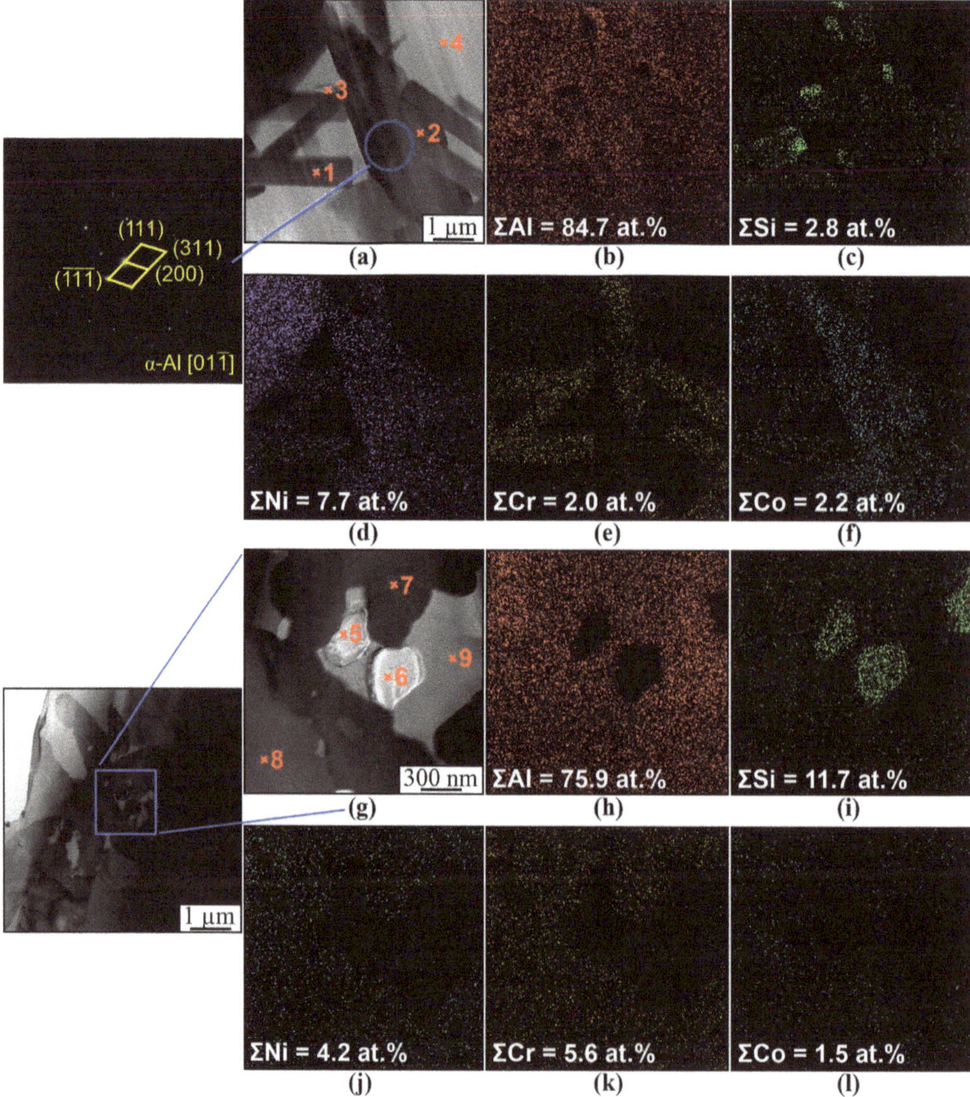

Figure 9. Microstructure of the composite material Al-5%Ni. Bright-field TEM image of the Al-5%Ni composite (**a**,**g**), and EDS element distribution maps (**b–f**,**h–l**).

Table 3. The chemical composition of the Al-5%Ni samples in the local areas shown in Figure 9.

Spectrum	Chemical Composition, at.%								Possible Phase
	Al	Si	Ti	Cr	Co	Ni	Fe	Mo	
1	77.8	5.6	0.6	7.5	1.9	5.2	0.1	0.3	$Al_3(Ni,Cr,Si)$
2	80.8	1.6	0.03	1.6	6.7	8.9	0.1	0.1	$Al(Co,Ni)$
3	2.0	98.0	-	-	-	-	-	-	β-Si
4	99.0	0.1	-	-	-	-	-	-	α-Al
5	1.7	98.3	-	-	-	-	-	-	β-Si
6	3.1	96.9	-	-	-	-	-	-	β-Si
7	76.1	7.2	0.2	7.3	1.9	6.3	-	-	$Al_3(Ni,Cr,Si)$
8	77.3	5.5	0.3	7.6	2.1	5.6	-	-	$Al_3(Ni,Cr,Si)$
9	99.9	0.1	-	-	-	-	-	-	α-Al

In addition to the formation of the particles of the above intermetallic compounds, intermetallic phases based on aluminum and cobalt are formed (Figure 10). According to the results of transmission microscopy, Al_9Co_2 particles are formed in the samples (Figure 10). The chemical analysis data of individual sections also show the formation of Al_9Co_2Ni. In this case, X-ray diffraction analysis shows the formation of intermetallic compounds of the complex composition, $Al_{75}Co_{22}Ni_3$. In this case, it is rather difficult to determine which intermetallic phases are formed. Presumably, due to the inhomogeneity of the distribution of chemical elements, a fairly wide range of intermetallic compounds is formed in local areas of the sample, with the predominance of $Al_{75}Co_{22}Ni_3$ in the main volume as a whole, which is shown by X-ray diffraction analysis.

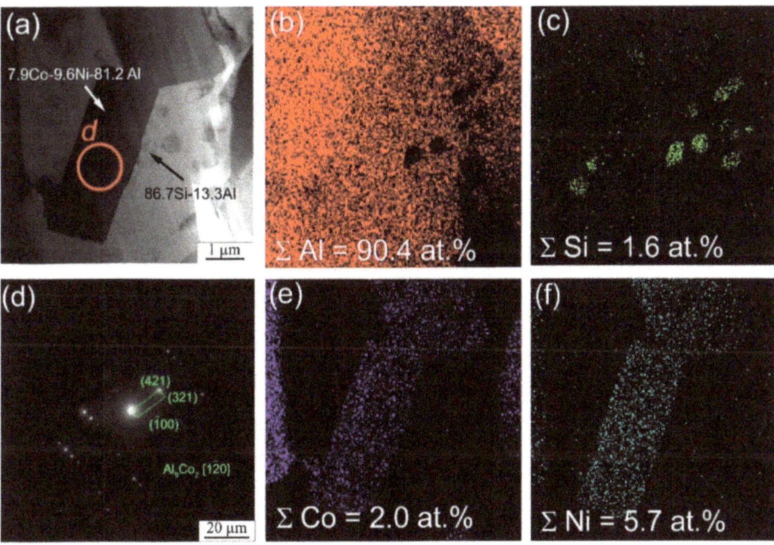

Figure 10. Bright-field TEM image of the intermetallic particles Al_9Co_2 (**a**), EDS element distribution maps (**b,c,e,f**) and SAED pattern from a particles Al_9Co_2 (**d**).

Increasing the content of nickel alloy in the matrix of the composite up to 10% leads to the formation of a structure similar to that described above. However, the proportion of intermetallic phases increases and the content of the α-Al solid solution decreases (Figure 11 and Table 4). In those areas where poor mixing of the components is observed, chromium enrichment is observed along the grain boundaries of AlNi (1, 2 in Figure 11 and Table 4). Presumably, AlCCr carbide is formed (3 in Figure 11b and Table 4). AlNi intermetallic phases based on aluminum and nickel also contain some dissolved chromium and cobalt.

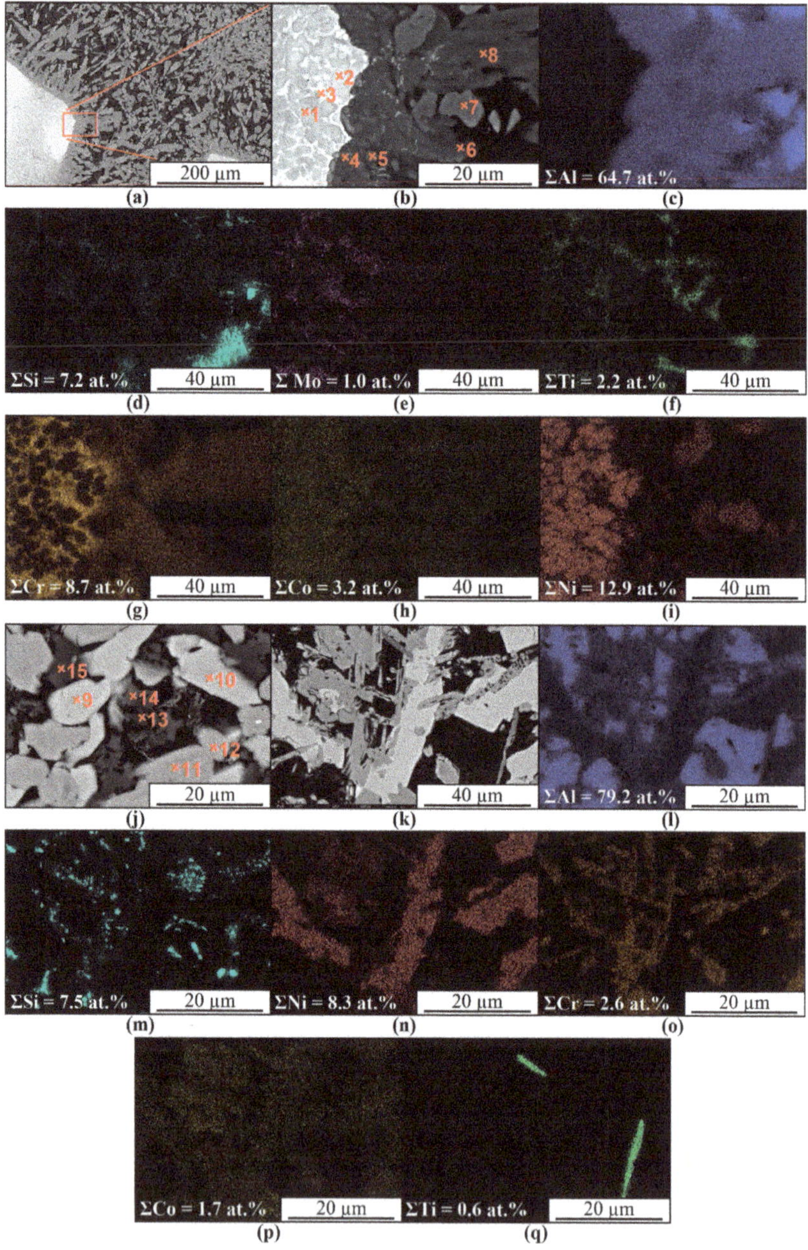

Figure 11. SEM images in BSE mode (**a,b,j,k**) and EDS maps of element distribution (**c–i,l–q**) obtained from the plot (**b,k**).

Table 4. The chemical composition of the Al-10%Ni sample in the local areas shown in Figure 11.

Spectrum	Chemical Composition, at.%							Possible Phase
	Al	Si	Ti	Cr	Co	Ni	Mo	
1	42.8	1.0	1.8	7.3	7.9	39.2	-	AlNi(Cr,Co)
2	40.0	0.4	1.8	6.2	8.0	43.7	-	AlNi(Cr,Co)
3	26.9	6.0	7.0	39.2	5.1	11.6	4.3	AlCCr$_2$
4	53.5	4.0	2.9	10.2	5.2	23.4	0.8	Al$_3$Ni
5	73.1	4.7	0.5	9.6	3.0	8.1	1.0	Al$_3$(Ni,Cr,Si)
6	73.7	5.4	0.5	9.9	2.8	6.6	1.1	Al$_3$(Ni,Cr,Si)
7	72.6	0.7	0.2	0.5	4.0	22.0	-	Al$_3$Ni
8	74.0	6.2	0.4	9.4	2.3	6.7	0.9	Al$_3$(Ni,Cr,Si)
9	73.6	0.6	0.1	0.2	3.4	22.1	-	Al$_3$Ni
10	74.0	0.5	0.2	0.3	2.6	22.4	-	Al$_3$Ni
11	75.4	4.5	1.0	9.4	2.8	5.9	1.0	Al$_3$(Ni,Cr,Si)
12	75.9	5.5	0.2	8.8	2.3	6.4	1.0	Al$_3$(Ni,Cr,Si)
13	32.5	58.8	0.3	2.2	0.8	5.3	-	β-Si
14	96.7	2.6	0.2	0.1	0.1	0.3	-	α-Al
15	96.4	1.5	0.3	0.8	0.2	0.8	-	α-Al

Therefore, in this case, the formation of AlNi(Cr,Co) and Al$_3$(Ni,Cr,Si) intermetallic compounds of complex composition also takes place (1, 2, 4, 5, 6, 8, 11, 12 at Figure 11 and Table 4). Al$_3$Ni grains with a number of impurities are also formed. The formation of Al$_3$Ti intermetallic phases in the form of thin plates or needles is also observed (Figure 11q). The microstructure of the aluminum matrix composite and globular particles also contains components of the original aluminum-silicon alloy (Figure 12). There is a solid solution of α-Al with dissolved Si (Figure 12, item 2). Microdiffraction images obtained from particles (1, 4, 5 in Figure 12a,c) indicate that they are silicon particles (Figure 12b,d). However, EDS analysis of a particle (item 3 in Figure 12a) containing 75.2 at.% Al and 22 at.% Ni indicates that it is an Al$_3$Ni particle. The particle also contains 1.1 at.% Si and 1.6 at.% Co.

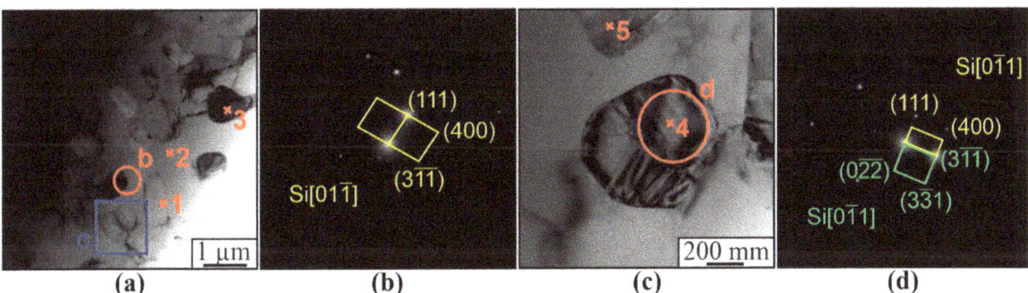

Figure 12. Bright field TEM images of a section of the microstructure of the Al-10% Ni composite (**a,c**), microdiffraction patterns (**b,d**) obtained from the area (**a,c**), respectively.

As shown by SEM, the composite structure contains lamellar particles enriched with titanium, mainly located in the eutectic (α+β) of the α-Al matrix (Figure 13a,b). This is also confirmed by TEM results with electron microdiffraction interpretation (Figure 13a,c). These particles are Al$_3$Ti; according to the EDS data, they also contain 6.9 at.% Si, 1.8 at.% Cr, and 2.7 at.% Mo. According to TEM data, the particle adjacent to Al$_3$Ti is Al$_3$Ni (Figure 13a,d).

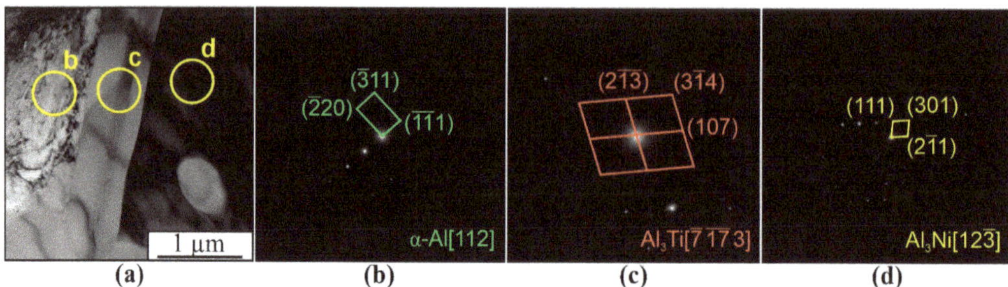

Figure 13. Bright field TEM image of a fragment of the microstructure of the Al-10% Ni composite (**a**) and microdiffraction patterns (**b–d**) obtained from the areas in figure (**a**).

The results of the uniaxial tensile tests show the brittle nature of the fracture of Al-Ni5% and Al-Ni10% composites caused by the formation of large intermetallic particles. Microcracking also occurs during the cooling of components with different crystallization temperatures and coefficients of thermal expansion. Cracks are formed mainly in intermetallic particles. When the additively produced AlSi5 aluminum alloy is modified with a nickel alloy during printing, the ultimate strength decreases from the initial 142–164 MPa to 55–123 MPa (Table 5 and Figure 14). Since the volume fraction of intermetallic phases is at a maximum in the Al-10%Ni composite, the least plastic and strong material is formed in this case.

Uniaxial compression tests show a significant increase in tensile strength due to multiphase hardening. Al-Ni5% composites have a tensile strength of 490–500 MPa in the horizontal direction and 530–570 MPa in the vertical loading direction (Figure 14 and Table 5). According to the results obtained, the tensile strength of Al-Ni10% composites is 905–945 MPa in the horizontal direction and 1150–1200 MPa in the vertical direction (Figure 14 and Table 5). The deformation behavior of the Al-5%Ni composite specimens is closest to that of the pure aluminum alloy specimens. At the same time, specimens of pure AlSi5 alloy practically do not break under compression and, although high plasticity is typical for Al-5%Ni specimens, at a strain higher than 0.075, the specimens break due to the formation of diagonal cracks. For Al-10%Ni specimens, due to the higher volume fraction of intermetallic compounds, higher strength is characteristic but fracture occurs without a pronounced plastic flow phase when the yield point is reached.

Table 5. Tensile strength values for uniaxial tension and compression in the horizontal and vertical directions, MPa.

Material	Test	Al		Al-Ni5%		Al-Ni10%	
		Top	Bottom	Top	Bottom	Top	Bottom
Vertical	Tension	142	143	74	121	57	123
Horizontal		145	164	79	59	55	88
Vertical	Compression	-	-	530	570	1150	1200
Horizontal		-	-	490	500	905	945

When the AlSi5 alloy is tested for wear resistance, no dependence of the coefficient of friction on the wear rate is observed (Figure 14e,f). The friction coefficient of the Al-5%Ni composite has a constant value of the friction coefficient of the order of 0.22–0.25 and does not depend on the sliding speed. The Al-10%Ni composite shows a decrease in the coefficient of friction from 0.46 to 0.36 with an increase in the friction speed from 0.06 to 0.24 m/s. The amount of wear, determined by the analysis of the cross profile of the friction marks, generally shows an increase with increasing friction velocity for all specimens

(Figure 14f). The highest wear is typical for pure aluminum alloy samples. Average values are typical for Al-5%Ni samples. The minimum wear value for Al-10%Ni samples is when the effect of the friction speed on the degree of wear of the material is also the smallest.

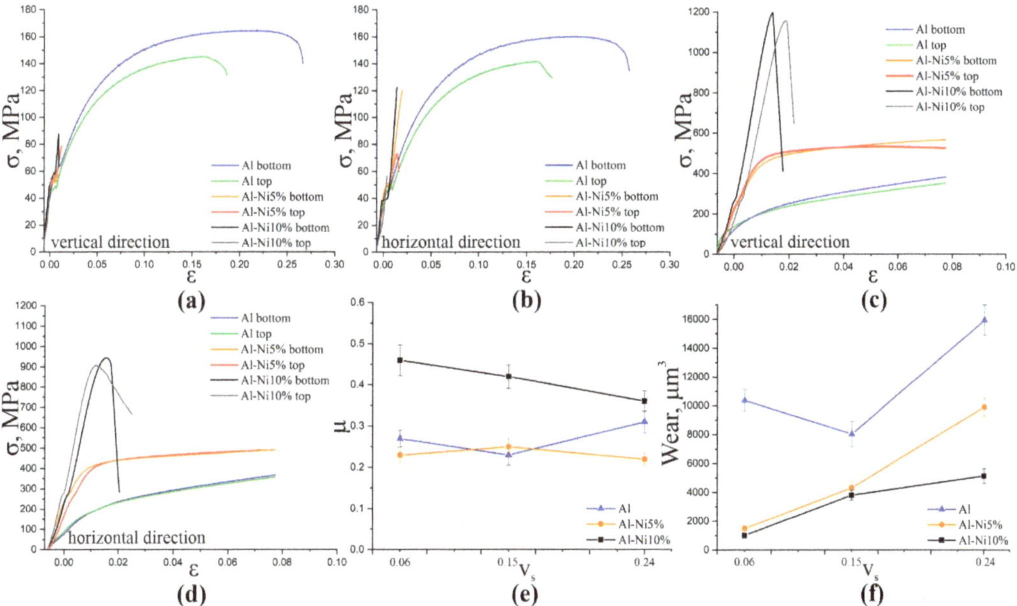

Figure 14. Engineering tensile stress–strain diagrams (**a**,**b**) and engineering compression stress-strain diagrams (**c**,**d**), friction coefficient diagrams (**e**), and wear rate diagrams (**f**) of aluminum alloy AlSi5 and composite materials Al-5%Ni, Al-10%Ni.

4. Discussion

Al_3Ni phase is the most favorable reaction in the aluminum-rich nickel region of the -Ni phase diagram [21]. The Gibbs free energy formations ΔGi (703 K) (kJ mol^{-1}) calculated by Qian et al. [32] connected with more low free energy at the Al_3Ni particles: Ni_3Al $(Ni_{0.75}Al_{0.25})$—35.86 kJ mol^{-1}; NiAl $(Ni_{0.50}Al_{0.50})$—66.92 kJ mol^{-1}; $Ni_2Al_3(Ni_{0.40}Al_{0.60})$—61.94 kJ mol^{-1}; $NiAl_3(Ni_{0.25}Al_{0.75})$—39.84 kJ mol^{-1}.

The existence of Al_3Ni XRD peaks (Figure 5) confirms that the intermetallic particles were formed due to the reaction between aluminum and nickel according to the equation $3Al + Ni \rightarrow Al_3Ni$. The height of the XRD peaks of Al_3Ni increases as the weight percentage of the nickel alloy increases. No nickel peaks were detected, suggesting that nickel was completely consumed in the reaction to form the intermetallic compound. This does not mean that the reaction time and holding period were sufficient for the complete synthesis of Al_3Ni, since in the structure of the composite we find areas with an aluminum and nickel ratio corresponding to the AlNi compound (Figures 8 and 9, Tables 2 and 3). The fact that we do not see the AlNi phase using XRD (Figure 5) means that its relative amount is not large compared to Al_3Ni.

As mentioned in the description of the results, the presence of Al_3Ni particles correlated with a decrease in the size of the dendrites of the α-solid solution in aluminum. The reduction in aluminum grain size indicates that the Al_3Ni particles act as a grain refiner. By all visibility synthesis, Al_3Ni particles changed the picture solidification. A nucleus is necessary for birthing new grains during solidification. The growth of α-grains of aluminum is unhindered until the grain boundary meets the boundary of another growing grain. The synthesis of Al_3Ni particles provides several grain nucleation regions for the production of new α-aluminum grains. The melting temperature of Al_3Ni is higher than that of the aluminum matrix. The regions surrounding the Al$_3$Ni particles are subjected to supercooling during solidification, which initiates the solidification of the aluminum grains. Growing aluminum grains encounter resistance to free growth due to the presence of Al$_3$Ni particles in the aluminum melt. Increasing the content of Al$_3$Ni particles increases the hindrance of free growth and increases the number of seed spots for α -aluminum grains.

The presence of pores in Figures 3 and 4 may be due to the accumulation of brittle intermetallic particles, which are partially chipped during the preparation of thin sections to study the microstructure. Since EBAM is a multi-layer hardfacing process, residual stress buildup increases with the height of the hardfacing, which exacerbates the initiation of cracks in intermetallic grains.

Observed micrograph features, such as particle clumps, pores, and sharp particle corners, do not improve tensile behavior, while compressive strength and wear resistance levels improve with increasing amounts of synthesized intermetallic compounds (Figure 14).

5. Conclusions

The conducted studies show that it is possible to obtain multicomponent multimetal composite materials based on aluminum alloy ER4043 and nickel alloy Udimet-500 using the method of wire additive electron beam technology. The structure resulting from the interaction of the main and alloying elements is represented by a wide range of phases of complex composition and different morphologies. The main drawbacks of the obtained materials are the presence of inhomogeneities of various scale levels and the formation of microcracks in large particles of intermetallic compounds. The reasons for the formation of such defects are significant differences in the density and temperature of nickel and aluminum alloys, the formation of thermal stresses during cooling, and differences in the values of the coefficients of thermal expansion of the intermetallic phases and the metal matrix.

The results of the study of the structural phase state show that a multicomponent structure is formed in the samples with the presence of solutions based on aluminum -Al or silicon β-Si, intermetallic phases Al$_3$Ni, Al$_3$Ni, Al$_{75}$Co$_{22}$Ni$_3$, and Al$_5$Co, as well as carbides of complex composition AlCCr, Al$_8$SiC$_7$, which is confirmed by the results of both the energy dispersive analysis and the X-ray diffraction method. In addition to these phases, intermetallic compounds and solid solutions of AlNi, Al$_3$Ti, Al$_9$Co$_2$, Al(Co,Ni), Al(Co,Ni,Si), Al$_3$(Ni,Cr,Si), and AlNi(Cr,Co) are also found, which form in smaller amounts and are not identified by the X-ray diffraction method.

A large amount of solid phases leads to the formation of a material with high hardness and low ductility. The fracture of composite specimens under tension and compression is brittle without revealing the stage of plastic flow. Tensile strength values are significantly reduced from the initial 142–164 MPa to 55–123 MPa. In compression, the strength limits increase to 490–570 MPa and 905–1200 MPa with the introduction of 5% and 10% nickel superalloys, respectively. The amount of wear in dry friction tests decreases significantly in composite specimens of both compositions. Its dependence on the nickel concentration is rather ambiguous, although the maximum wear resistance is observed for specimens with a nickel alloy content of 10%. Thus, in addition to the basic possibility of obtaining multiphase composite materials of complex composition using the method of two-wire electron beam technology, it is also possible to obtain relatively wear-resistant products based on aluminum alloys.

Author Contributions: Conceptualization, A.D. and A.C.; investigation, A.D., A.C., N.S. (Nikolay Savchenko), D.G., P.S., A.N., N.S. (Natalia Semenchuk), S.N. and A.Z.; resources, V.R. and E.K.; writing—original draft preparation, A.D., A.C. and A.Z.; translation, D.G.; guidance and corrections to the article, A.C., V.R. and A.Z. All authors have read and agreed to the published version of the manuscript.

Funding: The work was performed under State Assignment for ISPMS SB RAS (project No. FWRW-2021-0012).

Institutional Review Board Statement: Not applicable.

Informed Consent Statement: Not applicable.

Data Availability Statement: Data sharing is not applicable to this article.

Acknowledgments: The investigations have been carried out using the equipment of Share Use Centre "Nanotech" of the ISPMS SB RAS.

Conflicts of Interest: The authors declare no conflict of interest.

References

1. Lukach, I.; Shlesar, M.; Khrokh, P. Structure and mechanical properties of Silumin. *Met. Sci. Heat Treat.* **1976**, *18*, 624–626. [CrossRef]
2. Hernandez, F.C.R.; Ramírez, J.M.H.; Mackay, R. *Al-Si Alloys: Automotive, Aeronautical, and Aerospace Applications*; Springer: Berlin/Heidelberg, Germany, 2017.
3. Abdel-Jaber, G.T.; Omran, A.M.; Khalil, K.A.; Fujii, M.; Seki, M.; Yoshida, A. An investigation into solidification and mechanical properties behavior of Al-Si casting alloys. *Int. J. Mech. Mechatron. Eng. IJMME-IJENS* **2010**, *10*, 30–35.
4. Kobayashi, T. Strength and fracture of aluminum alloys. *Mater. Sci. Eng. A* **2000**, *280*, 8–16. [CrossRef]
5. Caceres, C. Microstructure Design and Heat Treatment Selection for Casting Alloys Using the Quality Index. *J. Mater. Eng. Perform.* **2000**, *9*, 215–221. [CrossRef]
6. Perrin, C.; Rainforth, W.M. Work hardening behaviour at the worn surface of Al-Cu and Al-Si alloys. *Wear* **1997**, *203*, 171–179. [CrossRef]
7. Sigworth, G.K. The Modification of Al-Si Casting Alloys: Important Practical and Theoretical Aspects. *Int. J. Met.* **2008**, *2*, 19–40. [CrossRef]
8. Heusler, L.; Schneider, W. Influence of alloying elements on the thermal analysis results of Al-Si cast alloys. *J. Light Met.* **2002**, *2*, 17–26. [CrossRef]
9. Morri, A.; Ceschini, L.; Messieri, S.; Cerri, E.; Toschi, S. Mo addition to the a354 (Al-Si-Cu-Mg) casting alloy: Effects on microstructure and mechanical properties at room and high temperature. *Metals* **2018**, *8*, 393. [CrossRef]
10. Shaha, S.K.; Czerwinski, F.; Kasprzak, W.; Friedman, J.; Chen, D.L. Effect of Zr, V and Ti on hot compression behavior of the Al–Si cast alloy for powertrain applications. *J. Alloys Compd.* **2014**, *615*, 1019–1031. [CrossRef]
11. Cho, Y.H.; Kim, H.W.; Lee, J.M.; Kim, M.S. A new approach to the design of a low Si-added Al-Si casting alloy for optimising thermal conductivity and fluidity. *J. Mater. Sci.* **2015**, *50*, 7271–7281. [CrossRef]
12. Rana, R.S.; Purohit, R.; Das, S. Reviews on the influences of alloying elements on the microstructure and mechanical properties of aluminum alloys and aluminum alloy composites. *Int. J. Sci. Res. Publ.* **2012**, *2*, 1–7.
13. Saini, N.; Pandey, C.; Thapliyal, S.; Dwivedi, D.K. Mechanical Properties and Wear Behavior of Zn and MoS2 Reinforced Surface Composite Al-Si Alloys Using Friction Stir Processing. *Silicon* **2018**, *10*, 1979–1990. [CrossRef]
14. Shaha, S.K.; Czerwinski, F.; Kasprzak, W.; Friedman, J.; Chen, D.L. Microstructure and mechanical properties of Al–Si cast alloy with additions of Zr-V-Ti. *Mater. Des.* **2015**, *83*, 801–812. [CrossRef]
15. El-Labban, H.F.; Abdelaziz, M.; Mahmoud, E.R. Preparation and characterization of squeeze cast-Al–Si piston alloy reinforced by Ni and nano-Al_2O_3 particles. *J. King Saud Univ.-Eng. Sci.* **2016**, *28*, 230–239. [CrossRef]
16. Mohan, R.R.; Venkatraman, R.; Raghuraman, S.; Kumar, P.M.; Rinawa, M.L.; Subbiah, R.; Arulmurugan, B.; Rajkumar, S. Processing of Aluminium-Silicon Alloy with Metal Carbide as Reinforcement through Powder-Based Additive Manufacturing: A Critical Study. *Scanning* **2022**, *2022*, 5610333. [CrossRef]
17. Rometsch, P.A.; Zhu, Y.; Wu, X.; Huang, A. Review of high-strength aluminium alloys for additive manufacturing by laser powder bed fusion. *Mater. Des.* **2022**, *219*, 10779. [CrossRef]
18. Hernández-Méndez, F.; Altamirano-Torres, A.; Miranda-Hernández, J.G.; Térres-Rojas, E.; RochaRangel, E. Effect of nickel addition on microstructure and mechanical properties of aluminum-based alloys. *Mater. Sci. Forum* **2011**, *691*, 10–14. [CrossRef]
19. Suwanpreecha, C.; Pandee, P.; Patakham, U.; Limmaneevichitr, C. New generation of eutectic Al-Ni casting alloys for elevated temperature services. *Mater. Sci. Eng. A* **2018**, *709*, 46–54. [CrossRef]
20. Deng, J.; Chen, C.; Liu, X.; Li, Y.; Zhou, K.; Guo, S. A high-strength heat-resistant Al−5.7Ni eutectic alloy with spherical Al_3Ni nano-particles by selective laser melting. *Scr. Mater.* **2021**, *203*, 114034. [CrossRef]
21. Gxowa-Penxa, Z.; Daswa, P.; Modiba, R.; Mathabathe, M.; Bolokang, A. Development and characterization of Al-Al_3Ni-Sn metal matrix composite. *Mater. Chem. Phys.* **2020**, *259*, 124027. [CrossRef]
22. Zuo, L.; Ye, B.; Feng, J.; Zhang, H.; Kong, X.; Jiang, H. Effect of ε-Al_3Ni phase on mechanical properties of Al-Si-Cu-Mg-Ni alloys at elevated temperature. *Mater. Sci. Eng. A* **2020**, *772*, 138794. [CrossRef]
23. Balakrishnan, M.; Dinaharan, I.; Kalaiselvan, K.; Palanivel, R. Friction stir processing of Al_3Ni intermetallic particulate reinforced cast aluminum matrix composites: Microstructure and tensile properties. *J. Mater. Res. Technol.* **2020**, *9*, 4356–4367. [CrossRef]

24. Pariyar, A.; Perugu, C.S.; Dash, K.; Kailas, S.V. Microstructure and mechanical behavior of high toughness Al-based metal matrix composite reinforced with in-situ formed nickel aluminides. *Mater. Charact.* **2020**, *171*, 110776. [CrossRef]
25. Khaki-Davoudi, S.; Nourouzi, S.; Aval, H.J. Microstructure and mechanical properties of AA7075/Al3Ni composites produced by compocasting. *Mater. Today Commun.* **2021**, *28*, 102537. [CrossRef]
26. Czerwinski, F.; Aniolek, M.; Li, J. Strengthening retention and structural stability of the Al-Al3Ni eutectic at high temperatures. *Scr. Mater.* **2022**, *214*, 114679. [CrossRef]
27. Wang, X.; Ran, Z.; Wei, Z.; Zou, C.; Wang, H.; Gouchi, J.; Uwatoko, Y. The formation of bulk β-Al3Ni phase in eutectic Al-5.69wt%Ni alloy solidified under high pressure. *J. Alloys Compd.* **2018**, *742*, 670–675. [CrossRef]
28. Gonzalez, G.; Sagarzazu, A.; Bonyuet, D.; D'angelo, L.; Villalba, R. Solid state amorphisation in binary systems prepared by mechanical alloying. *J. Alloys Compd.* **2009**, *483*, 289–297. [CrossRef]
29. Zhang, Z.; Akiyama, E.; Watanabe, Y.; Katada, Y.; Tsuzaki, K. Effect of α-Al/Al3Ni microstructure on the corrosion behaviour of Al-5.4wt% Ni alloy fabricated by equal-channel angular pressing. *Corros. Sci.* **2007**, *49*, 2962–2972. [CrossRef]
30. Uan, J.; Chen, L.; Lui, T. On the extrusion microstructural evolution of Al-Al3Ni in situ composite. *Acta Mater.* **2001**, *49*, 313–320. [CrossRef]
31. Song, C.-J.; Xu, Z.-M.; Li, J.-G. In-situ Al/Al3Ni functionally graded materials by electromagnetic separation method. *Mater. Sci. Eng. A* **2007**, *445–446*, 148–154. [CrossRef]
32. Qian, J.; Li, J.; Xiong, J.; Zhang, F.; Lin, X. In situ synthesizing Al3Ni for fabrication of intermetallic-reinforced aluminum alloy composites by friction stir processing. *Mater. Sci. Eng. A* **2012**, *550*, 279–285. [CrossRef]
33. Zhang, S.; Li, Y.; Frederick, A.; Wang, Y.; Wang, Y.; Allard, L.; Koehler, M.; Shin, S.; Hu, A.; Feng, Z. In-situ formation of Al3Ni nano particles in synthesis of Al 7075 alloy by friction stir processing with Ni powder addition. *J. Mater. Process. Technol.* **2023**, *311*, 117803. [CrossRef]
34. Dinaharan, I. Liquid metallurgy processing of intermetallic matrix composites. In *Intermetallic Matrix Composites: Properties and Applications*; Mitra, R., Ed.; Elsevier: Amsterdam, The Netherlands, 2018; pp. 167–202.
35. Mitra, R. *Composites of Nickel Aluminides, Intermetallic Matrix Composites: Properties and Applications*; Woodhead Publishing Limited: Sawston, UK, 2018; pp. 37–59.
36. Camagu, S.; Mathabathe, M.; Motaung, D.; Muller, T.; Arendse, C.; Bolokang, A. Investigation into the thermal behaviour of the B2–NiAl intermetallic alloy produced by compaction and sintering of the elemental Ni and Al powders. *Vacuum* **2019**, *169*, 108919. [CrossRef]
37. Fuchs, J.; Schneider, C.; Enzinger, N. Wire-based additive manufacturing using an electron beam as heat source. *Weld. World* **2018**, *62*, 267–275. [CrossRef]
38. Osipovich, K.S.; Astafurova, E.G.; Chumaevskii, A.V.; Kalashnikov, K.N.; Astafurov, S.V.; Maier, G.G.; Kolubaev, E.A. Gradient transition zone structure in "steel–copper" sample produced by double wire-feed electron beam additive manufacturing. *J. Mater. Sci.* **2020**, *55*, 9258–9272. [CrossRef]
39. Osipovich, K.; Kalashnikov, K.; Chumaevskii, A.; Gurianov, D.; Kalashnikova, T.; Vorontsov, A.; Zykova, A.; Utyaganova, V.; Panfilov, A.; Nikolaeva, A.; et al. Wire-Feed Electron Beam Additive Manufacturing: A Review. *Metals* **2023**, *13*, 279. [CrossRef]
40. Ding, D.; Pan, Z.; Cuiuri, D.; Li, H. Wire-feed additive manufacturing of metal components: Technologies, developments and future interests. *Int. J. Adv. Manuf. Technol.* **2015**, *81*, 465–481. [CrossRef]
41. Pu, Z.; Du, D.; Wang, K.; Liu, G.; Zhang, D.; Zhang, H.; Xi, R.; Wang, X.; Chang, B. Study on the NiTi shape memory alloys in-situ synthesized by dual-wire-feed electron beam additive manufacturing. *Addit. Manuf.* **2022**, *56*, 102886. [CrossRef]
42. Pu, Z.; Chang, B. Control of droplet transfer during in-situ synthesis of NiTi alloys by dual-wire electron beam additive manufacturing. *J. Phys. Conf. Ser.* **2022**, *2369*, 012011. [CrossRef]
43. Kolubaev, E.A.; Rubtsov, V.E.; Chumaevsky, A.V.; Astafurova, E.G. Micro-, Meso- and Macrostructural Design of Bulk Metallic and Polymetallic Materials by Wire-Feed Electron-Beam Additive Manufacturing. *Phys. Mesomech.* **2022**, *25*, 479–491. [CrossRef]
44. Li, Z.; Chang, B.; Cui, Y.; Zhang, H.; Liang, Z.; Liu, C.; Wang, L.; Du, D.; Chang, S. Effect of twin-wire feeding methods on the in-situ synthesis of electron beam fabricated Ti-Al-Nb intermetallics. *Mater. Des.* **2022**, *215*, 110509. [CrossRef]
45. Zykova, A.; Chumaevskii, A.; Vorontsov, A.; Kalashnikov, K.; Gurianov, D.; Gusarova, A.; Kolubaev, E. Evolution of microstructure and properties of Fe-Cu, manufactured by electron beam additive manufacturing with subsequent friction stir processing. *Mater. Lett.* **2021**, *307*, 131023. [CrossRef]
46. Lei, S.; Li, X.; Deng, Y.; Xiao, Y.; Chen, Y.; Wang, H. Microstructure and mechanical properties of electron beam freeform fabricated TiB2/Al-Cu composite. *Mater. Lett.* **2020**, *277*, 128273. [CrossRef]
47. Chen, G.; Shu, X.; Liu, J.; Zhang, B.; Feng, J. A new coating method with potential for additive manufacturing: Premelting electron beam-assisted freeform fabrication. *Addit. Manuf.* **2020**, *33*, 101118. [CrossRef]
48. Zykova, A.; Chumaevskii, A.; Panfilov, A.; Vorontsov, A.; Nikolaeva, A.; Osipovich, K.; Gusarova, A.; Chebodaeva, V.; Nikonov, S.; Gurianov, D.; et al. Aluminum Bronze/Udimet-500 Composites Prepared by Electron-Beam Additive Double-Wire-Feed Manufacturing. *Materials* **2022**, *15*, 6270. [CrossRef]

49. Kalashnikov, K.; Kalashnikova, T.; Semenchuk, V.; Knyazhev, E.; Panfilov, A.; Cheremnov, A.; Chumaevskii, A.; Nikonov, S.; Vorontsov, A.; Rubtsov, V.; et al. Development of a Multimaterial Structure Based on CuAl9Mn2 Bronze and Inconel 625 Alloy by Double-Wire-Feed Additive Manufacturing. *Metals* **2022**, *12*, 2048. [CrossRef]

50. Filippov, A.; Utyaganova, V.; Shamarin, N.; Vorontsov, A.; Savchenko, N.; Gurianov, D.; Chumaevskii, A.; Rubtsov, V.; Kolubaev, E.; Tarasov, S. Microstructure and Corrosion Resistance of AA4047/AA7075 Transition Zone Formed Using Electron Beam Wire-Feed Additive Manufacturing. *Materials* **2021**, *14*, 6931. [CrossRef]

Article

Influence of Different Build Orientations and Heat Treatments on the Creep Properties of Inconel 718 Produced by PBF-LB

Anke Kaletsch [1,2,*], Siyuan Qin [1] and Christoph Broeckmann [1,2]

[1] Institute for Materials Applications in Mechanical Engineering (IWM), RWTH Aachen University, Augustinerbach 4, 52062 Aachen, Germany
[2] Institute of Applied Powder Metallurgy and Ceramics at RWTH Aachen e.V. (IAPK), Augustinerbach 4, 52062 Aachen, Germany
* Correspondence: a.kaletsch@iwm.rwth-aachen.de

Abstract: Inconel 718 is a nickel-based superalloy with excellent creep properties and good tensile and fatigue strength. In the field of additive manufacturing, it is a versatile and widely used alloy due to its good processability in the powder bed fusion with laser beam (PBF-LB) process. The microstructure and mechanical properties of the alloy produced by PBF-LB have already been studied in detail. However, there are fewer studies on the creep resistance of additively manufactured Inconel 718, especially when the focus is on the build direction dependence and post-treatment by hot isostatic pressing (HIP). Creep resistance is a crucial mechanical property for high-temperature applications. In this study, the creep behavior of additively manufactured Inconel 718 was investigated in different build orientations and after two different heat treatments. The two heat treatment conditions are, first, solution annealing at 980 °C followed by aging and, second, HIP with rapid cooling followed by aging. The creep tests were performed at 760 °C and at four different stress levels between 130 MPa and 250 MPa. A slight influence of the build direction on the creep properties was detected, but a more significant influence was shown for the different heat treatments. The specimens after HIP heat treatment show much better creep resistance than the specimens subjected to solution annealing at 980 °C with subsequent aging.

Keywords: PBF-LB; Inconel 718; creep; HIP; post-processing; hot isostatic pressing

Citation: Kaletsch, A.; Qin, S.; Broeckmann, C. Influence of Different Build Orientations and Heat Treatments on the Creep Properties of Inconel 718 Produced by PBF-LB. *Materials* **2023**, *16*, 4087. https://doi.org/10.3390/ma16114087

Academic Editors: Bartłomiej Wysocki, Joseph Buhagiar and Tomasz Durejko

Received: 4 May 2023
Revised: 25 May 2023
Accepted: 29 May 2023
Published: 31 May 2023

1. Introduction

Powder bed fusion with laser beam (PBF-LB) is a widely researched and utilized additive manufacturing process that uses a high-power laser to selectively melt metal powder layerwise, thereby producing components with the highest geometric freedom in a layer-by-layer structure [1–3]. PBF-LB produced samples typically present an anisotropic microstructure due to epitaxial grain growth in the build direction, resulting in anisotropic mechanical properties [2,4]. Aerospace engineering has a strong interest in beam-based additive manufacturing processes like PBF-LB or powder bed fusion with electron beam (PBF-EB), focusing on processing titanium alloys, titanium aluminides, and nickel-based alloys [5–8].

Inconel 718, a precipitation hardening and solid solution strengthened nickel-based superalloy, is commonly used in high-temperature applications, such as gas turbine disks and combustors, due to its excellent creep properties, good tensile and fatigue strength, and corrosion resistance [9,10]. The matrix of Inconel 718 consists of Ni-Fe-Cr austenite (γ), with very fine γ'-($Ni_3(Al,Ti)$) and γ''-(Ni_3Nb) precipitates dispersed in the γ-matrix after heat treatment for strengthening. γ'' is metastable and has an identical chemical composition to the stable δ-phase. However, δ-phase cannot contribute to the strength of the matrix because, unlike the cubic γ''-phase, the orthorhombic δ-phase is incoherent with the γ-matrix. In addition, its needle-like shape might have a notch effect in the microstructure and, as a result, have a negative influence on the fatigue strength [11]. Additionally, when δ-phase forms, it consumes Nb from the matrix, leading to a loss of γ''-precipitates and strength [12]. However, if δ-phase precipitates at the grain boundaries, which is mostly the case, it suppresses grain growth [13] and affects the grain boundary creep fracture [14]. The grain growth inhibiting effect of the δ-phase is particularly desirable when forging Inconel 718.

Although plenty of publications exist on the PBF-LB manufacturing of nickel-based alloys, the creep strength of additively manufactured nickel superalloys has not yet been extensively researched. In this context, most publications exist on the creep of Inconel 718 [15–19]. The influence of the typical PBF-LB microstructure, mentioned above, on the creep behavior has already been described [20]. Moreover, some investigations have been made on the influence of heat treatment. Using the AMS 5662 standard heat treatment for additively manufactured Inconel 718, which includes solution annealing at 980 °C and double aging at 720 °C and 620 °C, results in a lower creep strength than a forged reference due to the high content of the Laves- and δ-phase [21]. The Laves phase is found in the as-built condition due to Nb segregation in the dendrite interstices. Solution annealing at 980 °C is not sufficient to completely dissolve the Laves phase and homogeneously dissolve and distribute Nb in the matrix. Therefore, a large number of acicular δ-phase forms within the crystallites in regions with increased Nb content. By consuming Nb, the δ-phase reduces the formation of the γ''-phase, which is needed for superior creep properties. The very coarse δ-phase at the grain boundaries can also have an unfavorable effect on the creep properties [22]. Therefore, increasing the solution temperature in the heat treatment can improve the creep properties [23,24].

Hot isostatic pressing (HIP) is a process in powder metallurgy that consolidates metal powder filled in a metallic capsule into solid materials by simultaneously applying high pressure and high temperature [25]. This technology can also be used to consolidate materials with pores or voids to full density. For example, cast turbine blades for safety-relevant aerospace applications are post-densified with HIP as standard to guarantee high mechanical properties and high reliability. The HIP process is also suitable for the post-densification of additively manufactured materials and, thus, optimizing their mechanical properties. In particular, fatigue strength can be significantly increased via HIP post-treatment [26–31], due to minimizing the fatigue critical defects like the lack of fusion (LOF), microcracks, and porosity. HIP can also have a positive effect on the creep properties of PBF-LB manufactured materials. Kuo et al. [23] and Qin et al. [32] showed that HIP can improve the creep life of PBF-LB manufactured Inconel 718. This can be attributed, on the one hand, to the fact that the solution annealing temperature in HIP treatment is higher than in conventional heat treatment. High HIP temperatures, for nickel-based superalloys between 1100 °C and 1280 °C [25], are generally required to achieve full densification of the material at the applied pressure. This leads to a different solution state of the alloying elements and possibly grain growth. On the other hand, the defects typical for PBF-LB, such as LOF, cracks, and pores, lower the creep rupture strain. Thus, densification of these defects can improve the creep properties.

Nonetheless, HIP usually also leads to a homogenization of the microstructure, which is maybe not advantageous in the case of creep properties. The elongated grains resulting from epitaxial grain growth could have a positive influence on the creep behavior in the build direction. Rickenbacher et al. [33] have shown in creep tests on IN738LC that the columnar grain structure resulting from epitaxial grain growth in the PBF-LB process leads to better creep properties for specimens built vertically compared to those built horizontally. Homogenization of the columnar grain structure during HIP could, thus, also have an effect on the orientation dependency of the creep properties.

Therefore, the aim of the present study was to investigate the orientation dependency of the creep properties of Inconel 718 produced via PBF-LB. Furthermore, it considered whether a HIP post-treatment influences the orientation dependency of the creep properties. For this purpose, two different build orientations were investigated to evaluate the influence of the anisotropic PBF-LB microstructure. Then, HIP heat treatment with integrated quenching and subsequent aging was performed on one half of the samples.

As a reference, the other half of the specimens were subjected to a conventional standard heat treatment (AMS 5662). In addition to the microstructure, the rupture characteristics of the fractured specimens were evaluated after the creep tests and correlated with the obtained creep properties.

2. Materials and Methods

2.1. Material and PBF-LB Process

Argon-atomized Inconel 718 (IN718, Alloy 718, EN NiCr19NbMo, 2.4668) powder was used for this study and supplied by Carpenter Technology Corporation. The chemical composition of the powder was in accordance with the ASTM B637-18 specification [34], as shown in Table 1. The powder particles had a spherical shape (see Figure 1) and a particle size in the range of 15 to 45 μm.

Table 1. Chemical composition of the IN718 powder in comparison to the standard range of chemical composition according to ASTM B637-18 [34].

	C	Mn	Si	Cr	Ni	Mo	Nb	Ti	Al	Cu	Fe
Investigated Material	0.04	0.01	0.02	18.73	54.24	2.95	4.81	0.97	0.46	0.01	Bal.
ASTM Standard	<0.08	<0.35	<0.35	17–21	50–55	2.8–3.3	4.75–5.5	0.65–1.15	0.2–0.8	<0.3	Bal.

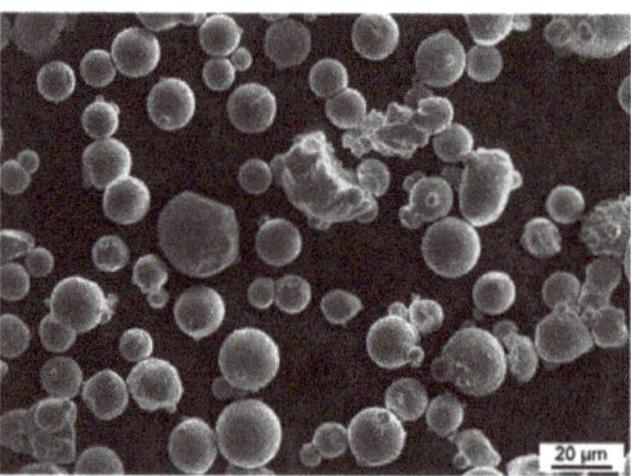

Figure 1. The morphology of the IN718 powder used in this study.

The creep specimens were produced under an argon atmosphere with a ReaLizer SLM 100 machine (ReaLizer GmbH, Borchen, Germany) utilizing a ytterbium fiber laser. The parameters for the PBF-LB fabrication of IN718 were taken from a previous study [35]: a laser power of 160 W, a laser scanning speed of 1000 mm/s, and a hatch distance of 100 μm were used. With these parameters, cylindrical specimens with a diameter of 11.5 mm were built up in two different orientations, either horizontally or at an angle of 30° with respect to the build direction, as depicted in Figure 2. For the analysis of the microstructure before creep testing, additional cubic specimens with an edge length of 10 mm were built.

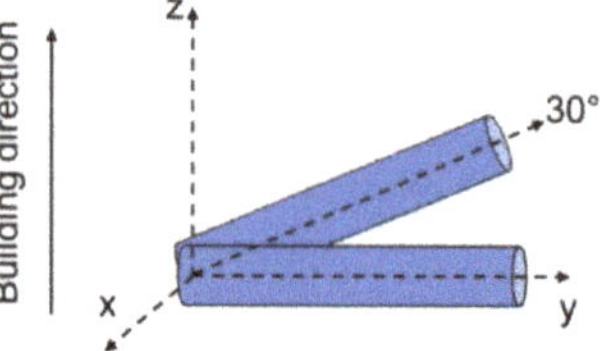

Figure 2. Schematic representation of the different build orientations. The cylindrical specimens were built either horizontally or at an angle of 30°.

2.2. Heat Treatment

Different heat treatment conditions were investigated for the specimens built in different orientations. In one variation, a usual standard heat treatment for IN718 was performed, which means that half of all the specimens were heat treated according to the AMS 5662 standard. This heat treatment included solution annealing at 980 °C for 1 h, followed by quenching in water. Subsequent double aging was performed at 720 °C for 8 h, followed by furnace cooling at 50 K/h and 620 °C for 8 h, followed by air cooling, as shown in Figure 3a.

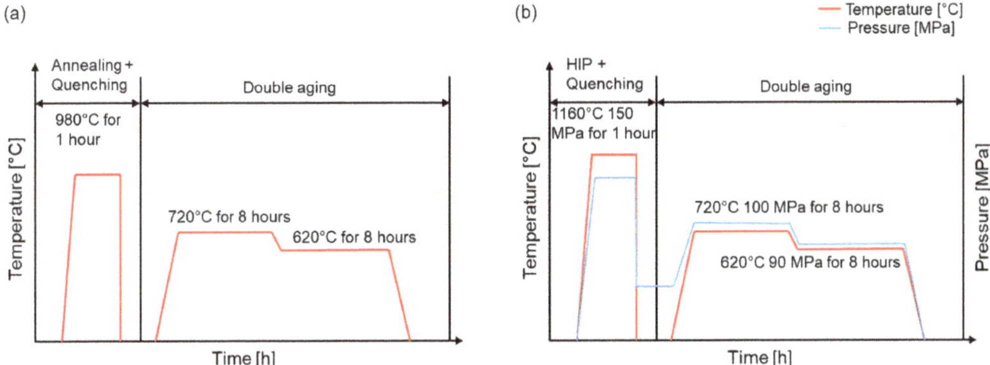

Figure 3. Graphical representation of the temperature profiles for the two heat treatments used in this study: (**a**) heat treatment for IN718 according to AMS 5662; (**b**) HIP heat treatment with integrated quenching and aging under pressure.

The second half of the specimens were heat treated under pressure in the HIP with a Uniform Rapid Cooling (URC©) furnace at Quintus Technologies Application Centre (Västerås, Sweden). Typical HIP parameters for IN718 were utilized, with a HIP temperature set to 1160 °C at 150 MPa for 4 h. This was followed by gas quenching and a double aging treatment at 710 °C and 100 MPa for 8 h, and 610 °C and 90 MPa for 8 h, as shown in Figure 3b.

2.3. Microstructure Characterization

To study the microstructure before the creep testing, the cubic specimens were mechanically ground, polished and then etched with Kalling's 2 reagents for 5 min. Microstructural analyses were performed using a light optical microscope (LOM) Zeiss 'Axio Imager M2m' (Carl Zeiss AG, Oberkochen, Germany) and Leica DM4000 (Leica Microsystems GmbH, Wetzlar, Germany) and a scanning electron microscope (SEM) Helios Nanolab G3 CX (FEI, Hillsboro, OR, USA). The porosity of the specimens was determined using image analysis on the unetched cross-sections.

2.4. Creep Testing and Fracture Analysis

The creep tests were performed at four different stress levels of 130 MPa, 170 MPa, 210 MPa, and 250 MPa, and a temperature of 760 °C. The specimens were machined to the geometry required for the creep test, with a gauge length of 50 mm, and a diameter of 5 mm, as shown in Figure 4. A lever-arm creep testing machine was used for the creep test, leading to loading with a constant axial force during the test, according to ASTM E139-2011 [36]. After testing, the fracture surfaces were examined, and the microstructures were analyzed using SEM. Due to the extended duration of creep tests, substantial noise is present in the raw data. MATLAB was employed to filter the noise and process the data effectively.

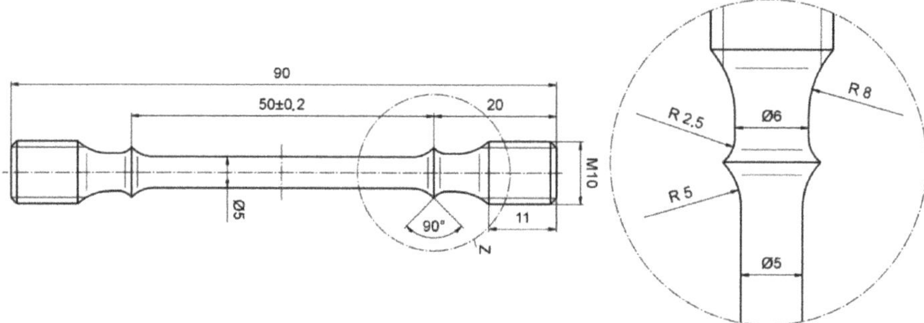

Figure 4. Specimen geometry for the creep tests.

3. Results and Discussion

3.1. Microstructure

The polished cross-sections of the unetched specimens revealed spherical, PBF-LB process-induced pores in the as-built condition, as depicted in Figure 5a. The as-built porosity was determined via image analysis to be 0.083%. After HIP treatment, almost all the pores were densified, resulting in a porosity of 0.04%. A few small pores can still be seen after the HIP process, see Figure 5b.

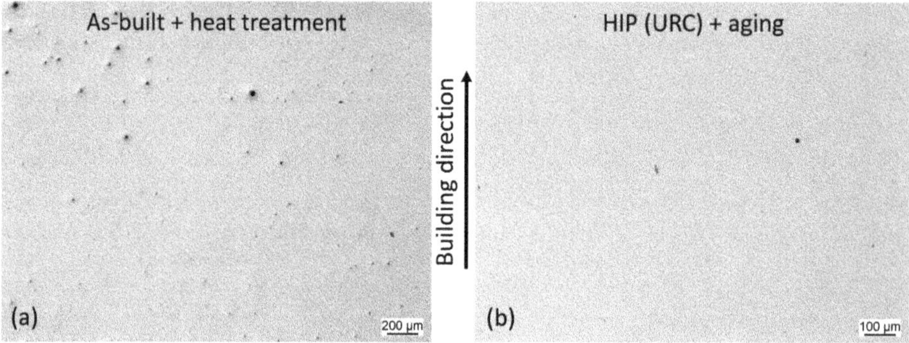

Figure 5. Cross-sections of the specimen: (**a**) as-built; (**b**) after HIP.

After chemical etching, the microstructure can be revealed in more detail. Figure 6a illustrates the melt pools and layer formation resulting from an alternating (x-y) layer-by-layer melt scan. The average layer thickness is approximately 50 μm, while the width of the melt pools varies between 70 and 120 μm. The grains exhibit an elongated shape in the build-up direction, resulting from the PBF-LB typical epitaxial grain growth [14]. The dendritic structures observed within the grains are a consequence of rapid cooling during the PBF-LB process. Segregations of the alloying elements can be seen in dark gray at the lower edges of the melt pools.

After the standard heat treatment with solution annealing and aging, the melt pools are dissolved, but the elongated grains remain (Figure 6b). In contrast, after HIP and aging, the microstructure resulting from the PBF-LB process is dissolved and homogenized, as shown in Figure 6c. The black dots that can be seen in Figure 6c are carbides that were not dissolved in the matrix during the HIP process. These carbides can be observed at the previous grain boundaries of the PBF-LB microstructure, where they formed during the PBF-LB process.

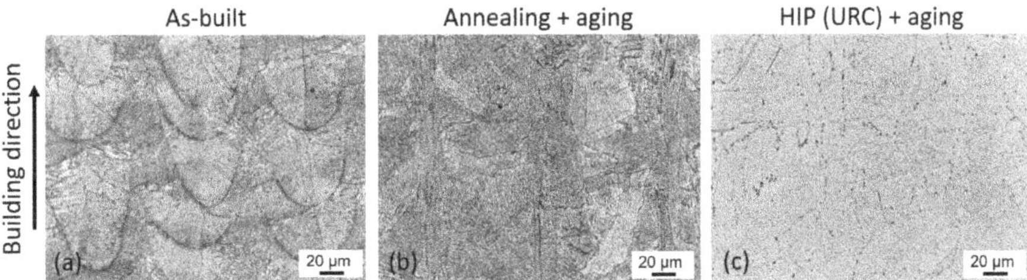

Figure 6. LOM images of the etched specimens: (**a**) as built; (**b**) after annealing at 980 °C and aging; (**c**) after HIP with integrated quenching and aging.

The microstructure of the specimens was further analyzed in detail using SEM. Figure 7 presents the different microstructural conditions: in the as-built condition (Figure 7a,d), a fine solidification structure can be observed inside and around the melt pools, while the structure outside the melt pools is coarser. The interdendritic phase in the PBF-LB samples, visible as white spots in Figure 7d, was assumed to be a Laves phase because this is consistent with descriptions from the literature [37,38]. The Nb-rich Laves phase is formed during the rapid solidification of the PBF-LB process, due to Nb segregation in the interdendritic regions. The standard AMS 5662 heat treatment cannot dissolve the Laves phase due to the too low solution annealing temperature of 980 °C and leads to the formation of intergranular acicular δ-phase (Ni_3Nb) from the Nb-rich regions of the Laves phase (Figure 7b,e). However, the HIP process can completely dissolve both the Laves- and the δ-phase due to the high solution heat treatment temperature of 1160 °C, resulting in a homogeneous distribution of Nb in the matrix. In addition, the rapid cooling during HIP prevents further δ-phase precipitation. As depicted in Figure 7c,f, the microstructure after HIP and aging does not display any δ-phase. Instead, small white spheres are observable at the grain boundaries, representing the carbides previously mentioned. Additionally, even smaller white dots can be observed in the matrix, which could be carbides or potentially γ' precipitates, as presented in Figure 7f. The γ'' precipitates, on the other hand, are too small to be seen with the SEM used in this study. However, it can be assumed that due to the dissolution of the Laves- and δ-phase and, thus, a high Nb content in the matrix, a high proportion of γ'' phase is precipitated during the heat treatment used in the HIP process. However, even though the material was recrystallized during HIP, the carbides still map the grain morphology of the PBF-LB microstructure, indicating that they did not completely dissolve during the HIP.

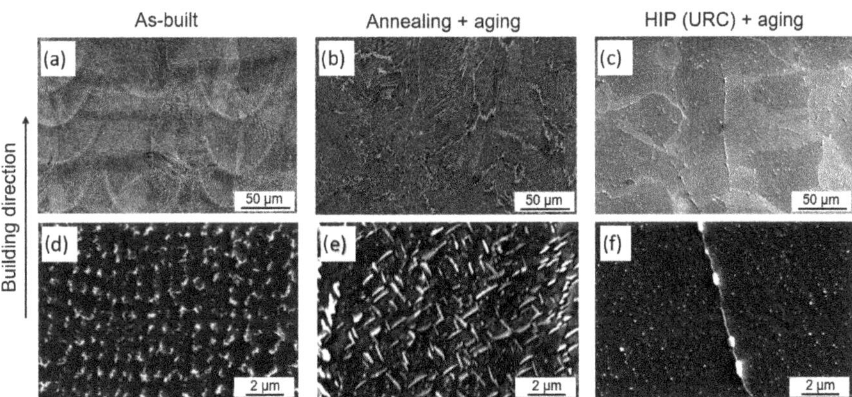

Figure 7. SEM images of the microstructure in different conditions: (**a**,**d**) as-built; (**b**,**e**) after standard heat treatment; (**c**,**f**) after HIP with integrated quenching and aging.

These results on the microstructure are in very good agreement with the results from a previous study [35], on additively manufactured Inconel 718 using the same PBF-LB and heat treatment parameters.

3.2. Tensile Creep Results

The creep performance for the different conditions was investigated at a constant temperature of 760 °C with four different stress levels from 130 MPa to 250 MPa. The creep curves in Figure 8 and the creep rates in Figure 9 exhibit a typical creep behavior for metallic material: after an initially unsteady plastic deformation, where the material starts to flow, a steady-state creep region with a constant deformation rate is reached and, finally, the deformation accelerate until failure.

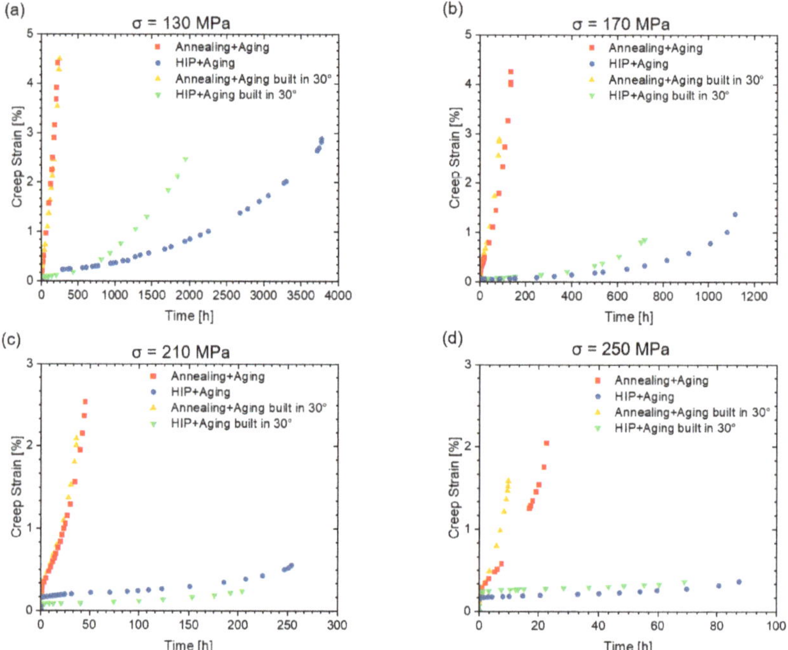

Figure 8. Creep curves of the specimens in different heat treatment conditions. The creep tests were conducted at 760 °C with four different stresses, specifically (**a**) 130 MPa, (**b**) 170 MPa, (**c**) 210 MPa, and (**d**) 250 MPa.

Figures 8a and 9a show the creep curves at a stress of 130 MPa. The specimens after HIP present a lower strain rate and a longer time to rupture than the specimen from the group with conventional heat treatment. The influence of build direction in the group with conventional heat treatment is not significant at the lower stress levels, while the HIP post-treated specimens built horizontally creep slower than those built at an angle of 30°. As the creep stress increases, the difference in creep rate, as indicated in Figure 9, between the two groups becomes more pronounced: the HIP post-treated specimens creep more slowly and last a longer time before fracture, suggesting a better creep resistance. At the stress level of 250 MPa (Figure 8d), the influence of build direction in the conventional heat treatment group becomes apparent, and the horizontally built specimens exhibit better creep resistance, consistent with the results of the HIP post-treated group.

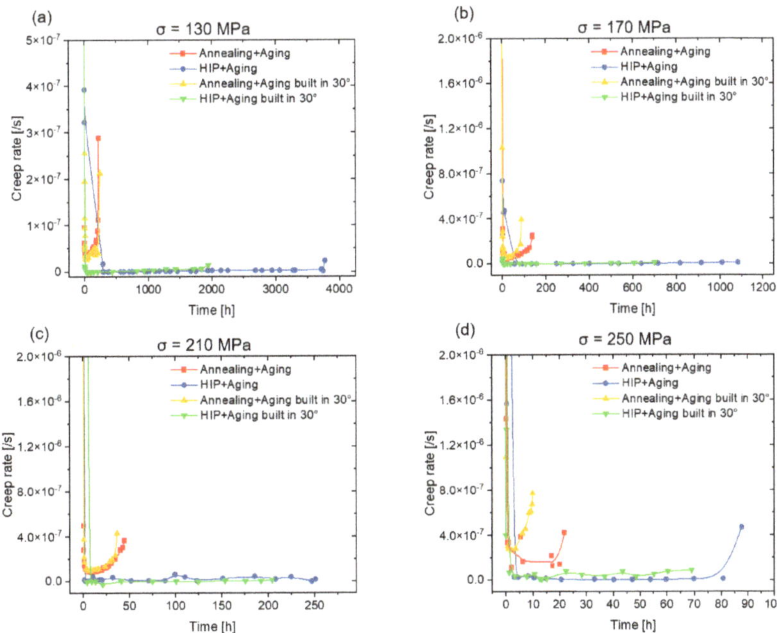

Figure 9. Creep rates of the specimens in different heat treatment conditions. The creep tests were conducted at 760 °C with four different stresses, specifically (**a**) 130 MPa, (**b**) 170 MPa, (**c**) 210 MPa, and (**d**) 250 MPa.

The minimum creep rate of the specimens is shown in Figure 10a. In general, the HIP-treated specimens, which contain fewer δ-phase and, thus, expectedly a larger amount of fine γ″ precipitates in the matrix, exhibit lower minimum creep rates at the applied stresses compared to the conventionally heat-treated specimens. The Norton creep law can be used to describe the minimum creep rate:

$$\dot{\varepsilon} = A \cdot \sigma^{n}$$

where $\dot{\varepsilon}$ is the minimum creep rate, A is a material constant, σ is the tensile stress, and n is the creep stress exponent. The stress exponent provides information about the mechanism that dominates creep deformation.

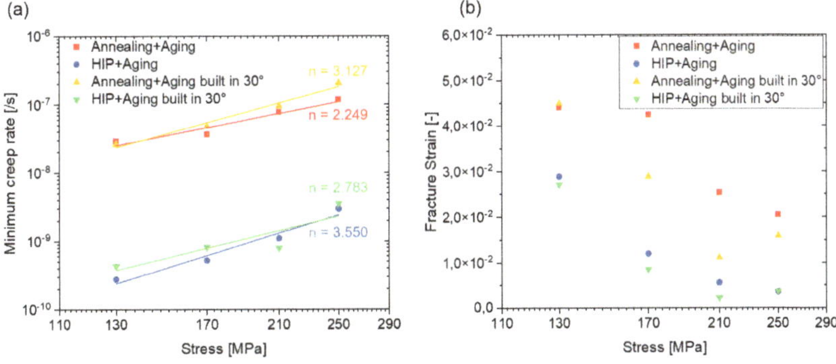

Figure 10. The dependence of the applied stress on the minimum creep rate (**a**) and fracture strain (**b**).

The stress exponents of the specimens in this study range from 2.245 to 3.55, indicating dislocation creep as the dominant mechanism. The higher proportions of the uniformly distributed fine γ'' precipitates, expected in the HIP specimens, limit the mobility of the dislocations and may, therefore, be one reason for the higher creep strength in the HIP heat treatment condition.

Figure 10b presents the fracture strain at different stresses. Higher stress levels result in lower elongation at fracture due to lower plastic deformation. This can be explained by the fact that at low stress levels, void nucleation happens significantly later leading to increased rupture time and allowing more time for creep deformation. Therefore, specimens tested under low stresses present higher fracture strain. Furthermore, the fracture strain is higher in the conventionally heat-treated group than in the HIP heat-treated group, which will be further discussed in the next chapter.

The fracture time for the specimens in various conditions in this study is compared with conventionally produced IN718 references [39,40], as shown in Figure 11. The HIP post-treated group shows a higher creep resistance, in particular a longer fracture time than the group with conventional heat treatment. In addition, the creep resistance of the specimens fabricated under 30° is worse than that of the horizontally fabricated specimens. Compared to conventionally fabricated IN718 [39,40], the HIP group exhibits even better creep resistance. The reasons will be discussed in more depth in the further sections, when analyzing the fracture characteristics.

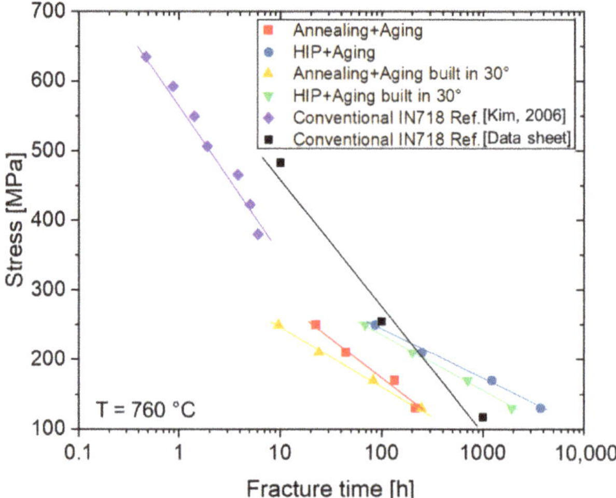

Figure 11. The fracture time in various conditions in this study compared with conventionally produced IN718 in references [39,40].

A few studies already describe the influence of the build direction on the creep strength in the horizontal and vertical directions [20,41–43]. These studies have consistently shown that specimens built vertically exhibit better creep resistance compared to horizontally built specimens. This is attributed to the <001> texture and elongated grains along the build direction, resulting also in a vertically aligned grain boundary δ-phase. When the specimens are loaded horizontally, the grain boundary δ-phase can act as sites for the rapid formation of creep voids, leading to accelerated rupture. Hilal et al. [44] also found in creep tests on the Ni-based superalloy CM247LC produced via PBF-LB that specimens built at an angle of 30° exhibit poorer creep properties than specimens built vertically. As a reason for the better creep properties in the vertically built samples, they also attribute the columnar grain shape resulting from epitaxial grain growth. However, these theories do not explain the lower creep resistance in the 30° build direction compared to horizontally built specimens. One possible explanation is that the largest shear stresses and, thus, the strongest grain boundary slip occurs in general at 45° grain boundaries in the direction of the applied stress. If grain boundary sliding is assumed to be one of the creep mechanisms present, then creep damage will occur at those grain boundaries with the most creep and stress superelevation. Since epitaxial grain growth occurs in the PBF-LB process, the specimens built horizontally have most of the grain boundaries at an angle of 90° in the direction of loading. In contrast, within the specimens built at an angle of 30°, most of the grain boundaries have an orientation of 60° to the loading direction. This may result in higher shear stresses and more grain boundary slip, for these specimens.

3.3. Failure Analysis

Figure 12 shows exemplary post-fracture cross-sections for two creep specimens tested at a stress of 250 MPa.

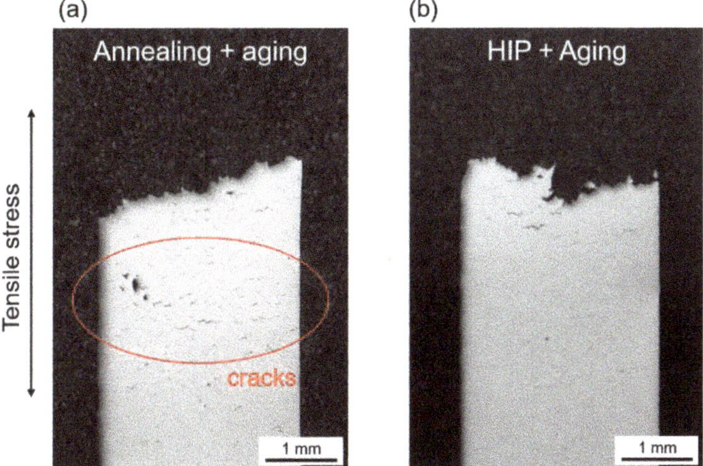

Figure 12. Representative cross-sections of the creep specimens after fracture. (**a**) Specimen built horizontally with standard heat treatment, tested at 250 MPa; (**b**) specimen built horizontally with HIP integrated heat treatment, tested at 250 MPa.

Both specimens were built in the horizontal orientation. Figure 12a shows a conventionally heat-treated specimen, while Figure 12b shows a specimen with HIP heat treatment. First, it should be mentioned that all samples show creep fractures without significant necking. Microcracks developed throughout the whole volume until void coalescence finally determined the main fracture.

Compared to the HIP heat-treated specimen, the conventionally heat-treated specimen exhibited a higher number of cracks and pores near the fracture surface. These cracks and pores presumably contributed to the final elongation of the specimens, resulting in a higher fracture strain in the conventional heat-treated group (Figure 10b). On the other hand, the lower number of cracks and pores in the HIP heat-treated group indicated a better creep resistance. The formation of cracks and pores can be attributed to the accumulation of plastic strains and the initiation and growth of microvoids, which are strongly influenced by the microstructure of the material. The microstructure of the specimens after creep testing is shown in Figure 13.

Due to the high temperature, the microstructure changed during the creep tests, resulting in the degradation of the material. Creep damage can generally be divided into several types, the two most important being: transgranular creep damage and intergranular creep damage. Which mechanism is dominant depends, on the one hand, on the temperature and the applied stress and, the other hand, on the creep properties of the material. In transgranular creep fracture, voids and cracks form within the matrix during the creep process due to the high temperature and stresses. Figure 13d shows that the HIP heat-treated sample contains nearly no cracks and voids within the grains compared to the conventionally heat-treated sample, but only along the grain boundaries or phase boundaries, respectively, and, thus, shows no transgranular creep damage.

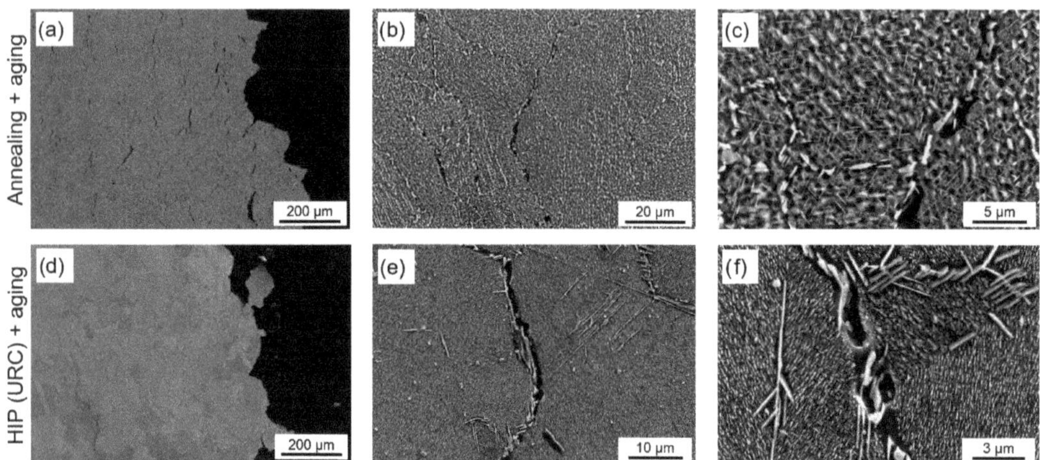

Figure 13. SEM images of the microstructure from the specimens tested at 250 MPa after creep testing. (**a–c**) Specimen built horizontally with standard heat treatment; (**d–f**) specimen built horizontally with HIP heat treatment.

Another type of failure is intergranular creep damage. In intergranular fracture, creep damage occurs at grain boundaries and triple points. In this case, the damage is often initiated at the crack nuclei, which during further development leads to microvoids and cavities, that later become cracks. Figure 13e shows that in HIP heat-treated specimens cracks are clearly visible at the grain boundaries (near δ-precipitates). In these specimens, long acicular δ-precipitates appear along the grain boundaries, which can act as crack nuclei and lead to intergranular fracture (Figure 14f). In contrast, in the conventionally heat-treated specimens, failure is evident not only at the grain boundaries, but especially within the grains leading to a certain amount of transgranular fracture.

In all specimens, a coarsening of the γ'' precipitates can be observed, as shown in Figure 13c,f, which generally reduces the strengthening effect of the γ'' precipitates and leads to a decrease in the creep strength of the material.

The morphology of the specimens' fracture surfaces after the different heat treatment conditions provides information about the deformation and failure mechanisms, as shown in Figure 14. The grain morphology observable in both fractures (Figure 14a,d) indicates the presence of intergranular fracture. The small dimples and pitting fracture surface of the conventionally heat-treated specimen in Figure 14b further indicate proportions of microscopic ductile fracture, while the fracture surface of the HIP post-treated specimen in Figure 14e is relatively smooth, indicating a lower degree of ductile deformation and a higher degree of intergranular fracture. Furthermore, coarsened precipitations can be observed on the fracture surface of both specimens (Figure 14c,f), indicating microstructure degradation.

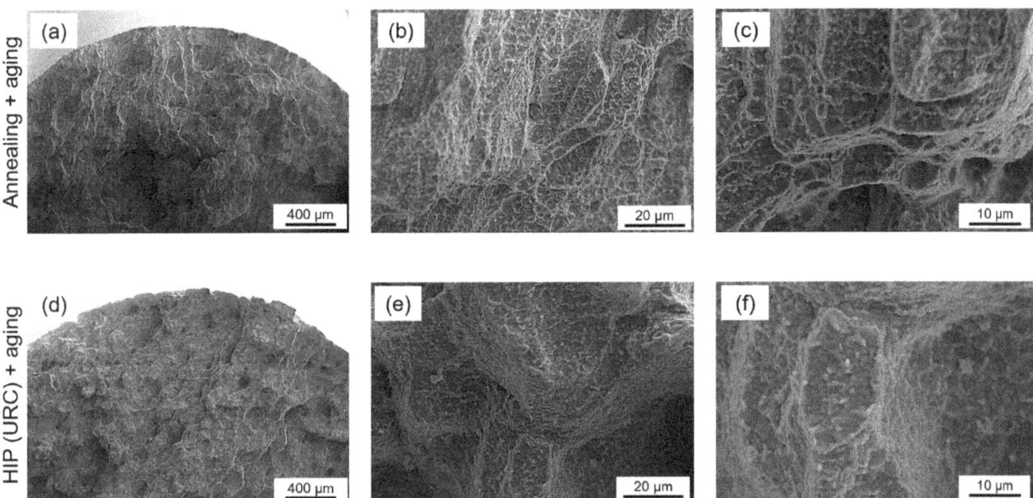

Figure 14. SEM images of the morphology of the fracture surfaces at different magnifications from specimens tested at 250 MPa. (**a**–**c**) Specimen built horizontally with standard heat treatment; (**d**–**f**) specimen built horizontally with HIP heat treatment.

4. Summary and Conclusions

In the present study, creep tests were performed on IN718 specimens built with PBF-LB in different orientations and subjected to two different heat treatments. The different build orientations were horizontal (0° to the build plate) and 30° inclined to the build plate. The heat treatment conditions were the conventional heat treatment according to AMS 5662 and the HIP heat treatment with integrated quenching and aging.

The study provides a description of the microstructural evolution in PBF-LB fabricated IN718. In the as-built initial condition, a large fraction of the Laves phase is visible, which is formed by Nb-segregations during rapid cooling during the PBF-LB process. The solution annealing temperature of 980 °C in the standard heat treatment was not sufficient to dissolve the Laves phase and homogenize the microstructure, resulting in a large amount of δ-phase within the matrix. In contrast, during the HIP treatment at 1160 °C, the matrix could be homogenized and both the Laves- and the δ-phase were dissolved. After HIP treatment with integrated quenching and aging, only the γ′ precipitates and carbides were observed. The γ″ precipitates present could not be resolved with the SEM used in this study, but it can be assumed that a high amount of γ″ precipitates is present after the HIP heat treatment.

The PBF-LB specimens exhibit lower minimum creep rates and longer life after HIP heat treatment than after conventional heat treatment, indicating better creep resistance. Microstructural fracture analysis shows that the specimens after HIP heat treatment exhibit mainly intergranular creep damage and little damage inside the grains. In contrast, the specimens with the conventional heat treatment exhibit a higher percentage of transgranular fracture, indicating that the creep damage was initiated also in the grain interior. Sites for crack initiation can be voids, such as pores, but also larger precipitates that are not coherent with the matrix, such as δ-phase. Both are present within the matrix of the specimens with conventional heat treatment. The samples after HIP heat treatment show a strongly reduced porosity and, also, no initial amount of δ-phase inside the grains.

In addition, it can be assumed that the specimens after HIP heat treatment have a much larger fraction of creep-strengthening γ″ precipitates. The reason for this assumption is that due to the dissolution of the Laves- and δ-phase at the higher solution annealing temperatures in HIP, more Nb is available to form the γ″ precipitates in the matrix. Thus, after HIP heat treatment, the specimens exhibit a more creep-resistant matrix and, at the same time, fewer defects in the microstructure, which can act as crack initiation points. That leads to superior creep properties in the samples with HIP treatment.

The specimens built horizontally present, for both heat treatment conditions, a better creep resistance than the specimens built at 30°. The reason for this is seen in the higher shear stresses at the grain boundaries for the 30° built samples. Due to the PBF-LB typical epitaxial grain growth in build direction, the orientation of the grain boundaries is less favorable to the load direction for a build direction of 30° than in a horizontal direction. The higher shear stresses at the grain boundaries lead to faster creep damage in this area.

Nevertheless, based on the results of the present study, it can be concluded that the influence of the PBF-LB build direction on the creep behavior is much smaller than the influence of the different heat treatments.

Author Contributions: A.K.: Conceptualization, Methodology, Validation, Writing—original draft; S.Q.: Methodology, Investigation, Formal analysis, Writing—original draft; C.B.: Supervision, Resources, Writing—Review & Editing. All authors have read and agreed to the published version of the manuscript.

Funding: This research received no external funding.

Data Availability Statement: The data presented in this study are available on request from the corresponding author.

Acknowledgments: The authors would like to thank Carpenter Technology Corporation for the powder supply and Quintus Technologies Application Centre for the HIP treatment.

Conflicts of Interest: The authors declare no conflict of interest.

References

1. Tan, C.; Weng, F.; Sui, S.; Chew, Y.; Bi, G. Progress and perspectives in laser additive manufacturing of key aeroengine materials. *Int. J. Mach. Tools Manuf.* **2021**, *170*, 103804. [CrossRef]
2. Bajaj, P.; Hariharan, A.; Kini, A.; Kürnsteiner, P.; Raabe, D.; Jägle, E.A. Steels in additive manufacturing: A review of their microstructure and properties. *Mater. Sci. Eng. A* **2020**, *772*, 138633. [CrossRef]
3. *DIN EN ISO/ASTM 52900:2022-03*; Additive Manufacturing—General Principles—Fundamentals and Vocabulary (ISO/ASTM 52900:2021). German Version EN ISO/ASTM 52900:2021; Beuth Verlag GmbH: Berlin, Germany, 2022. [CrossRef]
4. Zhang, X.; Yocom, C.J.; Mao, B.; Liao, Y. Microstructure evolution during selective laser melting of metallic materials: A review. *J. Laser Appl.* **2019**, *31*, 031201. [CrossRef]
5. Caiazzo, F.; Alfieri, V.; Corrado, G.; Argenio, P. Laser powder-bed fusion of Inconel 718 to manufacture turbine blades. *Int. J. Adv. Manuf. Technol.* **2017**, *93*, 4023–4031. [CrossRef]
6. Qi, H.; Azer, M.; Ritter, A. Studies of Standard Heat Treatment Effects on Microstructure and Mechanical Properties of Laser Net Shape Manufactured INCONEL 718. *Met. Mater. Trans. A* **2009**, *40*, 2410–2422. [CrossRef]
7. Guo, N.; Leu, M.C. Additive manufacturing: Technology, applications and research needs. *Front. Mech. Eng.* **2013**, *8*, 215–243. [CrossRef]
8. Blakey-Milner, B.; Gradl, P.; Snedden, G.; Brooks, M.; Pitot, J.; Lopez, E.; Leary, M.; Berto, F.; du Plessis, A. Metal additive manufacturing in aerospace: A review. *Mater. Des.* **2021**, *209*, 110008. [CrossRef]
9. Deng, D.; Moverare, J.; Peng, R.L.; Söderberg, H. Microstructure and anisotropic mechanical properties of EBM manufactured Inconel 718 and effects of post heat treatments. *Mater. Sci. Eng. A* **2017**, *693*, 151–163. [CrossRef]
10. Amato, K.N.; Gaytan, S.M.; Murr, L.E.; Martinez, E.; Shindo, P.; Hernandez, J.; Collins, S.; Medina, F. Microstructures and mechanical behavior of Inconel 718 fabricated by selective laser melting. *Acta Mater.* **2012**, *60*, 2229–2239. [CrossRef]
11. Belan, J.; Vaško, A.; Kuchariková, L.; Tillová, E.; Matvija, M. The High-Temperature Loading Influence on Orthorhombic Ni_3Nb DOa δ—Phase Formation and its Effect on Fatigue Lifetime in Alloy 718. *Manuf. Technol.* **2018**, *18*, 875–882. [CrossRef]
12. Chang, S.-H. In situ TEM observation of γ', γ'' and δ precipitations on Inconel 718 superalloy through HIP treatment. *J. Alloy Compd.* **2009**, *486*, 716–721. [CrossRef]
13. Azadian, S.; Wei, L.-Y.; Warren, R. Delta phase precipitation in Inconel 718. *Mater. Charact.* **2004**, *53*, 7–16. [CrossRef]
14. Chen, W.; Chaturvedi, M. Dependence of Creep Fracture of Inconel 718 on Grain Boundary Precipitates. *Acta Mater.* **1997**, *45*, 2735–2746. [CrossRef]
15. Nagahari, T.; Nagoya, T.; Kakehi, K.; Sato, N.; Nakano, S. Microstructure and Creep Properties of Ni-Base Superalloy IN718 Built up by Selective Laser Melting in a Vacuum Environment. *Metals* **2020**, *10*, 362. [CrossRef]
16. Kassner, M.E.; Son, K.T.; Lee, K.A.; Kang, T.-H.; Ermagan, R. The creep and fracture behavior of additively manufactured Inconel 625 and 718. *Mater. High Temp.* **2022**, *39*, 499–506. [CrossRef]

17. Wu, S.; Peng, H.; Gao, X.; Hodgson, P.; Song, H.; Zhu, Y.; Tian, Y.; Huang, A. Improving creep property of additively manufactured Inconel 718 through specifically-designed post heat treatments. *Mater. Sci. Eng. A* **2022**, *857*, 144047. [CrossRef]

18. Xu, Z.; Cao, L.; Zhu, Q.; Guo, C.; Li, X.; Hu, X.; Yu, Z. Creep property of Inconel 718 superalloy produced by selective laser melting compared to forging. *Mater. Sci. Eng. A* **2020**, *794*, 139947. [CrossRef]

19. Xu, Z.; Murray, J.; Hyde, C.; Clare, A. Effect of post processing on the creep performance of laser powder bed fused Inconel 718. *Addit. Manuf.* **2018**, *24*, 486–497. [CrossRef]

20. Sanchez, S.; Gaspard, G.; Hyde, C.; Ashcroft, I.; Ravi, G.A.; Clare, A. The creep behaviour of nickel alloy 718 manufactured by laser powder bed fusion. *Mater. Des.* **2021**, *204*, 109647. [CrossRef]

21. Xu, Z.; Guo, C.; Yu, Z.R.; Li, X.; Hu, X.G.; Zhu, Q. Tensile and Compressive Creep Behavior of IN718 Alloy Manufactured by Selective Laser Melting. *Mater. Sci. Forum* **2020**, *986*, 102–108. [CrossRef]

22. Chen, W.; Chaturvedi, M. The effect of grain boundary precipitates on the creep behavior of Inconel 718. *Mater. Sci. Eng. A* **1994**, *183*, 81–89. [CrossRef]

23. Kuo, Y.-L.; Nagahari, T.; Kakehi, K. The Effect of Post-Processes on the Microstructure and Creep Properties of Alloy718 Built Up by Selective Laser Melting. *Materials* **2018**, *11*, 996. [CrossRef]

24. Pröbstle, M.; Neumeier, S.; Hopfenmüller, J.; Freund, L.; Niendorf, T.; Schwarze, D.; Göken, M. Superior creep strength of a nickel-based superalloy produced by selective laser melting. *Mater. Sci. Eng. A* **2016**, *674*, 299–307. [CrossRef]

25. Atkinson, H.V.; Davies, S. Fundamental aspects of hot isostatic pressing: An overview. *Metall. Mater. Trans. A* **2000**, *31*, 2981–3000. [CrossRef]

26. Qin, S.; Herzog, S.; Kaletsch, A.; Broeckmann, C. Improving the Fatigue Strength of Laser Powder Bed-Fused AISI M3:2 by Hot Isostatic Pressing. *Steel Res. Int.* **2022**, *94*, 2200435. [CrossRef]

27. Kunz, J.; Herzog, S.; Kaletsch, A.; Broeckmann, C. Influence of initial defect density on mechanical properties of AISI H13 hot-work tool steel produced by laser powder bed fusion and hot isostatic pressing. *Powder Met.* **2021**, *65*, 1–12. [CrossRef]

28. Kunz, J.; Saewe, J.; Herzog, S.; Kaletsch, A.; Schleifenbaum, J.H.; Broeckmann, C. Mechanical Properties of High-Speed Steel AISI M50 Produced by Laser Powder Bed Fusion. *Steel Res. Int.* **2019**, *91*, 1900562. [CrossRef]

29. Masuo, H.; Tanaka, Y.; Morokoshi, S.; Yagura, H.; Uchida, T.; Yamamoto, Y.; Murakami, Y. Effects of Defects, Surface Roughness and HIP on Fatigue Strength of Ti-6Al-4V manufactured by Additive Manufacturing. *Procedia Struct. Integr.* **2017**, *7*, 19–26. [CrossRef]

30. Zhang, H.; Li, C.; Yao, G.; Zhang, Y. Hot isostatic pressing of laser powder-bed-fused 304L stainless steel under different temperatures. *Int. J. Mech. Sci.* **2022**, *226*, 107413. [CrossRef]

31. Kunz, J.; Boontanom, A.; Herzog, S.; Suwanpinij, P.; Kaletsch, A.; Broeckmann, C. Influence of hot isostatic pressing post-treatment on the microstructure and mechanical behavior of standard and super duplex stainless steel produced by laser powder bed fusion. *Mater. Sci. Eng. A* **2020**, *794*, 139806. [CrossRef]

32. Qin, S.; Herzog, S.; Kaletsch, A.; Broeckmann, C. Effects of HIP on microstructure and creep properties of Inconel 718 fabricated by laser powder-bed fusion. In Proceedings of the Euro PM 2019, Maastricht, The Netherlands, 13–16 October 2019; European Powder Metallurgy Association: Shrewsbury, UK, 2019; ISBN 978-1-899072-51-4.

33. Rickenbacher, L.; Etter, T.; Hövel, S.; Wegener, K. High temperature material properties of IN738LC processed by selective laser melting (SLM) technology. *Rapid Prototyp. J.* **2013**, *19*, 282–290. [CrossRef]

34. *ASTM B637-18*; Standard Specification for Precipitation-Hardening and Cold Worked Nickel Alloy Bars, Forgings, and Forging Stock for Moderate or High Temperature Service. ASTM: West Conshohocken, PA, USA, 2018.

35. Kaletsch, A.; Qin, S.; Herzog, S.; Broeckmann, C. Influence of high initial porosity introduced by laser powder bed fusion on the fatigue strength of Inconel 718 after post-processing with hot isostatic pressing. *Addit. Manuf.* **2021**, *47*, 102331. [CrossRef]

36. *ASTM E 139:2011*; Standard Test Methods for Conducting Creep, Creep-Rupture, and Stress-Rupture Tests of Metallic Materials. Beuth Verlag GmbH: Berlin, Germany, 2011. [CrossRef]

37. Fayed, E.M.; Shahriari, D.; Saadati, M.; Brailovski, V.; Jahazi, M.; Medraj, M. Influence of Homogenization and Solution Treatments Time on the Microstructure and Hardness of Inconel 718 Fabricated by Laser Powder Bed Fusion Process. *Materials* **2020**, *13*, 2574. [CrossRef]

38. Zhang, D.; Niu, W.; Cao, X.; Liu, Z. Effect of standard heat treatment on the microstructure and mechanical properties of selective laser melting manufactured Inconel 718 superalloy. *Mater. Sci. Eng. A* **2015**, *644*, 32–40. [CrossRef]

39. Kim, D.-H.; Kim, J.-H.; Sa, J.-W.; Lee, Y.-S.; Park, C.-K.; Moon, S.-I. Stress rupture characteristics of Inconel 718 alloy for ramjet combustor. *Mater. Sci. Eng. A* **2008**, *483–484*, 262–265. [CrossRef]

40. Internet Data, Creep and Stress-Rupture Strengths HAYNES® 718 Data Sheet, Age-Hardened. Available online: https://www.haynesintl.com/alloys/alloy-portfolio_/High-temperature-Alloys/haynes718-alloy/creep-and-stress-rupture-strengths.aspx (accessed on 15 May 2023).

41. Hautfenne, C.; Nardone, S.; De Bruycker, E. Influence of heat treatments and build orientation on the creep strength of additive manufactured IN718. In Proceedings of the 4th International ECCC Conference, Düsseldorf, Germany, 10–14 September 2017.

42. Kuo, Y.-L.; Horikawa, S.; Kakehi, K. Effects of build direction and heat treatment on creep properties of Ni-base superalloy built up by additive manufacturing. *Scr. Mater.* **2017**, *129*, 74–78. [CrossRef]

43. Strondl, A.; Palm, M.; Gnauk, J.; Frommeyer, G. Microstructure and mechanical properties of nickel based superalloy IN718 produced by rapid prototyping with electron beam melting (EBM). *Mater. Sci. Technol.* **2011**, *27*, 876–883. [CrossRef]
44. Hilal, H.; Lancaster, R.; Jeffs, S.; Boswell, J.; Stapleton, D.; Baxter, G. The Influence of Process Parameters and Build Orientation on the Creep Behaviour of a Laser Powder Bed Fused Ni-based Superalloy for Aerospace Applications. *Materials* **2019**, *12*, 1390. [CrossRef]

*materials*

MDPI

Review

The State of the Art in Machining Additively Manufactured Titanium Alloy Ti-6Al-4V

Chen Zhang [1], Dongyi Zou [2], Maciej Mazur [1], John P. T. Mo [1], Guangxian Li [3,*] and Songlin Ding [1,*]

[1] School of Engineering, RMIT University, Melbourne, VIC 3083, Australia;
s3458355@student.rmit.edu.au (C.Z.); maciej.mazur@rmit.edu.au (M.M.); john.mo@rmit.edu.au (J.P.T.M.)

[2] Faculty of Engineering and Information Technology, The University of Melbourne,
Melbourne, VIC 3010, Australia; dzzo@student.unimelb.edu.au

[3] College of Mechanical Engineering, Guangxi University, Nanning 530004, China

* Correspondence: 20210037@gxu.edu.cn (G.L.); songlin.ding@rmit.edu.au (S.D.)

Abstract: Titanium alloys are extensively used in various industries due to their excellent corrosion resistance and outstanding mechanical properties. However, titanium alloys are difficult to machine due to their low thermal conductivity and high chemical reactivity with tool materials. In recent years, there has been increasing interest in the use of titanium components produced by additive manufacturing (AM) for a range of high-value applications in aerospace, biomedical, and automotive industries. The machining of additively manufactured titanium alloys presents additional machining challenges as the alloys exhibit unique properties compared to their wrought counterparts, including increased anisotropy, strength, and hardness. The associated higher cutting forces, higher temperatures, accelerated tool wear, and decreased machinability lead to an expensive and unsustainable machining process. The challenges in machining additively manufactured titanium alloys are not comprehensively documented in the literature, and this paper aims to address this limitation. A review is presented on the machining characteristics of titanium alloys produced by different AM techniques, focusing on the effects of anisotropy, porosity, and post-processing treatment of additively manufactured Ti-6Al-4V, the most commonly used AM titanium alloy. The mechanisms resulting in different machining performance and quality are analysed, including the influence of a hybrid manufacturing approach combining AM with conventional methods. Based on the review of the latest developments, a future outlook for machining additively manufactured titanium alloys is presented.

Keywords: additive manufacturing; machining; cutting force; surface integrity; tool wear; porosity; anisotropy; post-processing processes; hybrid manufacturing

Citation: Zhang, C.; Zou, D.; Mazur, M.; Mo, J.P.T.; Li, G.; Ding, S. The State of the Art in Machining Additively Manufactured Titanium Alloy Ti-6Al-4V. *Materials* **2023**, *16*, 2583. https://doi.org/10.3390/ma16072583

Academic Editors: Bartlomiej Wysocki, Joseph Buhagiar and Tomasz Durejko

Received: 30 January 2023
Revised: 16 March 2023
Accepted: 17 March 2023
Published: 24 March 2023

1. Introduction

Titanium and its alloys are applied in a wide range of industries due to their outstanding mechanical and chemical properties, such as high strength/weight ratio, high elastic modulus, and excellent corrosion resistance. Specifically, the remarkable biocompatibility and good mechanical properties at elevated service temperatures make titanium alloys especially desirable for use in biomedical and aerospace industries, respectively [1,2]. However, the machining of titanium alloys is challenging due to the high strength, low thermal conductivity, and high chemical reactivity of the material [3–5]. Due to these properties, it is common to machine titanium alloys with a low cutting speed of less than 90 m/min, which significantly reduces productivity and increases machining costs, when compared to other alloys, such as aluminium, that can be machined with cutting speeds as high as 2500 m/min. The comparatively high feedstock cost of titanium also imposes further economic challenges when machining parts for cost-sensitive applications with high buy-to-fly ratios, which result in large amounts of waste [6,7].

Additive manufacturing (AM) is a new class of manufacturing processes which can produce complex parts directly from a 3D digital model using a layer-by-layer approach with little material waste. Compared to conventional machining processes, such as milling, turning, and grinding in which parts are produced by subtracting redundant materials from feedstock, additively manufactured parts are produced by sequentially consolidating layers of material [8–10]. The layered manufacturing approach effectively decomposes a complex and difficult-to-shape 3D part geometry into a series of comparatively simple to manufacture 2D layers. In comparison to conventional machining processes, AM imposes few design constraints on part geometry and requires a low setup effort without the requirement for custom tooling. This allows for the manufacture of complex parts that are highly customisable. AM processes were first commercialised in 1986 through the stereolithography (SLA) process and were only suitable for the manufacture of plastic prototype components [11]. The commercialisation of AM technology for the manufacture of metal components began in the 1990s and has seen extensive development since then [12]. Through continuous development over the past 30 years, it is now possible to build fully dense, functional metallic components with complex geometries for use in demanding industrial applications [13]. For example, as of August 2021, GE Aviation has produced 100,000 additively manufactured fuel nozzle tips for its LEAP jet engines used in Boeing 737 and Airbus 320 [14]. By replacing 855 conventional parts with 12 consolidated components manufactured by AM, GE Aviation has significantly reduced manufacturing process complexity, while also increasing performance, which contributed to a 20% increase in engine fuel efficiency [15].

Despite the advantages of additive manufacturing, metal parts produced by AM processes can exhibit inferior properties compared to wrought machined parts, including higher surface roughness, porosity, dimensional variability, and residual stresses [16,17]. In particular, the high-surface roughness can significantly reduce fatigue life for components subjected to dynamic loading, thereby limiting suitable application areas. As such, additively manufactured metal parts may need to be subjected to various post-processing processes, including machining, in order to obtain acceptable surface quality [18,19], although chemical and electrochemical methods can be applied to improve surface finish of parts with unique geometrical structures, for example, the scaffolds for bone tissue engineering and cellular foams [20–22]. A significant number of studies have been carried out on the machining of wrought Ti alloys in the past decades. These studies have focused either on the development of specially designed cutting tools and new tool materials [23,24], optimisation of cutting parameters, application of new cooling media and tribology theories [25], or on the investigation of hybrid machining processes such as thermal-assisted machining (TAM) [26–30]. Since Ti-6Al-4V is the most widely used titanium alloy constituting 50% of total titanium alloy production worldwide, the machining of Ti-6Al-4V has become one of the most widely studied subjects [31–33].

The microstructure and mechanical properties of additively manufactured titanium alloys are notably different from their wrought counterparts due to recurring rapid heating and cooling thermal cycles that occur during common metal AM processes (Section 2) [34]. The processing conditions can also vary between specific AM process, resulting in different part mechanical performance, microstructure characteristics, and surface integrity even for the same feedstock material composition [35]. For example, it has been found that Ti-6Al-4V manufactured by Laser Powder Bed Fusion (PBF-LB) (Section 2), the most widely used AM technology, exhibits higher tensile and yield strength than traditionally wrought Ti-6Al-4V due to the presence of higher residual stress and refined α' martensite resulting from rapid PBF-LB heating and cooling cycles [36]. Similarly, PBF-LB-processed Ti-6Al-4V alloy has been reported to also exhibit higher Young's modulus [37]. Another factor that drastically affects the mechanical strength of Ti alloys is the oxygen content. Controlling the addition of oxygen, which induces the solution-strengthening phenomenon in the manufacturing process, can improve the mechanical properties of additively manufactured parts [38–40]. In additional to differences in machinal properties, the surface roughness, Ra, of PBF-LB manufactured parts (typically 5–40 μm [41]) also differs compared to their machined counterparts (typically 0.4–6.3 μm [42]). The unique properties of PBF-LB parts subsequently affect tool wear, cutting force, cutting temperature, and formation of chips, resulting in significant differences between the machining of additively manufactured and wrought titanium alloys [43,44].

A significant amount of research has been conducted on the microstructure and mechanical properties of additively manufactured Ti-6Al-4V alloys [44–49], including studies on their machining characteristics [50,51]. However, there is very limited published literature summarising the influence of mechanical/microstructure properties on the machinability of additively manufactured Ti alloys. Generally, the machinability of materials is influenced by factors including surface integrity, cutting force, cutting temperature, tool wear/life, and chip formation [32,52,53]. Understanding the influence of these factors on the machinability of Ti alloys produced by various AM technologies requires a comprehensive review of the state of the art in several research areas. This paper aims to provide such a review by focusing on the machining of Ti-6Al-4V alloys produced by several Powder Bed Fusion and Direct Energy Deposition additive manufacturing technologies (Section 2). The working principles of these AM processes and how process differences can influence the machinability of additively manufactured Ti-6Al-4V alloys are first discussed. The effects of anisotropy, porosity, and post-processing treatment on additively manufactured Ti-6Al-4V are subsequently reviewed. The mechanisms resulting in different machining performance and quality are analysed. The effects of post-processing and hybrid manufacturing approach on additively manufactured Ti alloys are also reviewed. Finally, based on a summary of the latest achievements, future development trends in machining additively manufactured titanium alloys are presented.

2. Additive Manufacturing Processes

The AM process workflow starts with a digital 3D model of the part file, formulated as surface mesh (commonly a stereolithography (STL) file), which is digitally subdivided into individually processable layers for which material deposition or fusion tool paths are generated. During the build process, the tool paths are consolidated in a layer-by-layer approach to produce parts without the need for any custom tooling. The layer-by-layer manufacturing approach effectively deconstructs complex and difficult-to-shape 3D geometry into a series of relatively simpler to shape 2D layers. This key characteristic allows AM to produce parts with unprecedented levels of geometric complexity and design freedom [54–56].

A range of AM processes have been developed, which vary in their material feedstock form and material consolidation approach. The most common metal AM approaches are based on Powder Bed Fusion (PBF) and Direct Energy Deposition (DED) techniques [57–59]. In a PBF process, a thermal point source scans and selectively melts and fuses the top layer of a bed of atomised metal powder feedstock according to the pre-programmed toolpaths and layers which constitute the desired part. The melt pool rapidly solidifies, the build substrate is lowered by one layer thickness, new powder is deposited, and the next layer is selectively melted. The process repeats until the part build is completed. Following build completion, the unfused powder surrounding the solidified part is removed for reuse in subsequent builds. The ability to rapidly reuse unfused powder in PBF processes provides a significant advantage in material use efficiency compared to conventional subtractive machining processes, which may generate substantial amounts of waste swarf.

In the case of PBF-LB (Figure 1), the irradiating heat source is a high powered, mirror-actuated laser (typically 100–2000 W and ~1 μm wavelength) that is sufficient to melt and fuse the metal powder layers (typically 30–90 μm thick) [60,61]. The representative lasers applied in the PBF-LB process include CO2 laser, Nd: YAG laser (neodymium-doped yttrium aluminum garnet laser), and Yb-fiber laser (ytterbium-doped fiber laser) [60–62]. The build chamber is filled with an inert gas (such as argon), and the build substrate onto which the part is built is preheated (preheating temperature can be up to 1200 C [63,64]) to reduce the thermal gradients between the solidifying layers and the substrate. However, the thermal gradients remain high and PBF-LB parts are prone to high levels of residual stress which, in extreme cases, can cause severe part distortion or fracture. Post-build heat treatments are often applied to relieve stresses [65]. The PBF-LB process is also commonly referred to by synonymous trademark terms, such as Selective Laser Melting (SLM) and Direct Metal Laser Sintering (DMLS), that have been popularised by machine manufacturers. However, the terminology standardised by the ASTM International is laser powder bed fusion [66].

The Electron Beam-Powder Bed Fusion (PBF-EB) variant of the PBF process (also referred to as Electron Beam Melting (EBM), Figure 1b) uses an electromagnetically actuated electron beam (typically 3–6 kW) to melt the powder layers (typically 50–150 µm) [67,68]. Compared to PBF-LB, the electron beam allows for higher scanning speeds, powder preheat temperatures, and scan speeds. The higher process temperatures also result in lower thermal gradients and thereby lower residual stress in PBF-EB manufactured parts. However, unlike the PBF-LB, the PBF-EB process needs to take place in a vacuum to avoid E-beam interaction with gas molecules, and this requirement can increase machine cost and setup times [69]. Furthermore, the resolution of manufactured parts is lower than with PBF-LB [70].

Both PBF-LB and PBF-EB technologies allow the creation of parts with small features, high precision, and complex shapes using a variety of alloys. However, these processes involve a range of complex physical phenomena that can impact part manufacturability, quality, and cost, including powder rheology, laser/E-beam absorption in the powder bed, heat transfer, melting and solidification, melt pool dynamics, microstructure development, and thermomechanical stress. The number of influential parameters associated with these phenomena can be large, with estimates suggesting 130 parameters of relevance to PBF-LB [71]. A comprehensive understanding of such influential parameters is the subject of extensive ongoing research [72].

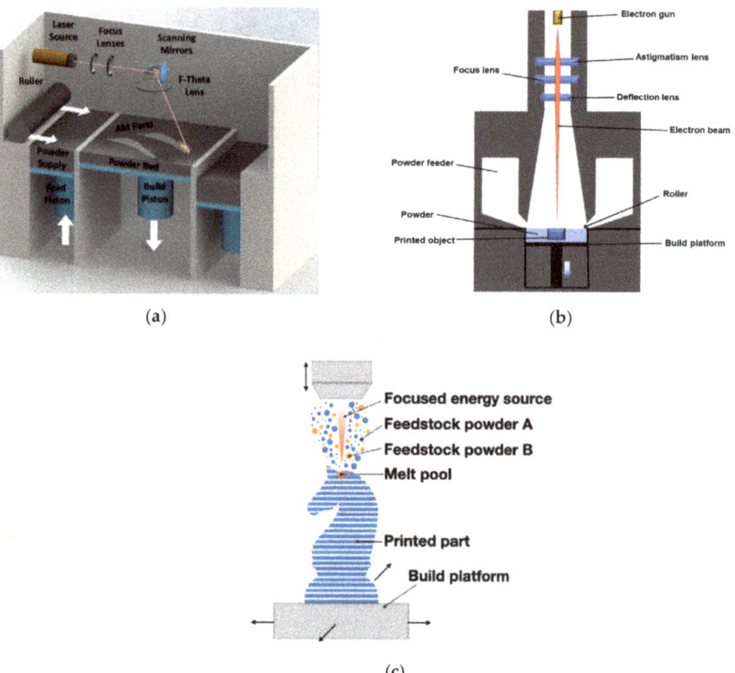

Figure 1. (**a**) Laser Powder Bed Fusion (PBF-LB) process [73] (Reprinted with permission from [73]. Copyright 2018, Elsevier). (**b**) Electron Beam-Powder Bed Fusion (PBF-EB) process [74] (Reprinted with permission from [74]. Copyright 2020, Springer Nature). (**c**) Direct Energy Deposition (DED) process [75] (Reprinted with permission from [75]. Copyright 2022, Elsevier).

In the DED AM process, feedstock material in the form of powder or wire is fed concurrently with an inert oxidisation-shielding gas (such as argon or helium) through a nozzle, which is coaxial with a thermal source, that melts the feedstock material onto a substrate (Figure 1c). The nozzle is mounted on an articulated robotic arm or gantry system that controls the deposition location of the feedstock material and the layer thickness (typically 200–5000 μm) [76–78]. The DED process is generally operated in an inert environment where the oxygen level is controlled to be within less than 5–10 ppm [79,80]. However, some oxidation of the deposited material can occur once the shielding gas is removed from the weld if the part temperature remains high. Unlike PBF processes, the DED process allows for the AM of larger parts or for the repair or refurbishment of existing parts with localised material addition to damaged areas [81,82]. Powder-fed DED processes can concurrently feed different types of powder materials via multiple hoppers, enabling the manufacture of composite materials including functionally graded material (FGM). Variants of DED technology include Laser Metal Deposition (LMD), which uses a laser heat source and can operate with wire and powder feedstock, Electron Beam Additive Manufacturing (EBAM), which uses an E-beam heat source and operates with wire in a vacuum build chamber, and Wire Arc Additive manufacturing (WAAM), which uses an electric arc heat source and operates with wire feedstock. A detailed review of the characteristics of these technologies is available in [83].

3. Machinability of Additively Manufactured Ti-6Al-4V

Additively manufactured metal parts typically have a complex, spatially varying thermal history due to the rapid melting and solidification inherent to metal AM processes. This leads to significant differences in residual stresses, surface roughness (Figure 2) and microstructural characteristics (such as grain size and morphology), which results in different mechanical properties compared to conventionally manufactured parts [35,84,85]. Surface roughness is a decisive factor that impacts the mechanical properties of the final products. A high surface roughness can induce stress concentration on the surface and adversely affect the fatigue properties [86]. In general, the surface roughness of additively manufactured Ti alloys is in the range of 10 to 70 μm (Ra), which is significantly larger than that achieved by machining [39,87–89]. These differences can also influence the machinability of additively manufactured parts by affecting the cutting force, surface integrity, and tool wear. In order to improve the machinability of Ti-6Al-4V, a comprehensive understating of these influential factors is critically important.

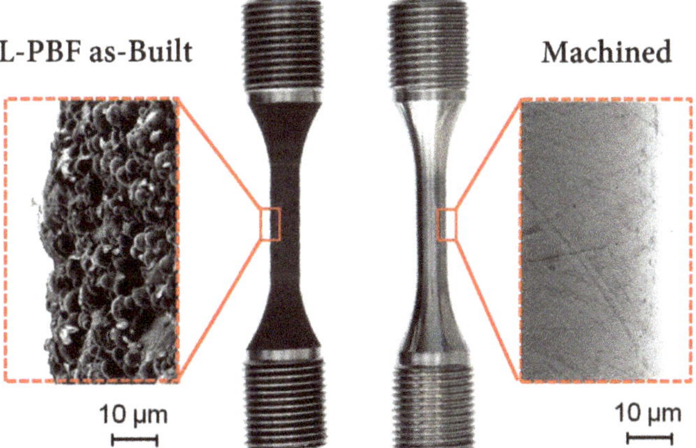

Figure 2. Example PBF-LB surface finish before and after machining process for a Ti-6Al-4V part [90] (Reprinted with permission from [90]. Copyright 2015, Elsevier).

3.1. Cutting Forces in Machining Additively Manufactured Ti-6Al-4V

In order to achieve high machining precision, a cutting tool has to follow specially designed tool paths to minimize dynamic changes in cutting load on the tool [91,92]. A sudden change of cutting force can cause poor surface integrity of the workpiece, severe tool wear, and premature tool failure. The cutting force generated by the cutting edge is a significant factor in characterising the machinability of titanium alloys [93]. During the cutting process, the force is exerted on the contact interface between the cutting tool and the chip [94]. Since additively manufactured Ti alloys have a unique acicular microstructure, their strength and hardness can be higher than those of wrought Ti alloys. The higher yield strength can result in a higher cutting force and a high cutting temperature in the cutting zone, which, in turn, will accelerate tool wear rate. The higher hardness can lead to less plastic flow and induce a lower surface roughness [95]. The interaction between the cutting tool and workpiece material, as well as the plastic deformation process during the formation of chips, is affected by the enhanced mechanical properties, the changes of which are finally reflected on the cutting force and surface quality.

Various experimental studies have been conducted to explore cutting force in machining additively manufactured Ti alloys. Polishetty et al. [96] investigated the effect of machining parameters on cutting force and surface roughness of SLM Ti-6Al-4V during the turning process. It was found that higher cutting forces were generated when machining the SLM Ti-6Al-4V alloy due to higher hardness and yield strength. The surface roughness of SLM Ti-6Al-4V was lower because of the high hardness and brittle properties of the material. Shunmugavel et al. [97] found that higher cutting forces when turning SLM Ti-6Al-4V led to higher cutting temperature and tool/chip wear, resulting in significant adhesive/abrasive wear. By comparing the machinability of SLM and wrought Ti-6Al-4V, they found that the cutting force was closely related to the mechanical strength and hardness of the material [18], and the change in these factors was the reason that large cutting forces were recorded when machining SLM Ti [96]. At a cutting speed of 60 m/min, the cutting force was not affected by the tool wear; at 120 m/min, the cutting force increased a little bit when machining both types of parts. However, it rapidly increased when machining SLM Ti alloy at 180 m/min due to the large concentration of wear on the flank face of the insert.

Ming et al. [98] compared the mechanical properties of SLM titanium alloy using different machining processes under dry and minimum quantity lubrication (MQL) conditions. It was found that chip curling was more obvious when milling forged Ti-6Al-4V due to its better plasticity, which indicated that the forged Ti-6Al-4V resulted in a larger main cutting force compared to SLM Ti-6Al-4V. The main cutting force when machining SLM components decreased when the cutting speed was at 80 mm/min due to the thermal-softening effect. However, when machining the forged Ti-6Al-4V, the main cutting force increased slowly. For additively manufactured products, the grain size/morphology is also an indicator of their performance. Coarse columnar grains may be generated in the AM process due to the ultra-high temperature gradient, which results in the anisotropic and poor mechanical properties of additively manufactured products [99,100]. An optimisation of the manufacturing process and alloy composition control has to be conducted to obtain finer equiaxed grains to achieve stronger mechanical properties in the additively manufactured components [101]. Kallel et al. [102] studied the machinability of Ti-6Al-4V produced by Laser Metal Deposition (LMD). Compared to the wrought Ti-6Al-4V parts, the cutting force when machining LMD-manufactured Ti-6Al-4V was 10–40% higher due to their finer grains, which provided higher yield stress and resistance to plastic deformation. The difference in ductility and microstructure (equiaxed grains and lamellar ones) of the wrought parts and the AM parts [35] led to the surface roughness of the LMD parts being 18–65% rougher than that of the conventional samples. The compressive residual stress of the LMD parts was 11–30% higher than that of the wrought samples.

3.2. Surface Integrity

Surface integrity of machined parts is influenced by factors such as machining parameters, cooling conditions, microstructure of the materials, tool wear, and so on. Compared with wrought parts, the surface quality of as-built additively manufactured parts is poor. Lizzul et al. [103] investigated the influence of microstructure on the surface integrity of additively manufactured Ti-6Al-4V fabricated by PBF-LB. According to the analysis of AM-induced microstructure, the material structure, including the α phase layer and the β grains, have a significant effect on the surface roughness of the samples. The AM parameters can impact microstructural anisotropy, which further affects the surface integrity of the as-built parts. The scanning strategy affects the size of prior β grain correlated to the microhardness of the workpiece [104]. Prior β grain boundaries can impact the crack propagation in AM Ti alloy, which has a significant influence on the tensile and fatigue properties of AM Ti parts [105]. The different sizes of β grains of the material are induced by the different thermal histories that have occurred on the fused powder layers. The wider β grains are generated by a lower cooling rate, which leads to thicker α lamellae and lower microhardness. The fracture phenomena occur in correspondence to the α-phase layers that are formed along the β grain boundaries upon heat treatment [38,105]. The α-phase layers formed along the β grain boundaries represent the AM-produced titanium material discontinuity that weakens the material integrity, which benefits material removal during the cutting process (Figure 3). The samples with a minor β grain width exhibit the highest density of discontinuity in the α-phase layers and show a lower cutting force, resulting in a lower surface roughness. The application of the cryogenic cooling strategy can reduce the surface roughness of the final additively manufactured components.

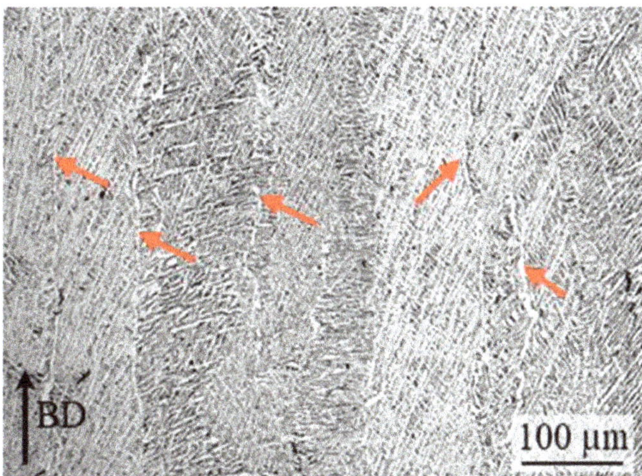

Figure 3. Microstructure along the build-up direction of the PBF-LB Ti-6Al-4V. The red arrows show the α-phase layers along the prior β grains boundaries [103] (Reprinted with permission from [103]. Copyright 2020, Elsevier).

By studying the machinability of wrought, EBM, and DMLS Ti-6Al-4V, Rotella et al. [35] found that the surface roughness of the wrought parts was 10–20% lower than that of the additively manufactured parts. A layer of plastically deformed grains was formed below the machined surface and the thickness of the affected layer, which increased with an increase in cutting speed. The thickness of the affected layer was largest when the cutting speed was 110 m/min. Compared to the wrought workpieces, the EBM parts showed better corrosion resistance and higher surface roughness, and they exhibited higher sensitivity to hardening and compressive residual stress, which was induced by machining, because of their higher hardness. It had been found that cryogenically machined Ti-6Al-4V parts fabricated by EBM had deeper affected layers [106]. The effect of cryogenic cooling was apparent as the thickness of the deformed layer increased on the workpiece surface. However, the plasticity of the material decreased under the cryogenically cooled condition due to the lower cutting temperature, which led to grooved and irregular feed marks on the machined surface. The amount of surface defects increased with an increase in feed rate regardless of the process route. Bordin et al. [107] investigated the surface integrity of Ti-6Al-4V alloy produced by EBM with different cooling strategies, including dry, wet, and cryogenic cooling. It was observed that the cooling condition did not affect the surface quality regardless of the cutting speed when the feed rate was set at the lowest. The use of liquid nitrogen (LN2) reduced the crater and flank wear as well as the surface roughness of the workpiece. Flaws in the machined surfaces, such as side flow, adhesion, tearing, and jagged feed marks, were found; these flaws were affected by the ploughing action between the cutting tool and the workpiece surface. The side flow of the material was caused by the plastic deformation of the surface materials induced by the tool motion. The long straight grooves were caused by the small fragments of the build-up edge (BUE), owing to the rubbing action between the cutting tool and the workpiece. When these fragments were machined through below the flank face of the cutting tool, the grooves were generated on the underlying material; these small fragments left long straight grooves on the material surface when they passed the bottom of the tool flank face. Additionally, adhered materials were left and welded on the surface during the turning process when the adhered particles were formed (Figure 4). Similar surface defects, including side flow and adhered material, can also be found when turning wrought Ti alloy [108]. Cleaner and unbroken surface morphology, as well as randomly oriented micro-scratches induced by chip entanglements, can be observed when the feed rate is 0.1 mm/rev during the wet cutting process (Figure 5).

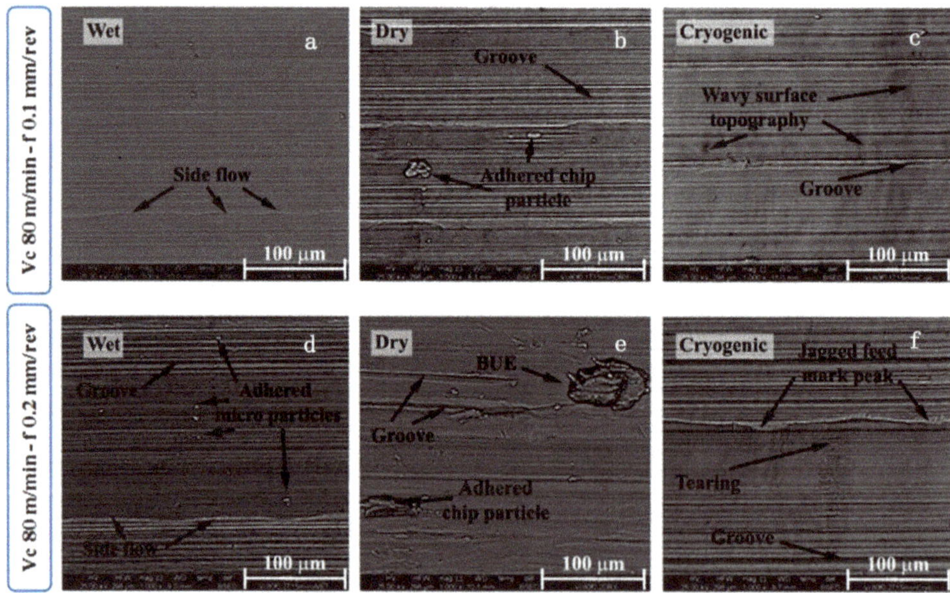

Figure 4. Main surface defects after 8 min of turning when adopting the wet, dry, and cryogenic cooling strategies: (**a,d**) material side flow, double feed marks, long grooves, and micro-particles adhered on the machined surface; (**b,e**) wider adhered chip fragments and BUE are attached to the machined surfaces; (**c,f**) the wavy surface topography along the cutting speed direction and the jagged feed mark peak [107] (Reprinted with permission from [107]. Copyright 2017, Elsevier).

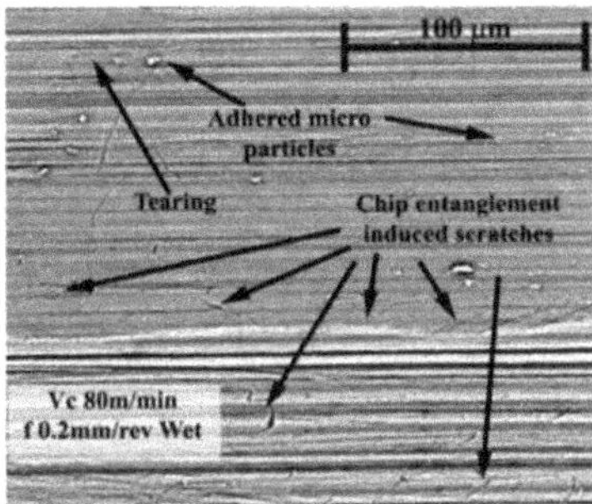

Figure 5. Scratches and adhered particles provoked by chip entanglements after 8 min of wet cutting [107] (Reprinted with permission from [107]. Copyright 2017, Elsevier).

Sartori et al. [109] studied the machining of DMLS-produced Ti-6Al-4V by using cryogenic cooling during the turning process to improve surface integrity (Figure 6). Compared to dry cutting, a more irregular and jagged surface was produced because of the application of LN2 (Figure 7). The tearing phenomenon was presented on the DMLS part irrespective of the use of cooling strategy. The application of LN2 resulted in the most significant residual stress in the axial direction and induced a thickening of the layer with compressive residual stress.

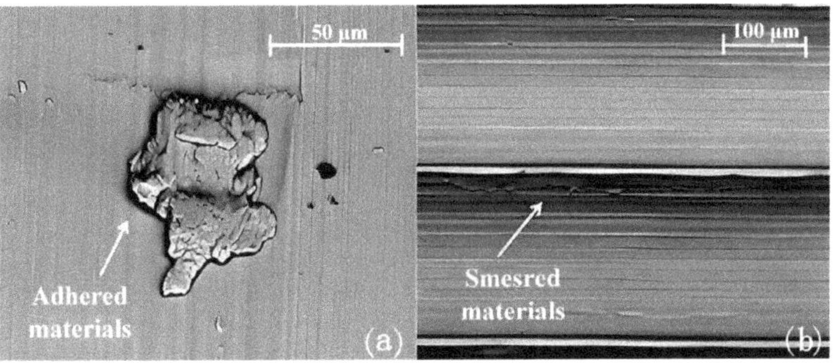

Figure 6. The surface defect generated during dry cutting: (**a**) adhered material and (**b**) smeared material [109] (Reprinted with permission from [109]. Copyright 2016, Elsevier).

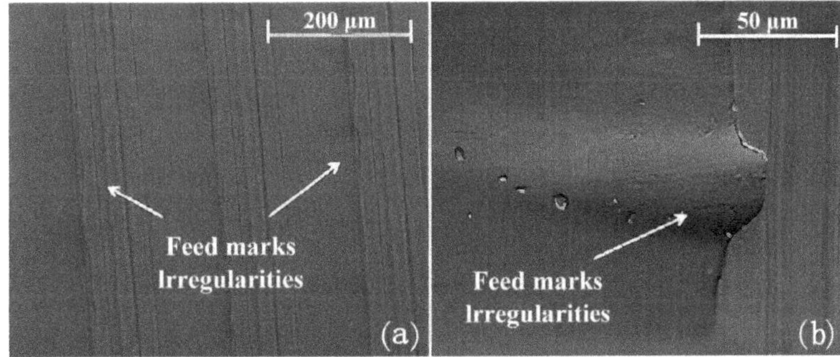

Figure 7. The feed marks and irregularities formed during turning when adopting different cryogenic cooling strategies: (**a**) 500× and (**b**) 2000× [109] (Reprinted with permission from [109]. Copyright 2016, Elsevier).

During the machining process, an ubiquitous compressive stress along the surface of additively manufactured parts could be observed [110]. The cutting region in front of the cutting tool experienced compressive plastic deformation. The compressive stress was generated when the machined surface was squeezed and plastically deformed during the machining process [111]. Oyelola et al. [110] adopted X-ray diffraction to determine the compressive stress in the two measurement directions (axial and circumferential directions) of an additively manufactured Ti workpiece. According to the comparison of the compressive stress in these two directions, it was found that the sample machined with a coated insert incurred higher compressive stresses than the sample machined with an uncoated insert in the parallel direction of the circumference. This difference was induced by the thermal effects associated with chip formation as well as the interactions between the edge of the cutting tool and the machined surface. Furthermore, heat treatment is able to homogenise the microstructure of additively manufactured parts. Coarse microstructures and larger grain size induced by heat treatment lead to a reduction in yield stress, ultimate tensile strength, and hardness of the workpiece, which in turn results in a lower machining force and decreased subsurface deformation, eventually improving the surface roughness [112].

3.3. Tool Wear

Compared to other metallic materials, titanium alloys are difficult to machine due to their lower thermal conductivity and higher chemical reactivity with the cutting tool. During the machining process, the high cutting temperature and the significant adhesion at the tool–workpiece and tool–chip interfaces can lead to severe tool wear and seriously impact tool life [113–117]. Three dominant types of tool wear mechanisms exist during the machining of titanium alloys: adhesion wear, abrasion wear, and diffusion wear [115,118,119]. Since additively manufactured titanium alloys have different mechanical properties and microstructure in comparison to their wrought counterparts, the effect caused by changes in strength, hardness, and microstructure has to be considered when analysing tool wear when machining additively manufactured Ti alloys. Su et al. [120] studied tool wear when milling SLM Ti-6Al-4V and found that even though the SLM Ti alloy has higher hardness and brittleness as well as lower plasticity than the wrought Ti alloy, severe chip adhesion still can be observed on the rake face and flank face of the cutting tool (Figure 8). Similarly, Al-Rubaie et al. [121] found that regardless of the material features, the major tool wear mechanisms are adhesion, coating delamination, and abrasion. The adhered material, which is caused by the high chemical reactivity of titanium material with the cutting tool, as well as the chips welded on the cutting tool at the interface between the tool and the chips, can be observed on the cutting tool in all machining processes (Figures 8 and 9). The built-up layer of titanium alloy can be seen on all cutting tools. Coating delamination, i.e., the removal of the coating and the exposure of the tool substrate material, can also be seen in the figure.

Minimum quantity lubrication is a new micro-lubrication machining technique that uses an oil mist, rather than a flood coolant, to increase the lubrication capability and reduce the cutting temperature. During the machining process, a mixture flow of cutting fluid and compressed air is atomised and jetted into the cutting area. It can enhance the penetration ability of the cutting fluid and provide significant machining advantages when compared to conventional flood cooling and dry machining in terms of tool life and surface quality [122]. When MQL was used during the machining of SLM Ti-6Al-4V, it was found that abrasive wear was the main reason leading to tool wear due to the adhesion of the cutting tool and the part material [123]. A low cutting speed was unsuitable for machining SLM titanium alloy in the micro-milling process. Specifically, a higher flank tool wear rate was found while using a cutting speed of no more than 55 m/s, and the use of MQL could decrease the flank wear and increase tool life. However, the material adhesion on the tool surface can take off the hard coating of the cutting tool and decrease tool hardness, leading to premature failure of the tool edges, as shown in Figure 10.

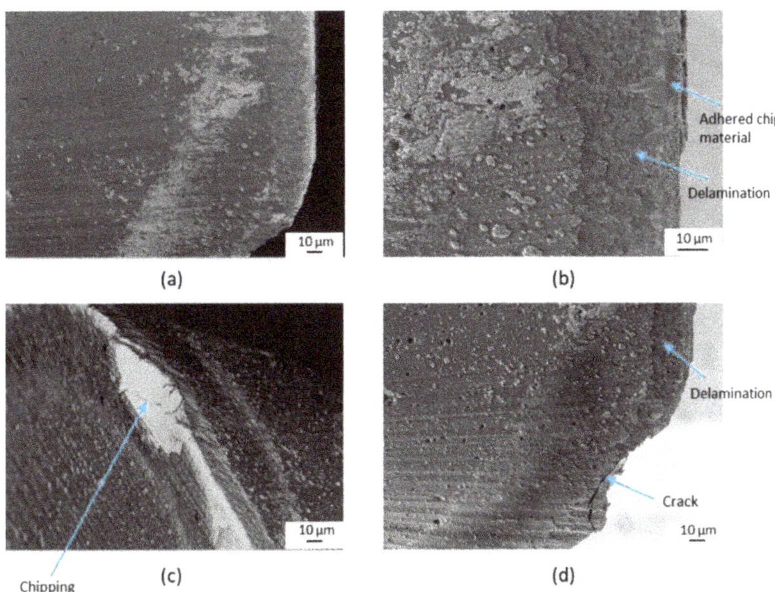

Figure 8. Tool wear mechanisms when machining wrought Ti-6Al-4V alloy: (**a**) SEM micrographs of the cutting tool after a cutting length of 500 m; (**b**) built-up layer of titanium material and coating delamination; (**c**,**d**) fractures and chipping of the cutting edge [121] (Reprinted with permission from [121]. Copyright 2020, Elsevier).

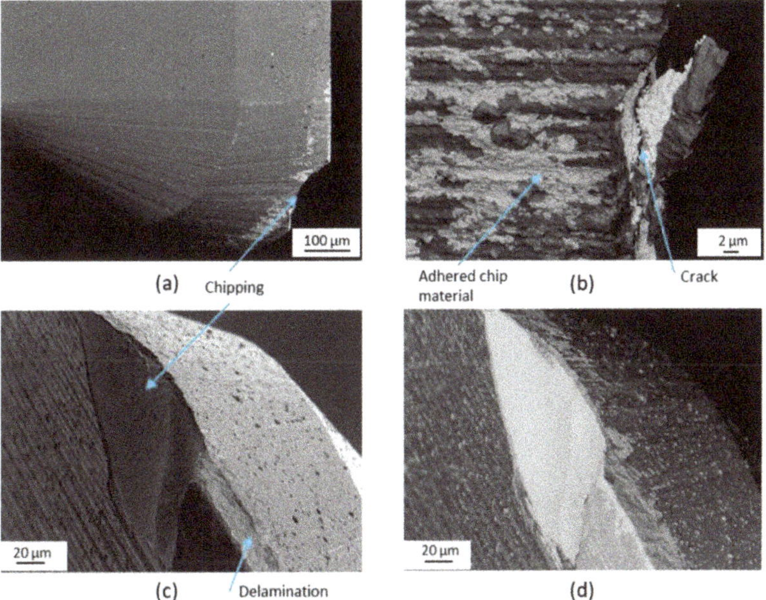

Figure 9. Tool wear mechanisms when machining as-built SLM Ti-6Al-4V alloy: (**a–c**) the built-up layer of titanium material, coating delamination, fracture, and chipping of the cutting edge; (**d**) SEM micrograph of the cutting tool after a cutting length of 500 m [121] (Reprinted with permission from [121]. Copyright 2020, Elsevier).

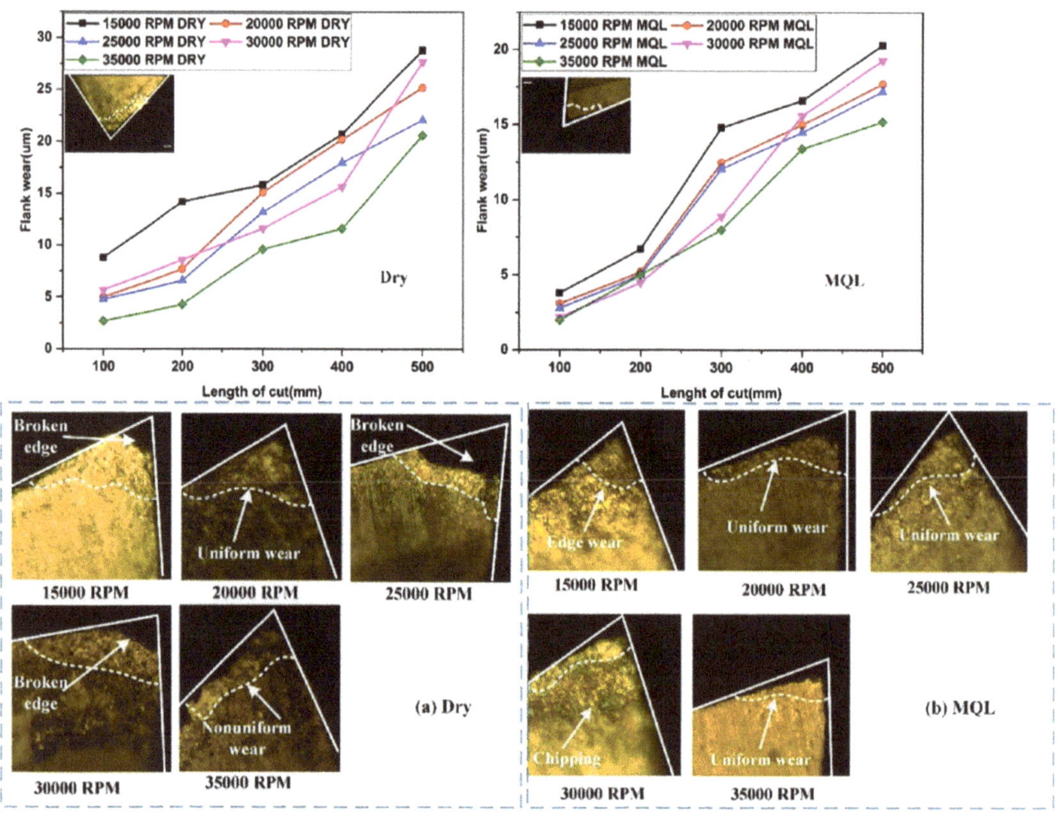

Figure 10. Tool flank wear with increasing feed rate under (**a**) dry conditions and (**b**) MQL coolant conditions [123] (Reprinted with permission from [123]. Copyright 2020, Elsevier).

During high-speed milling (HSM) of DMLS-built Ti-6Al-4V using ceramic cutting tools, the cutting edge experienced heavy mechanical and thermal loading with the cutting force being up to about 400–500 N and the cutting temperature being approximately 1000 °C [124]; intense friction existed between the chip and the tool, which caused a high temperature around the edge of the cutting tool [125]. Grave wear can be caused on the rake face by a rise in sliding velocity and friction stress [126,127]. The main wear mode of the rake face is adhesion, while both abrasion wear and adhesive wear exist on the flank face (Figure 11). The adhesion of the workpiece material on the rake face and the abrasion on the flank face are adjacent to the tool edge corner, which indicates that diffusion has occurred under the repeated thermal and mechanical loads. The wear is ascribed to the powerful extrusion between the workpiece and the cutting tool, as well as the cyclical robust friction produced by the hard particles of the workpiece. By analysing the degree of tool wear, it has been found that, compared to coated carbide tools, solid ceramic tools can provide better cutting performance during high-speed machining processes [128].

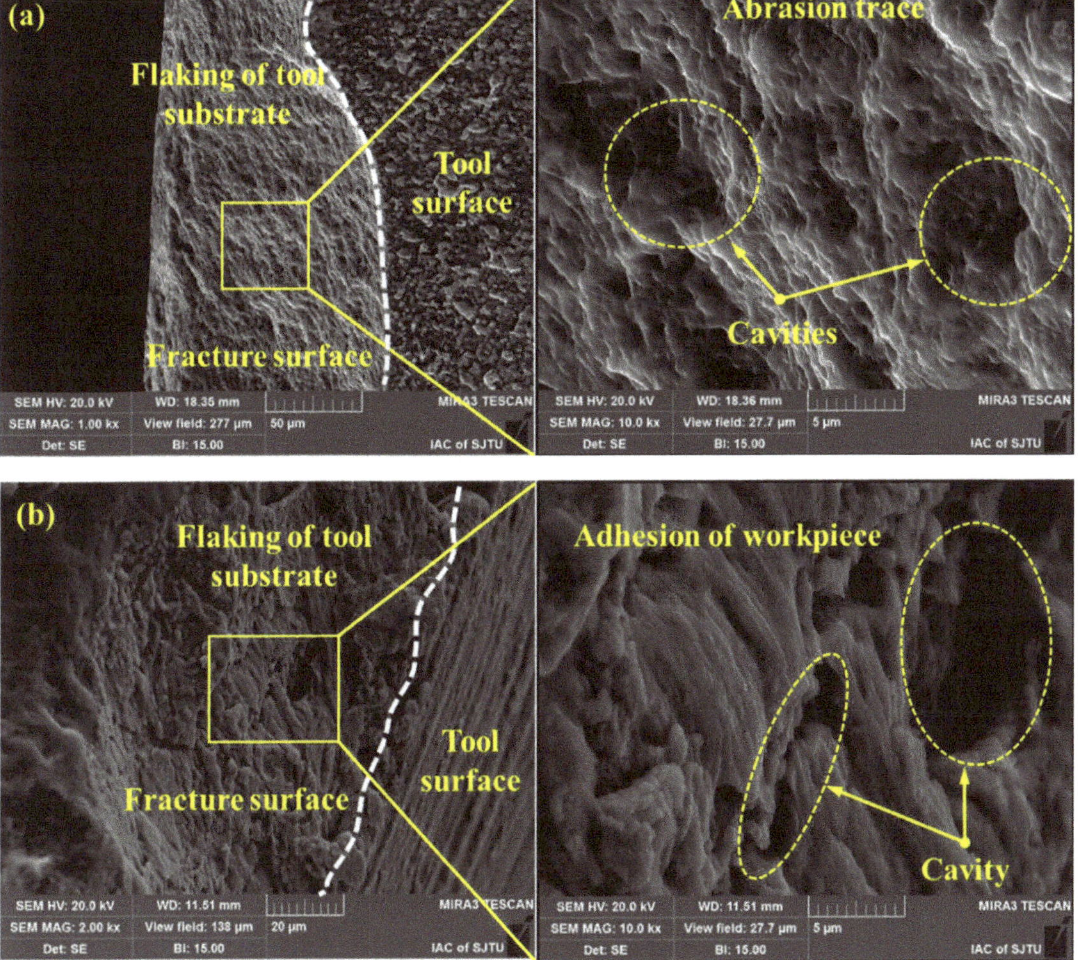

Figure 11. Wear types: (**a**) abrasion marks and (**b**) workpiece adhesion [125] (Reprinted with permission from [125]. Copyright 2020, Elsevier).

Bordin et al. [129] evaluated tool wear when machining EBM Ti-6Al-4V by adopting coated carbide tools under dry and cryogenic cooling conditions in a semi-finish turning process. It was found that the main wear mechanisms of the flank face are abrasion, chipping, and adhesion of workpiece material (Figure 12a,b). The BUE, built-up layer (BUL), and crater wear can also be seen on the rake face of the cutting tool (Figure 12c,d). Material adhesion is the most crucial wear mechanism during dry and cryogenic machining processes. Chipping is induced by unstable cutting edge fragment under the adhered workpiece material which is welded on the tool surface since the cutting begins. The supply of LN2 in the cutting zone can reduce material adhesion on the cutting edge and prevent the formation of crater wear.

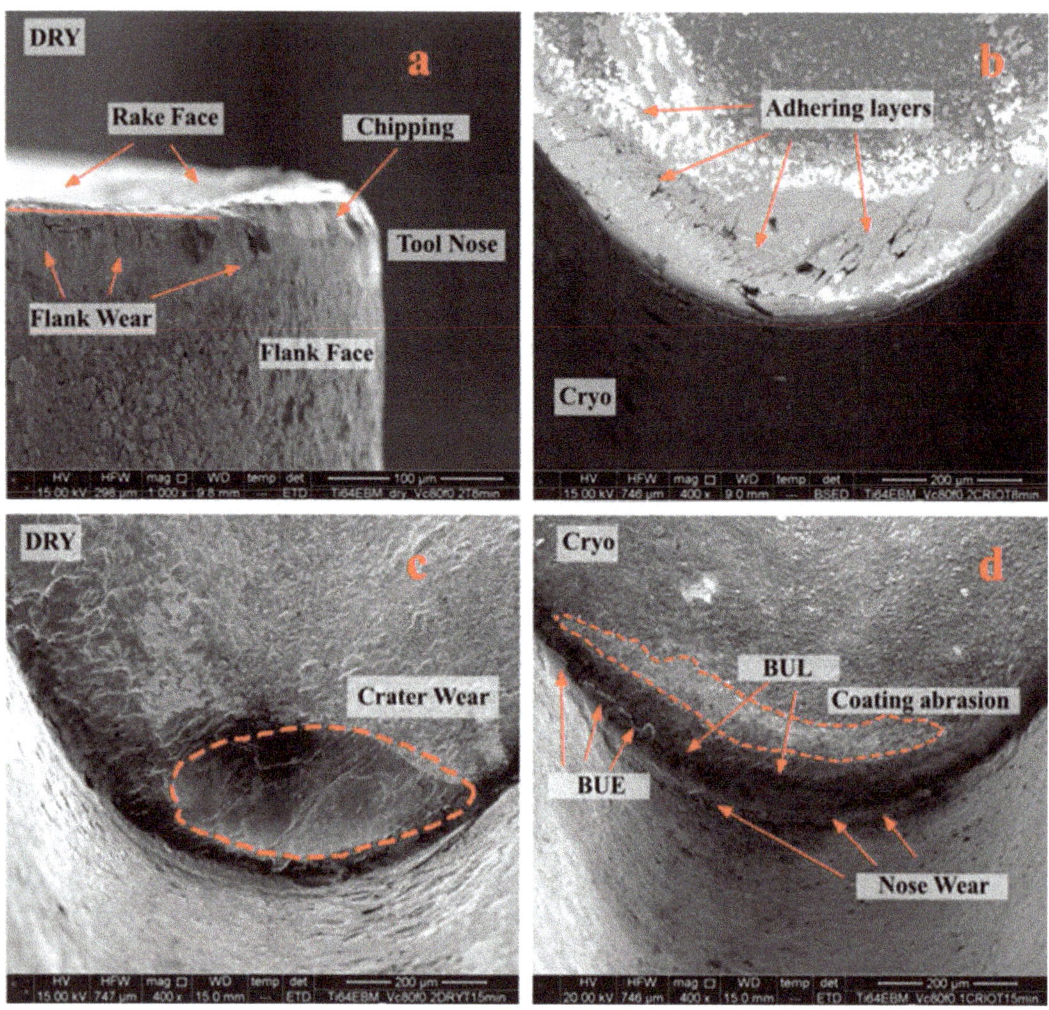

Figure 12. The main tool wear modes observed in dry and cryogenic turning processes: (**a**,**b**) abrasion, chipping, and adhesion on the flank face and (**c**,**d**) crater wear, BUE and BUL on the rake face [129] (Reprinted with permission from [129]. Copyright 2015, Elsevier).

Abrasive wear and adhesive wear are the primary wear mechanisms when turning EBM Ti-6Al-4V under dry and cryogenic conditions, irrespective of the as-delivered conditions of Ti-6Al-4V, cutting parameters, and cooling strategies [130]. Bruschi et al. [130] found that abrasive wear is presented when wear scars with feed mark peak on the workpiece surface. Due to the high contact pressure, these abrasions lead to the removal of material flakes from the surface together with some titanium oxide debris (Figure 13a,b). These titanium oxide fragments may induce three-body abrasion and increase the wear rate. As shown in Figure 13c,d, adhesive wear appears on the workpiece surface due to the transfer of material, and the appearance of micro-cracks perpendicular to the sliding direction is induced when these layers are subjected to serious deformations. Compared to dry cutting, the cryogenic cooling strategy provides a lower friction coefficient and fewer releases of titanium debris because of abrasive wear.

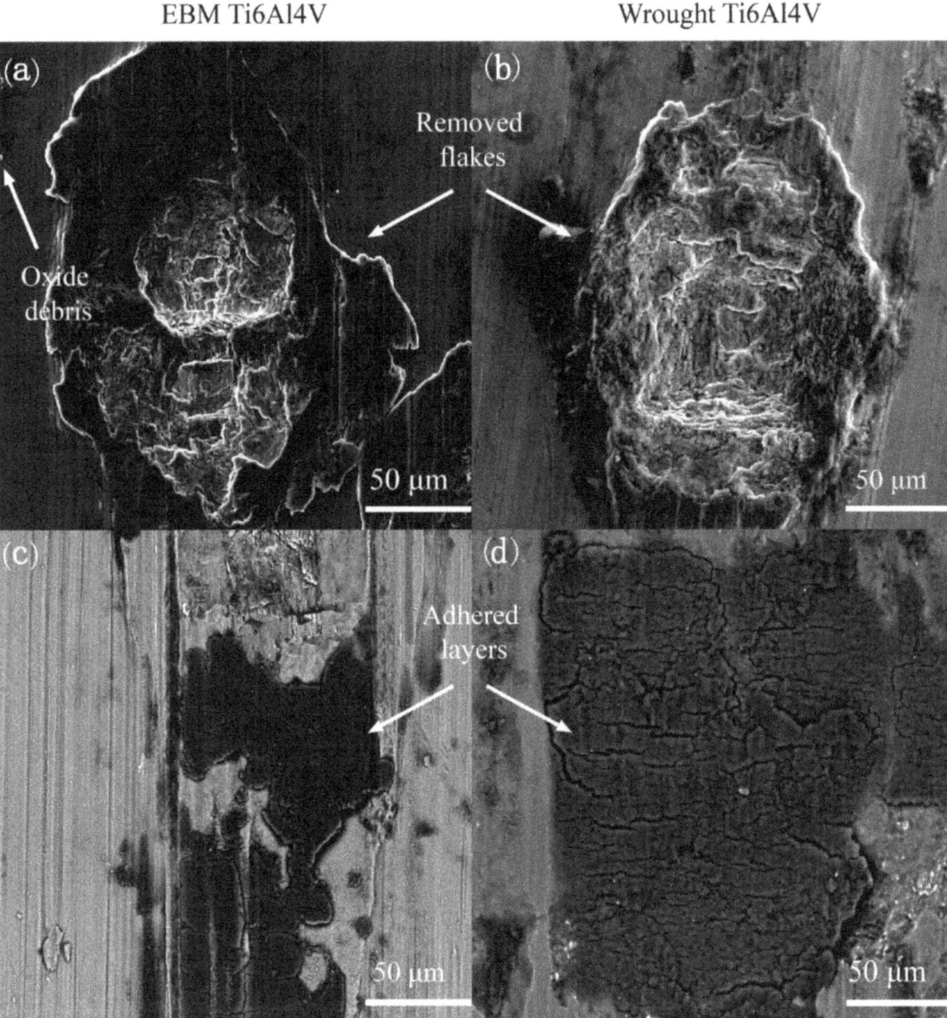

Figure 13. Main wear mechanisms for the EBM and wrought Ti-6Al-4V parts: (**a**,**b**) abrasive wear and (**c**,**d**) adhesive wear [130] (Reprinted with permission from [130]. Copyright 2016, Elsevier).

When machining Ti-6Al-4V samples produced by wrought, EBM, DMLS, and heat-treated DMLS, Sartori et al. [119] observed the deepest crater on the tool in the dry-cutting DMLS titanium samples because of their highest hardness, while in cryogenic cooling, the maximum crater depth was reduced to 58% of that in dry cutting due to the reduction in cutting temperature. Figure 14 shows the SEM images of the rake faces of the cutting tool when turning the Ti-6Al-4V under the dry cutting and cryogenic cooling conditions. It can be observed that adhesion and abrasion are the most critical wear mechanisms, irrespective of the as-received conditions and the adopted cutting strategies. In addition, the adoption of LN2 can reduce abrasive wear and flank wear. Based on an analysis of tool wear as well as mechanical and thermal properties, it has been found that EBM parts exhibit the best machinability due to their highest thermal conductivity (Figure 15) and the smallest hardness in comparison to wrought, DMSLed, and heat-treated DMLS parts.

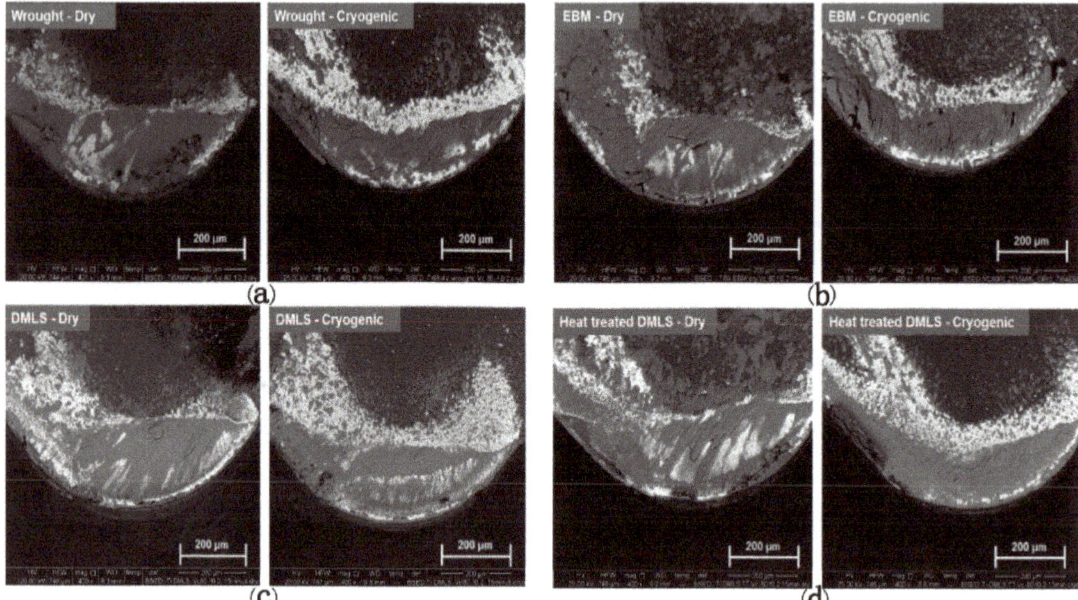

Figure 14. SEM images of worn tool rake faces after 15 min of turning under dry cutting and cryogenic cooling for different Ti-6Al-4V material: (**a**) wrought Ti-6Al-4V; (**b**) EBM Ti-6Al-4V; (**c**) DMLS Ti-6Al-4V; (**d**) Heat treated DMLS Ti-6Al-4V [119] (Reprinted with permission from [119]. Copyright 2017, Elsevier).

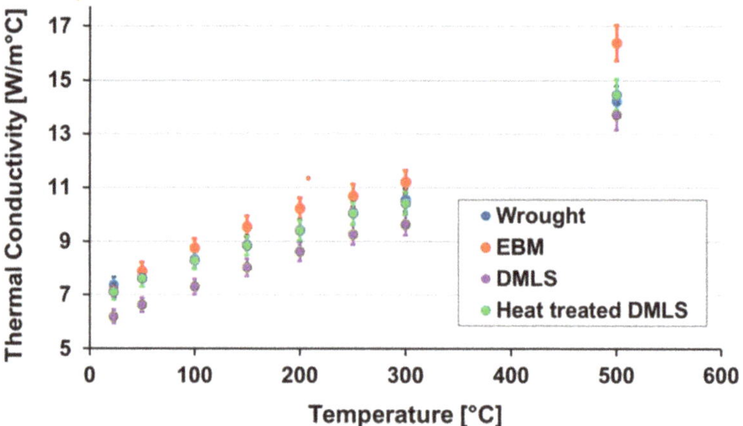

Figure 15. Effect of temperature on the thermal conductivity of Ti-6Al-4V samples under different cutting conditions [119] (Reprinted with permission from [119]. Copyright 2017, Elsevier).

According to Oyelola et al. [131], the rigid reinforcements and the heterogeneous microstructure in Ti-6Al-4V/WC metal matrix composite (MMC) produced via DED resulted in accelerated tool wear when turning the material with polycrystalline diamond (PCD), cubic boron nitride (CBN) and carbide tools. However, with the outermost layer being removed by subsequent passes, workpiece outrun decreased and tool life was improved. Because less material was pulled out from the MMC surface, the cutting tool experienced less shock impact when the machining process entered the steady state, and abrasion and chipping were found to be the main tool wear modes [132]. The surface integrity was affected by two-body and three-body abrasions induced by the interactions at the cutting region and the pull out of materials. The PCD tool provided the best performance after the outermost layer was removed. When machining with coated carbide tools, WC particles in the MMC acted like tiny cutting edges and grinded the flank face of the tool, resulting in cutting edge abrasion. For both the CBN and PCD cutting tools, the wear around the MMC region resulted in the pull out of material fragments from the cutting tool. These tiny fragments from the cutting tool and those from the workpiece led to accelerated tool wear as the self-abrasive media on the flank face of the cutting edge. On the other hand, the formation of the built-up edge at the cutting edge not only led to adhesion, but also promoted flank wear [133,134]. This phenomenon is most obvious in the CBN tool, as shown in Figure 16.

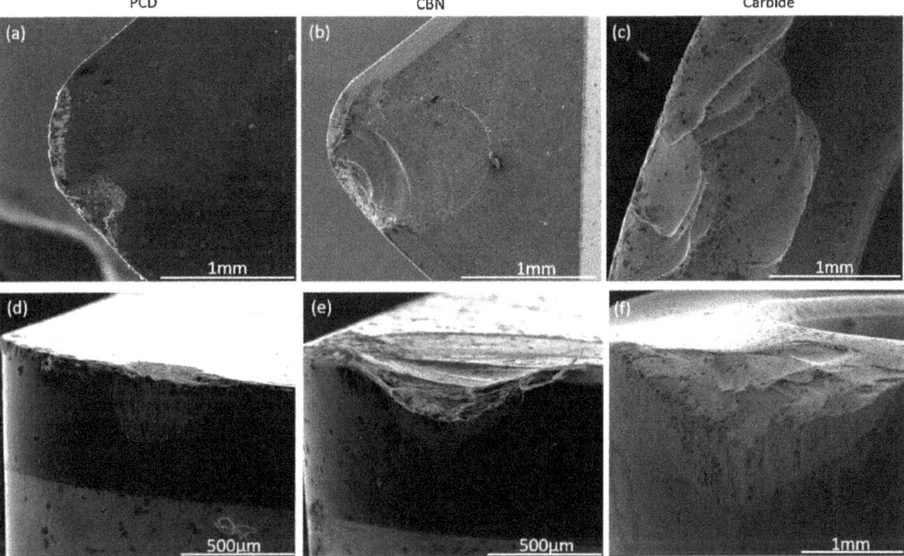

Figure 16. (**a**) Rake face of PCD tool; (**b**) rake face of CBN tool; (**c**) rake face of coated carbide tool; (**d**) flank face of PCD insert; (**e**) flank face of CBN insert; and (**f**) flank face of coated carbide tool [131] (Reprinted with permission from [131]. Copyright 2018, Elsevier).

3.4. Chip Morphology

When machining titanium alloys, the morphology and formation of chips are remarkable indicators of tool wear and level of machinability [28]. The formation of segmented chips is because of the growth of cracks on the chip's outer surface, or the formation of an adiabatic shear band that is generated by the localised shear deformation as a consequence of the predominance of thermal softening over strain hardening [94,135]. The shear localisation leads to a cyclic change of force (cutting force/thrust) with a sizable magnitude variation. The chatter/vibrations in the machining process can affect the chip and tool wear and restrict the material removal rate (MRR) [94,136]. By examining the chip morphology of wrought and SLM Ti-6Al-4V during the turning process, Coz et al. [137] found that the chip morphologies were similar for the two material states. Helical chips were generated due to the lower uncut chip thickness at a constant feed rate (60 m/min), and a higher cutting speed induced long chips. Additionally, the cutting speed had more significant effect on chip formation during the orthogonal cutting process, which mainly impacted the shear angles, segmentation frequency, and crack length. In micro-scale observations, chips showed irregularly serrated shape in form/frequency, and no particular crack was present.

Similarly, Al-Rubaie et al. [121] investigated the generation of chips from conventionally and SLM-produced Ti alloys during the milling processes and found that, despite all the generated chips being fragmented and discontinuous, the chips produced from the SLM workpiece exhibit an increase in the curling degree (Figure 17). This suggests that the chip flow is accelerated in the cutting zone due to minimal frictions. Specifically, the high hardness of the SLM parts tends to reduce the workpiece plasticity and the extent of lateral plastic flow. Therefore, it can reduce the friction at the tool–chip interface and accelerate the chip flow.

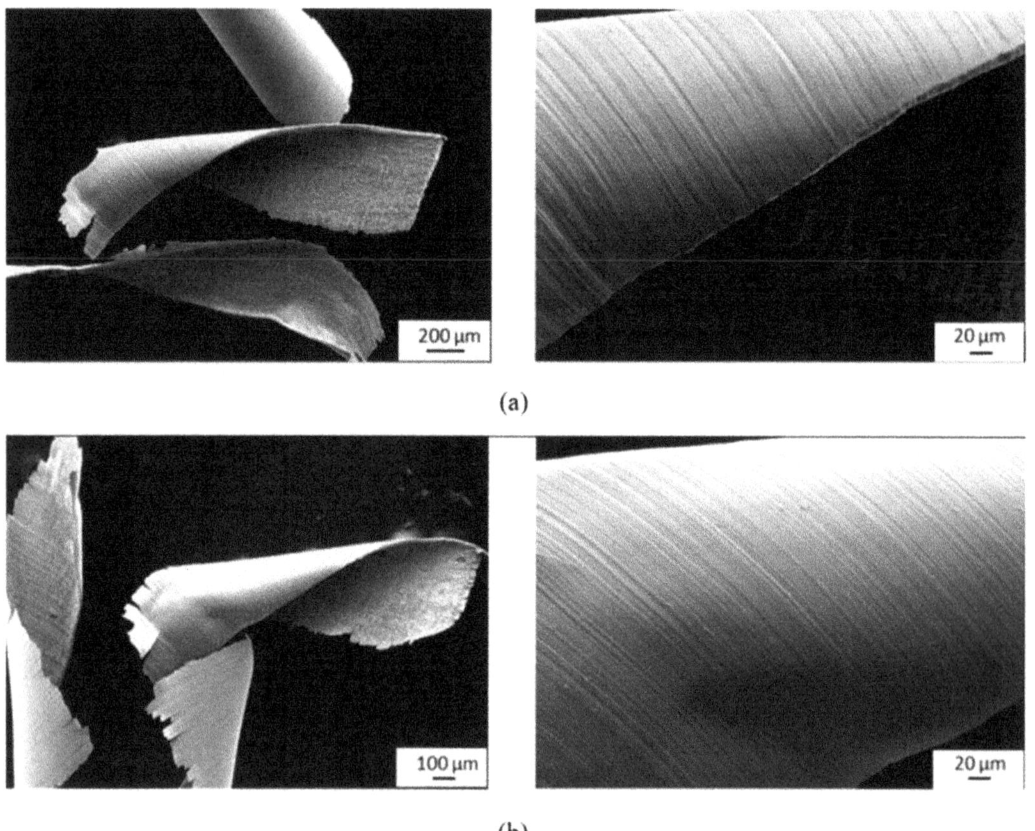

Figure 17. SEM micrographs showing the morphology of chips during the milling process: (**a**) conventional Ti-6Al-4V alloy and (**b**) SLM-formed Ti-6Al-4V alloy [121] (Reprinted with permission from [121]. Copyright 2020, Elsevier).

Zhang et al. [125] observed the chip morphologies of DMLS Ti-6Al-4V in high-speed milling tests. It was found that the chips showed continuous ribbon and broken pieces, and the chips in some positions presented a deep purple colour as a result of the super high cutting temperature causing the burn of the material. According to Figure 18, the typical serrated morphology on the free surface and some ripples that appear on the back surface of the chips can be observed. In the aggressive cutting processes, the high cutting speed generates enough strain to result in the formation of serrated chips.

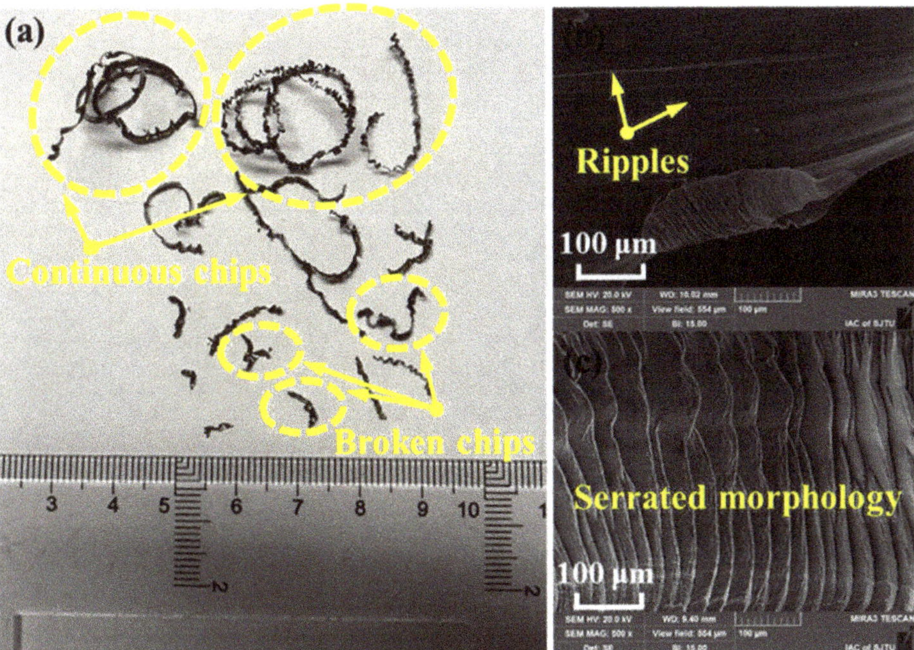

Figure 18. Chip morphologies of DMLS-fabricated Ti-6Al-4V obtained after high-speed milling: (**a**) chip present continuous ribbons and broken pieces, and chip color changes to deep purple in some positions; (**b,c**) ripples on the back surface and typically serrated morphology on the free surface of chips [125] (Reprinted with permission from [125]. Copyright 2020, Elsevier).

Furthermore, better control of chips can be conducted when adopting a cryogenic cooling strategy. This is due to the decreased material plasticity because the cutting temperature becomes lower, and the bending capacity and ductility of the material involved in chip formation are reduced [107].

3.5. Chatter Vibration

Chatter vibrations are the most frequently encountered problem in the machining of Ti alloys. It is caused by the violent relative motion between the workpiece and the tool because of the instability of the cutting system. Compared to the machining of conventionally shaped Ti workpieces, chatter vibration is more possibly generated during the machining of thin-walled Ti parts due to their low rigidity, time-varying tool–workpiece engagement conditions, and dynamic characteristics [138,139]. Chatter vibration can lead to dynamic interactions, high noise levels, unwanted residual stress, reduced tool life, and poor workpiece surface quality [140]. The mechanism of chatter vibration is still not fully understood because of its complex nature even though it is significantly detrimental to the machining process. The vibrations can affect the fatigue load on the cutting tool and deteriorate tool life. The high vibration magnitude exhibited by additively manufactured Ti parts can be mitigated by post-stress-relief heat treatment, as found by Raval et al. [141]

3.6. Mechanical Properties

Although post-processing by using finish machining is required for AM Ti parts to obtain the desired geometrical specifications and surface quality, it is worth noticing that the machining operations can cause machining-induced changes in the mechanical properties of the workpieces. Bertolini et al. [106] found that a highly deformed deformation layer always characterised the machined alloy parts, regardless of the cooling condition and machining route. The cryogenically cooled components showed a deeper affected layer. The EBM-produced samples exhibited a higher sensitivity for machined-induced hardening, and the feed rate significantly affected the compressive residual stresses. In addition, the low cutting temperature reduced material plasticity and induced more grooves on the cryogenically machined surfaces. However, the corrosion resistance grew with the adoption of cryogenic cooling. Ming et al. [98] studied the mechanical properties of SLM Ti-6Al-4V during milling tests and found that high strain rate and temperature could significantly decrease the material flow stress. More energy was consumed during the cutting of SLM parts owing to their low plasticity. The effect of the milling parameter on the residual stress of the SLM parts was not noticeable, but the residual stress of the SLM parts exhibited anisotropy. Furthermore, Oyelola et al. [110] found compressive residual stress on the machined surface of AM Ti parts produced by direct metal deposition (DMD). Umbrello et al. [142] proposed a finite element model to predict the variation in microstructure and nano-hardness of EBM Ti alloy during dry and cryogenic cooling machining processes. Their results showed that a higher cutting speed induced more plastic deformation due to the growth of the shear friction forces at the machined surface. The cryogenic cooling inhibited the strain softening of the surface layers caused by the dynamic recovery, and a higher hardness of the cryogenically machined components could be observed.

3.7. Influence of Microstructural Anisotropy

Titanium alloys manufactured by AM technology generally present microstructural anisotropy, which affects their macroscopic mechanical properties [143–147]. The microstructural anisotropy is induced by the unique thermal history at each location during the layer-by-layer building process. Ti-6Al-4V fabricated by the AM process is sensitive to thermal history because the material structure is affected by the temperature and cooling rate [148–150]. In general, Ti-6Al-4V manufactured by PBF-LB or PBF-EB exhibits fine acicular, Widmanstätten α–β or martensite alpha′ grains within the prior β phase boundaries [38], which are oriented in the build direction to grow across the building layers during solidification [151,152]. During the AM process, the large columnar prior β grains tend to grow along with the $<001>\beta$, which is in parallel with the build direction. The change in the growth direction of β grains leads to the directional solidification texture and anisotropic mechanical properties [79,89,152]. It was reported that horizontally oriented titanium samples provided lower ductility and higher strength than vertically orientated samples [153,154]; however, no apparent difference in tensile strength/yield strength was found [155]. In general, material anisotropy of AM parts is unwanted as it can lead to a fluctuation in cutting force and affect the performance of the final components [156,157].

Lizzul et al. [158] investigated the influence of anisotropy of PBF-LB Ti-6Al-4V parts on the surface quality of the workpiece and chip morphology. The parts were heat treated before the machining was carried out. It was found that α-phase layers (αGB) were present on the prior β grain boundaries. The effect of anisotropy on the machined surface was attributed to the interactions between the α-phase layers and the cutting edge of the tool. The 0 deg and 90 deg samples exhibited similar microhardness (maximum difference was 2%). The αGB layers are positively oriented in relation to the cutting edge when milling the 0 deg Ti samples, as shown in Figure 19. On the contrary, there is no favourable interaction between the αGB layers and the cutting edge in the 90 deg Ti samples. The αGB layers favourably reduced the dislocation movements and the formation of burrs. By analysing chip morphology, it was found a lower cutting force/power was generated when machining the 0 deg Ti samples instead of the 90 deg Ti samples. This phenomenon was more accentuated under the higher cutting speed condition, which prompted the plastic flow of the material and resulted in chips with higher curl radii.

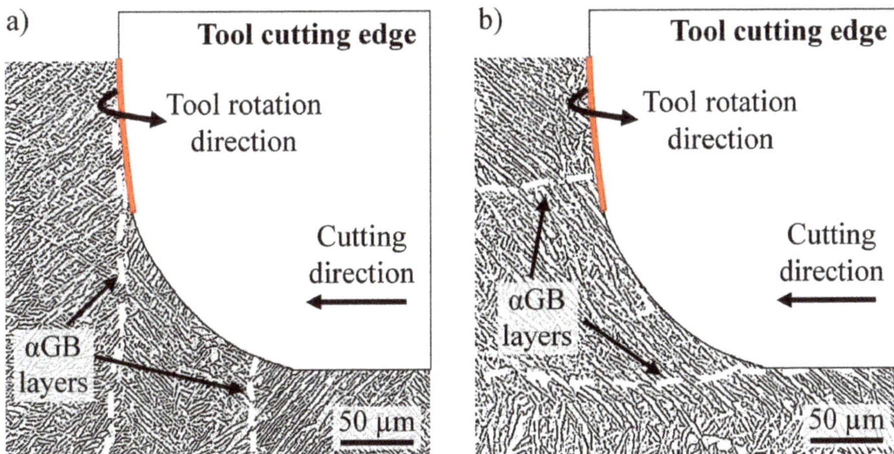

Figure 19. Microstructural interaction when milling 0 deg (**a**) and 90 deg (**b**) oriented PBF-LB Ti-6Al-4V samples [158] (Reprinted with permission from [158]. Copyright 2021, Elsevier).

In another paper, Lizzul et al. [159] evaluated the effect of anisotropy on tool wear when machining Ti-6Al-4V parts produced by PBF-LB with four different build orientations (0°, 36°, 72° and 90°) of αGB layers. Within 1 m of cutting length, the tool diameter was reduced by 3.7% and 7.5%, respectively, when machining the 0 deg Ti sample and 90 deg Ti sample. The tool diameter reduction when machining the 36 deg Ti sample and the 72 deg Ti sample was 5.5% and 6.5%, respectively. During the cutting process, the αGB layer developed along the prior β grains is characteristic of the discontinuity in the microstructure and the weak point along which cracks may develop (Figure 20) [105,150,160]. In addition, when the cutting tool rotates, it gradually engages the workpiece with the registration angle κ (approaches 90°) (Figure 20b). In this case, the orientation angle of the αGB layers corresponds to the registration angle κ for the 0 deg sample, which is favourable to material removal and chip formation, thereby decreasing the cutting force and improving the tool life.

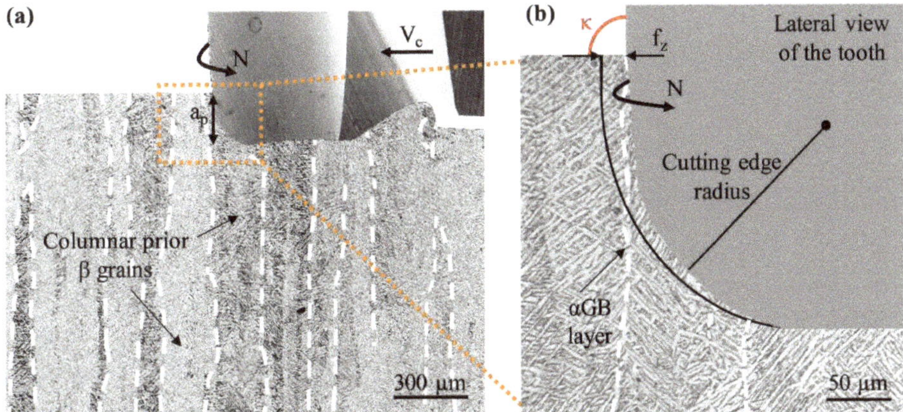

Figure 20. (**a**) Tool engagement with respect to the prior β grain orientations for the 0° sample, and (**b**) zoomed picture showing the orientation angle of the αGB layers relative to the cutting tool registration angle κ [159] (Reprinted with permission from [159]. Copyright 2020, Elsevier).

3.8. Influence of Porosity

Although additively manufactured Ti parts present excellent surface finish after machining or post-processing, their mechanical characteristics are still affected by the presence of porosity, which results in the deterioration of product quality and a reduction in ductility and strength [45,161,162]. Porosity defects are unavoidable in metal additive manufacturing due to the instability and complexity of the layer-by-layer manufacturing process. The defects can be caused by improper processing parameters, insufficient energy input, and gas entrapped in the powder particles [163], and they can be classified as keyhole pores, lack of fusion pores, and gas pores [164].

During the machining of porous parts, with an increase in machining distance, the spherical pore near the machining path can become tiny and closed, revealing the machining-induced pore closure phenomenon. The presence of pores promotes the dislocation slips and results in the work-hardening effect on the machined surface, in addition to causing changes in cutting force [165]. Ahmad et al. [166] investigated the effect of different porosity on the machineability of additively manufactured Ti-6Al-4V parts through micro-milling experiments. The cutting force was found to have a strong linear correlation with the porosity of additively manufactured parts, and the tool wear increased with an increase in porosity. Micro burrs caused by interrupted chip formation could be observed on the machined surface of the additively manufactured Ti-6Al-4V samples, and the surface roughness increased with an increase in porosity. A similar experiment was conducted by Varghese et al. [167], who investigated the influence of porosity on the cutting force and surface integrity of Ti-6Al-4V alloy produced by DMLS using a micro-milling process. The results demonstrated that the cutting force when machining additively manufactured Ti-6Al-4V with 0% porosity was maximum. It decreased with an increase in porosity and reached the minimal level at 46% porosity. On the other hand, with an increase in cutting depth, the cutting force when machining porous additively manufactured Ti parts increased due to the heterogeneous nature of the porous workpiece material and the non-uniformity in pore size. The surface roughness of the additively manufactured Ti-6Al-4V alloy decreased with the increase in cutting depth. With an increase in porosity, the surface roughness of the additively manufactured part began to increase first and then decreased with the continuous increase in porosity because of the smearing/porosity closure phenomena. Shunmugavel et al. [95] found that there was a significant difference in cutting force when machining conventional Ti parts and additively manufactured Ti parts at a high cutting speed of 180 m/min because of the thermal softening effects and the existence of pores. When machining the SLM Ti component, the pores resulted in micro impacts on the cutting tool, and the frequency of these micro impacts increased with an increase in the cutting speed. Hence, a higher cutting force could be induced during the high-speed machining of the SLM Ti-6Al-4V sample.

3.9. Influence of Post-Processing Processes

Ti-6Al-4V parts produced by AM show comparatively high yield stress (about 1000 Mpa) and high tensile strength (around 1150 MPa), but relatively low ductility (smaller than 10%) [168,169]. The high cooling rate during the AM process results in significant internal thermal stresses in the material structure. During the building process, the scanning by the heat resource (laser or electron beam) may lead to instabilities in the melt pool, which, consequently, causes a rise in porosity and surface roughness [55,168]. These features will deteriorate the fatigue performance of the additively manufactured Ti components and impact their reliability in engineering applications. Hence, to improve mechanical properties, reduce porosity, and alter the microstructure of additively manufactured Ti components so as to meet application requirements, besides machining, post-processing methods such as heat treatment and hot isostatic pressing (HIP) are usually applied after the parts are built [79,168,170–173]. These processes can relieve stresses, minimise porosity [171,174–176], and affect the response of subsequent machining processes. Oyelola et al. [112] studied the influence of heat treatment on the machining of

DED-fabricated Ti parts. They found that heat treatment can improve the machinability of DED parts due to the homogenising of microstructure. The microstructural homogenisation results in a reduction in the average machining force. The larger grain size generated by the post-heat treatment leads to a reduction in hardness, which improves surface roughness and substantially decreases subsurface deformation. Similarly, Al-Rubaie et al. [121] found that stress-relieved additively manufactured Ti parts exhibited a larger cutting force compared to wrought and as-built parts because the stress-relief heat treatment increased the compressive stress. In another study, Bruschi et al. [177] investigated the influence of heat treatment on tool wear behaviours when machining EBM-built Ti components. They found that the heat treatment had positive influences on tool wear behaviours, and all the heat-treated samples exhibited a lower friction coefficient/tool wear rate because the heat treatment increased the sub-surface hardness of the heat-treated parts through changing the microstructure.

4. Hybrid Manufacturing of Additively Manufactured Ti-6Al-4V

Additively manufactured metal parts have poor dimensional/geometric accuracies and poor surface quality because of the "staircase-effect" and non-uniform powder deposition [178,179]. In order to solve these issues, subtractive machining including milling and turning has been extensively applied when post-processing quality-critical components, such as low-pressure turbine blades and blade disks. However, the machining of internal surfaces, which are unreachable by cutting tools, is not possible. This issue has significantly limited the high-level design flexibility of AM and restricted its application in manufacturing functional components with complex shapes.

To solve this bottleneck problem, a new type of "hybrid manufacturing" machine has been developed recently by integrating the machining and AM process together [29,30]. The machining of internal surfaces is conducted during the AM process before the surfaces become inaccessible after the AM process is finished (Figure 21) [180]. Nevertheless, due to the lack of in-depth knowledge of the hybrid process, the strategies used in in situ machining are still those developed for conventional machining, and as a result, serious quality problems, such as defective surfaces and excessive tool wear, frequently occur [181–183].

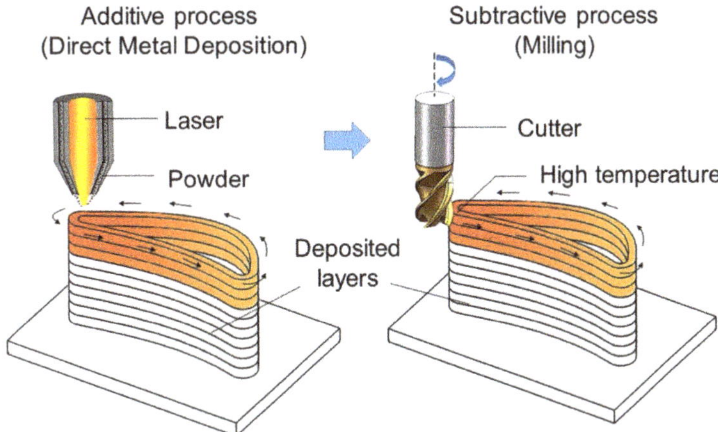

Figure 21. Additive and subtractive hybrid manufacturing processes [180] (Reprinted with permission from [180]. Copyright 2020, Springer Nature).

Ye et al. [184] developed a hybrid manufacturing machine tool consisting of high-speed milling and direct metal deposition to acquire additively manufactured products with good surface quality and high precision. The surface quality can be significantly improved due to the removal of the rough layers by the milling process. Jeng et al. [185] presented a hybrid process that combines selective laser cladding (SLC) and the milling process to prove the feasibility of modifying/repairing the parts by applying the hybrid process.

As a hybrid solution that combines laser and conventional machining, laser-assisted machining (LAM) can improve the machinability of metal parts by preheating materials using a laser beam as an external heat source to soften the workpiece material [186,187]. The research conducted by Gao, Bermingham, Dargusch, and Matthew et al. [188–191] provided a better understanding of the influence of cutting parameters on tool wear and surface quality and offered valuable insights into the complexities of machining titanium alloys at elevated temperature. Navas et al. [192] investigated the machinability of Inconel 718 and found that laser heating led to a significant reduction in the work-hardening effect, and the tensile and yield strength of Inconel 718 dropped when the temperatures on the shear zone were around 600–650 °C. Laser heating can result in reductions in cutting force if proper feed rates are applied. Similarly, Dandekar et al. [27] investigated machining characteristics during the hybrid manufacturing of a titanium workpiece with laser-assisted machining and cryogenic cooling of tools. They found that the specific cutting energy was reduced by up to 20% and the surface roughness was reduced by 30% compared to conventional machining. In the hybrid manufacturing process, cryogenic cooling led to lower temperatures in the tool–chip interface to reduce the cutting temperature and tool wear. The surface quality was also improved because of the lower friction between the tool flank face and the workpiece.

In LAM, the softening effect of the machined material by local heating is a critical factor that reduces the cutting force and tool wear. However, during the additive and subtractive hybrid manufacturing (ASHM) process, the laser source leads to a build-up of temperature, and the whole workpiece is heated. It results in a temperature field which is different from local heating, and the machining response (such as machining force, surface integrity, and tool wear) is impacted during successive machining processes [193,194]. Li et al. [180] investigated the influence of temperature build-up on the machinability of DMD-produced Ti-6Al-4V during the ASHM process. The heating device was designed to conduct the milling experiments at a specific temperature. The cutting temperature rose with the rise in the preheating temperature, which reduced the material flow stress. The thermal softening effect was counteracted by the work-hardening effect when the preheating temperature was less than 300 °C. With the further rise in the preheating temperature, the enhanced thermal softening effect led to significantly reduced cutting forces.

In another study, Moritz et al. [194] analysed the influence of cryogenic milling on the machinability of Ti-6Al-4V components fabricated by LMD by adopting carbon dioxide as the coolant to evaluate the surface integrity and tool wear. It was found that cryogenic machining could cause lower surface roughness and smaller tool wear, leading to contamination-free surfaces compared to other machining methods. The result shows that hybrid manufacturing consisting of LMD and cryogenic milling provides significant advantages for the final parts when machining additively manufactured Ti-6Al-4V. Likewise, to further improve the machinability of additively manufactured parts using ASHM, Du et al. [195] presented a creative method that integrated eddy current detection (ECD) into the ASHM process for defect detection/removal of additively manufactured metal parts during the manufacturing process. This method provides the ability for in-process defect detection/removal without the needs for advanced equipment and offers an effective way to further improve the quality of additively manufactured parts of complex shapes.

5. Future Development and Conclusions

PBF and DED processes have been progressively used to produce complex alloy parts with near-net shapes. As shown in Table 1, a lot of research has been conducted in the past ten years on powder bed fusion (both SLM and EBM) and directed energy deposition processes. In comparison, there is limited research on wire-based AM processes, such as wire arc additive manufacturing process.

Table 1. Recent publications on machining of additively manufactured Ti-6Al-4V alloy.

Reference	AM Process	Machining Process	Investigated Machining Characteristics	Year of Publication
[18]	SLM	Turning	Cutting force, tool wear, and surface roughness	2016
[97]	SLM	Turning	Tool wear and surface integrity	2016
[96]	SLM	Turning	Cutting force and surface integrity	2017
[98]	SLM	Milling	Cutting force and surface integrity	2019
[121]	SLM	Milling	Cutting force and tool wear	2020
[120]	SLM	Milling	Cutting force, tool wear, and surface roughness	2022
[107]	EBM	Turning	Tool wear, surface integrity, and chip morphology	2017
[106]	EBM	Turning	Surface integrity	2019
[35]	EBM	Milling	Surface integrity	2018
[196]	EBM	Milling	Cutting force and surface quality	2020
[112]	DED	Turning	Surface integrity and tool wear	2018
[131]	DED	Turning	Cutting force, surface finish, and tool wear	2018
[197]	DED	Milling	Subsurface deformation	2022
[198]	EBM/DMLS	Turning	Tool wear	2016
[119]	EBM/DMLS	Turning	Tool wear	2017
[110]	DMD	Turning	Surface integrity	2016
[109]	DMLS	Turning	Surface integrity	2016
[129]	EBM	Turning	Tool wear	2015
[137]	SLM	Turning	Cutting force, chip morphology, and microstructural characteristics	2017
[159]	PBF-LB	Milling	Tool wear, surface roughness, and chip morphology	2020
[123]	SLM	Milling	Tool wear, residual stresses, and surface quality	2020
[180]	DMD	Milling	Cutting force and tool wear	2020
[199]	SLM	Milling	Cutting force, tool wear, and chip morphology	2020
[125]	DMLS	Milling	Cutting force, tool wear, chip morphology, and surface integrity	2020
[200]	EBM	Milling	Surface finish, tool wear, and microstructure	2020
[201]	PBF-LB	Milling	Cutting force, surface topography, and chip morphology	2021
[202]	PBF-LB	Turning	Microstructure and surface topography	2021
[203]	PBF-LB	Milling	Surface quality and chip morphology	2021
[204]	SLM	Milling	Surface integrity	2022
[205]	SLM	Turning	Tool wear and surface characteristics	2022
[206]	SLM	Turning	Cutting force, surface integrity, and tool wear	2022
[207]	SLM	Turning	Cutting force, surface quality, and tool wear	2022
[208]	SLM	Milling	Cutting force and surface roughness	2022
[209]	DMLS	Turning	Tool wear, surface roughness, and chip morphology	2022
[210]	DMLS	Milling	Milling force, residual stresses, and subsurface plastic deformation	2022
[211]	EBM	Turning	Cutting force and surface roughness	2022

Since some mechanical properties of AM-produced Ti alloys, such as strength and hardness, are significantly larger than those of wrought Ti alloys, the machining of additively manufactured Ti-6Al-4V alloy is a difficult process. The higher cutting force caused by the higher hardness and higher yield strength of additively manufactured parts leads to a higher cutting temperature and severe tool wear, which not only impact the final surface quality of the additively manufactured parts, but also result in premature tool failure. The surface roughness of additively manufactured Ti parts is larger than that of wrought Ti parts. Similar cooling methods that are suitable for machining wrought Ti alloys, such as cryogenic cooling and MQL, are used when machining additively manufactured Ti alloy to reduce the cutting temperature and tool wear. Since the material properties of additively manufactured parts, including microstructural anisotropy and porosity, change with the building orientations, post-processing processes, such as HIP and heat treatment, can be used to alter the microstructural characteristics of additively manufactured components. Microstructural anisotropy of additively manufactured parts results in an unstable manufacturing process because of the fluctuation in cutting force, which further leads to unpredictable tool wear and poor surface integrity. It has been reported that the machining direction has a close relationship with the build-up orientation. The cutting force is affected by different build directions due to the orientation of α-grain boundaries. The best machining performances can be achieved when the machining processes are perpendicular to the build-up direction. However, the relationship between microstructural characteristics and machinability, as well as the fundamental mechanisms, is still not clearly understood. Further exploration of the relationship between machining characteristics and printing strategies is needed.

The additive and subtractive hybrid manufacturing process combines the benefits of both precision subtractive machining and additive manufacturing. It is a promising method with significant potentials for manufacturing additively manufactured Ti parts with complex geometries, which previously have not been possible. In the hybrid machining process, the heat source such as the laser beam, which can be applied to preheat and soften the material, reduces the hardness, yield strength, and tensile strength of the material, leading to an easier machining process and significant enhancement in the machinability of additively manufactured Ti alloy workpieces. In LAM, the softening effect of the machined material by local heating is a critical factor that decreases the cutting force and reduces tool wear. However, during the ASHM process, the whole workpiece is heated, thus inducing a different temperature field in comparison to the local heating method, and may affect the machining characteristics of additively manufactured Ti workpieces.

Additionally, machining strategies for conventional turning and milling processes are used in situ in the hybrid additive and subtractive manufacturing process. Nevertheless, these strategies may not be directly suitable for actual machining applications, for example, additively manufactured Ti-6Al-4V parts with specialised geometry, such as thin walls, and specially shaped overhang features. Hence, more studies are expected in this field. Additionally, conventional cutting tools are used in most machining processes. There is limited research on the design of new cutting tools for the cutting of additively manufactured Ti components. To obtain minimal tool wear and better surface quality, there is a pressing need for more in-depth research on hybrid manufacturing additively manufactured Ti alloys, including machining-induced defects on workpieces and design of new tools that are customised for the in situ machining of additively manufactured parts.

Author Contributions: Conceptualization, C.Z., D.Z., M.M., G.L., J.P.T.M. and S.D.; methodology, G.L. and M.M.; resources, J.P.T.M. and S.D.; writing—original draft preparation, C.Z. and D.Z.; writing—review and editing, M.M., G.L. and S.D.; visualization, C.Z.; supervision, J.P.T.M. and S.D.; project administration, M.M. and J.P.T.M.; funding acquisition, S.D. All authors have read and agreed to the published version of the manuscript.

Funding: This research was partially funded by the Australian Research Council under the Discovery Project scheme (Grant Numbers: DP180100762 and DP210103278).

Data Availability Statement: Not applicable.

Conflicts of Interest: The authors declare no conflict of interest.

References

1. Williams, J.C.; Boyer, R.R. Opportunities and Issues in the Application of Titanium Alloys for Aerospace Components. *Metals* **2020**, *10*, 705. [CrossRef]
2. Bartolo, P.; Kruth, J.-P.; Silva, J.; Levy, G.; Malshe, A.; Rajurkar, K.; Mitsuishi, M.; Ciurana, J.; Leu, M. Biomedical production of implants by additive electro-chemical and physical processes. *CIRP Ann.* **2012**, *61*, 635–655. [CrossRef]
3. Veiga, C.; Davim, J.; Loureiro, A. Properties and applications of titanium alloys: A brief review. *Rev. Adv. Mater. Sci.* **2012**, *32*, 133–148.
4. Veiga, C.; Davim, J.P.; Loureiro, A. Review on machinability of titanium alloys: The process perspective. *Rev. Adv. Mater. Sci.* **2013**, *34*, 148–164.
5. Ezugwu, E.O.; Wang, Z.M. Titanium alloys and their machinability—A review. *J. Mater. Process. Technol.* **1997**, *68*, 262–274. [CrossRef]
6. Liu, Y.; Dong, H.; Wang, H.; Xiao, G.; Meng, F. Multi-objective titanium alloy belt grinding parameters optimization oriented to resources allocation and environment. *Int. J. Adv. Manuf.* **2021**, *113*, 449–463. [CrossRef]
7. Muhammad, R.; Maurotto, A.; Demiral, M.; Roy, A.; Silberschmidt, V.V. Thermally enhanced ultrasonically assisted machining of Ti alloy. *CIRP J. Manuf. Sci. Technol.* **2014**, *7*, 159–167. [CrossRef]
8. Murr, L.E.; Quinones, S.A.; Gaytan, S.M.; Lopez, M.I.; Rodela, A.; Martinez, E.Y.; Hernandez, D.H.; Martinez, E.; Medina, F.; Wicker, R.B. Microstructure and mechanical behavior of Ti-6Al-4V produced by rapid-layer manufacturing, for biomedical applications. *J. Mech. Behav. Biomed. Mater.* **2009**, *2*, 20–32. [CrossRef]
9. Pruncu, C.I.; Hopper, C.; Hooper, P.A.; Tan, Z.; Zhu, H.; Lin, J.; Jiang, J. Study of the Effects of Hot Forging on the Additively Manufactured Stainless Steel Preforms. *J. Manuf. Process.* **2020**, *57*, 668–676. [CrossRef]
10. Kaur, I.; Singh, P. State-of-the-art in heat exchanger additive manufacturing. *Int. J. Heat Mass Transf.* **2021**, *178*, 121600. [CrossRef]
11. Vithani, K.; Goyanes, A.; Jannin, V.; Basit, A.W.; Gaisford, S.; Boyd, B.J. An Overview of 3D Printing Technologies for Soft Materials and Potential Opportunities for Lipid-based Drug Delivery Systems. *Pharm. Res.* **2018**, *36*, 4. [CrossRef] [PubMed]
12. Pratheesh Kumar, S.; Elangovan, S.; Mohanraj, R.; Ramakrishna, J.R. Review on the evolution and technology of State-of-the-Art metal additive manufacturing processes. *Mater. Today Proc.* **2021**, *46*, 7907–7920. [CrossRef]
13. Duda, T.; Raghavan, L.V. 3D Metal Printing Technology. *IFAC-Pap.* **2016**, *49*, 103–110. [CrossRef]
14. MANUFACTUR3D. GE Aviation Produced Its 100,000th 3D Printed Fuel Nozzle Tip. Available online: https://manufactur3dmag.com/ge-aviation-produced-its-100000th-3d-printed-fuel-nozzle-tip/ (accessed on 19 August 2022).
15. Electric, G. GE9X Additive Parts. Available online: https://www.ge.com/additive/sites/default/files/2020-08/GE9X%20Additive%20parts.pdf (accessed on 19 August 2022).
16. Mishurova, T.; Artzt, K.; Rehmer, B.; Haubrich, J.; Ávila, L.; Schoenstein, F.; Serrano-Munoz, I.; Requena, G.; Bruno, G. Separation of the impact of residual stress and microstructure on the fatigue performance of LPBF Ti-6Al-4V at elevated temperature. *Int. J. Fatigue* **2021**, *148*, 106239. [CrossRef]
17. Yadroitsev, I.; Smurov, I. Surface Morphology in Selective Laser Melting of Metal Powders. *Phys. Procedia* **2011**, *12*, 264–270. [CrossRef]
18. Shunmugavel, M.; Polishetty, A.; Nomani, J.; Goldberg, M.; Littlefair, G. Metallurgical and machinability characteristics of wrought and selective laser melted Ti-6Al-4V. *J. Metall.* **2016**, *2016*, 7407918. [CrossRef]
19. Yang, L.; Patel, K.V.; Jarosz, K.; Özel, T. Surface integrity induced in machining additively fabricated nickel alloy Inconel 625. *Procedia CIRP* **2020**, *87*, 351–354. [CrossRef]
20. Basha, M.M.; Basha, S.M.; Jain, V.K.; Sankar, M.R. State of the art on chemical and electrochemical based finishing processes for additive manufactured features. *Addit. Manuf.* **2022**, *58*, 103028. [CrossRef]
21. Wysocki, B.; Idaszek, J.; Szlązak, K.; Strzelczyk, K.; Brynk, T.; Kurzydłowski, K.J.; Święszkowski, W. Post Processing and Biological Evaluation of the Titanium Scaffolds for Bone Tissue Engineering. *Materials* **2016**, *9*, 197. [CrossRef]
22. Pyka, G.; Kerckhofs, G.; Papantoniou, I.; Speirs, M.; Schrooten, J.; Wevers, M. Surface Roughness and Morphology Customization of Additive Manufactured Open Porous Ti-6Al-4V Structures. *Materials* **2013**, *6*, 4737–4757. [CrossRef]
23. Li, G.; Rahim, M.Z.; Pan, W.; Wen, C.; Ding, S. The manufacturing and the application of polycrystalline diamond tools—A comprehensive review. *J. Manuf. Process.* **2020**, *56*, 400–416. [CrossRef]
24. Li, G.; Wu, G.; Pan, W.; Rahman Rashid, R.A.; Palanisamy, S.; Ding, S. The Performance of Polycrystalline Diamond (PCD) Tools Machined by Abrasive Grinding and Electrical Discharge Grinding (EDG) in High-Speed Turning. *J. Manuf. Mater. Process.* **2021**, *5*, 34. [CrossRef]
25. Yi, S.; Li, N.; Solanki, S.; Mo, J.; Ding, S. Effects of graphene oxide nanofluids on cutting temperature and force in machining Ti-6Al-4V. *Int. J. Adv. Manuf. Technol.* **2019**, *103*, 1481–1495. [CrossRef]
26. Niknam, S.A.; Khettabi, R.; Songmene, V. Machinability and Machining of Titanium Alloys: A Review. In *Machining of Titanium Alloys*; Davim, J.P., Ed.; Springer: Berlin/Heidelberg, Germany, 2014; pp. 1–30.

27. Dandekar, C.R.; Shin, Y.C.; Barnes, J. Machinability improvement of titanium alloy (Ti-6Al-4V) via LAM and hybrid machining. *Int. J. Mach. Tool Manuf.* **2010**, *50*, 174–182. [CrossRef]

28. Hourmand, M.; Sarhan, A.A.D.; Sayuti, M.; Hamdi, M. A Comprehensive Review on Machining of Titanium Alloys. *Arab. J. Sci. Eng.* **2021**, *46*, 7087–7123. [CrossRef]

29. Sommer, D.; Götzendorfer, B.; Esen, C.; Hellmann, R. Design Rules for Hybrid Additive Manufacturing Combining Selective Laser Melting and Micromilling. *Materials* **2021**, *14*, 5753. [CrossRef]

30. Sommer, D.; Pape, D.; Esen, C.; Hellmann, R. Tool Wear and Milling Characteristics for Hybrid Additive Manufacturing Combining Laser Powder Bed Fusion and In Situ High-Speed Milling. *Materials* **2022**, *15*, 1236. [CrossRef]

31. Lütjering, G.; Williams, J.C. (Eds.) Titanium Matrix Composites. In *Titanium*; Springer: Berlin/Heidelberg, Germany, 2007; pp. 367–382.

32. Arrazola, P.J.; Garay, A.; Iriarte, L.M.; Armendia, M.; Marya, S.; Le Maître, F. Machinability of titanium alloys (Ti-6Al-4V and Ti555.3). *J. Mater. Process. Technol.* **2009**, *209*, 2223–2230. [CrossRef]

33. Palanisamy, S.; McDonald, S.D.; Dargusch, M.S. Effects of coolant pressure on chip formation while turning Ti-6Al-4V alloy. *Int. J. Mach. Tool Manuf.* **2009**, *49*, 739–743. [CrossRef]

34. Baufeld, B.; Biest, O.V.d.; Gault, R. Additive manufacturing of Ti-6Al-4V components by shaped metal deposition: Microstructure and mechanical properties. *Mater. Des.* **2010**, *31*, S106–S111. [CrossRef]

35. Rotella, G.; Imbrogno, S.; Candamano, S.; Umbrello, D. Surface integrity of machined additively manufactured Ti alloys. *J. Mater. Process. Technol.* **2018**, *259*, 180–185. [CrossRef]

36. Ren, X.P.; Li, H.Q.; Guo, H.; Shen, F.L.; Qin, C.X.; Zhao, E.T.; Fang, X.Y. A comparative study on mechanical properties of Ti-6Al-4V alloy processed by additive manufacturing vs. traditional processing. *Mater. Sci. Eng. A* **2021**, *817*, 141384. [CrossRef]

37. Mierzejewska, Ż.A.; Hudák, R.; Sidun, J. Mechanical Properties and Microstructure of DMLS Ti-6Al-4V Alloy Dedicated to Biomedical Applications. *Materials* **2019**, *12*, 176. [CrossRef]

38. Wysocki, B.; Maj, P.; Sitek, R.; Buhagiar, J.; Kurzydłowski, K.J.; Święszkowski, W. Laser and Electron Beam Additive Manufacturing Methods of Fabricating Titanium Bone Implants. *Appl. Sci.* **2017**, *7*, 657. [CrossRef]

39. Wysocki, B.; Maj, P.; Krawczyńska, A.; Rożniatowski, K.; Zdunek, J.; Kurzydłowski, K.J.; Święszkowski, W. Microstructure and mechanical properties investigation of CP titanium processed by selective laser melting (SLM). *J. Mater. Process. Technol.* **2017**, *241*, 13–23. [CrossRef]

40. Sitek, R.; Szustecki, M.; Zrodowski, L.; Wysocki, B.; Jaroszewicz, J.; Wisniewski, P.; Mizera, J. Analysis of Microstructure and Properties of a Ti–AlN Composite Produced by Selective Laser Melting. *Materials* **2020**, *13*, 2218. [CrossRef]

41. Jiménez, A.; Bidare, P.; Hassanin, H.; Tarlochan, F.; Dimov, S.; Essa, K. Powder-based laser hybrid additive manufacturing of metals: A review. *Int. J. Adv. Manuf. Technol.* **2021**, *114*, 63–96. [CrossRef]

42. Kalpakjian, S. *Manufacturing Processes for Engineering Materials*; Pearson Education India: New Delhi, India, 1984.

43. Hamza, H.M.; Deen, K.M.; Khaliq, A.; Asselin, E.; Haider, W. Microstructural, corrosion and mechanical properties of additively manufactured alloys: A review. *Crit. Rev. Solid State Mater. Sci.* **2022**, *47*, 46–98. [CrossRef]

44. Zhai, Y.; Lados, D.A.; Brown, E.J.; Vigilante, G.N. Understanding the microstructure and mechanical properties of Ti-6Al-4V and Inconel 718 alloys manufactured by Laser Engineered Net Shaping. *Addit. Manuf.* **2019**, *27*, 334–344. [CrossRef]

45. Nguyen, H.D.; Pramanik, A.; Basak, A.K.; Dong, Y.; Prakash, C.; Debnath, S.; Shankar, S.; Jawahir, I.S.; Dixit, S.; Buddhi, D. A critical review on additive manufacturing of Ti-6Al-4V alloy: Microstructure and mechanical properties. *J. Mater. Res. Technol.* **2022**, *18*, 4641–4661. [CrossRef]

46. Lee, Y.; Kim, E.S.; Park, S.; Park, J.M.; Seol, J.B.; Kim, H.S.; Lee, T.; Sung, H.; Kim, J.G. Effects of Laser Power on the Microstructure Evolution and Mechanical Properties of Ti-6Al-4V Alloy Manufactured by Direct Energy Deposition. *Met. Mater. Int.* **2022**, *28*, 197–204. [CrossRef]

47. Chen, B.; Wu, Z.; Yan, T.; He, Z.; Sun, B.; Guo, G.; Wu, S. Experimental study on mechanical properties of laser powder bed fused Ti-6Al-4V alloy under post-heat treatment. *Eng. Fract. Mech.* **2022**, *261*, 108264. [CrossRef]

48. Yan, Q.; Chen, B.; Kang, N.; Lin, X.; Lv, S.; Kondoh, K.; Li, S.; Li, J.S. Comparison study on microstructure and mechanical properties of Ti-6Al-4V alloys fabricated by powder-based selective-laser-melting and sintering methods. *Mater. Charact.* **2020**, *164*, 110358. [CrossRef]

49. Tong, J.; Bowen, C.R.; Persson, J.; Plummer, A. Mechanical properties of titanium-based Ti-6Al-4V alloys manufactured by powder bed additive manufacture. *Mater. Sci. Technol.* **2017**, *33*, 138–148. [CrossRef]

50. Khanna, N.; Zadafiya, K.; Patel, T.; Kaynak, Y.; Rahman Rashid, R.A.; Vafadar, A. Review on machining of additively manufactured nickel and titanium alloys. *J. Mater. Res. Technol.* **2021**, *15*, 3192–3221. [CrossRef]

51. Li, G.; Chandra, S.; Rahman Rashid, R.A.; Palanisamy, S.; Ding, S. Machinability of additively manufactured titanium alloys: A comprehensive review. *J. Manuf. Process.* **2022**, *75*, 72–99. [CrossRef]

52. Lalbondre, R.; Krishna, P.; Mohankumar, G.C. An Experimental Investigation on Machinability Studies of Steels by Face Turning. *Procedia Mater. Sci.* **2014**, *6*, 1386–1395. [CrossRef]

53. Kechagias, J.D.; Aslani, K.-E.; Fountas, N.A.; Vaxevanidis, N.M.; Manolakos, D.E. A comparative investigation of Taguchi and full factorial design for machinability prediction in turning of a titanium alloy. *Measurement* **2020**, *151*, 107213. [CrossRef]

54. Pegues, J.W.; Shao, S.; Shamsaei, N.; Sanaei, N.; Fatemi, A.; Warner, D.H.; Li, P.; Phan, N. Fatigue of additive manufactured Ti-6Al-4V, Part I: The effects of powder feedstock, manufacturing, and post-process conditions on the resulting microstructure and defects. *Int. J. Fatigue* **2020**, *132*, 105358. [CrossRef]
55. Gu, D.D.; Meiners, W.; Wissenbach, K.; Poprawe, R. Laser additive manufacturing of metallic components: Materials, processes and mechanisms. *Int. Mater. Rev.* **2012**, *57*, 133–164. [CrossRef]
56. Frazier, W.E. Metal Additive Manufacturing: A Review. *J. Mater. Eng. Perform.* **2014**, *23*, 1917–1928. [CrossRef]
57. Li, C.; Liu, Z.Y.; Fang, X.Y.; Guo, Y.B. Residual Stress in Metal Additive Manufacturing. *Procedia CIRP* **2018**, *71*, 348–353. [CrossRef]
58. Beese, A.M.; Carroll, B.E. Review of Mechanical Properties of Ti-6Al-4V Made by Laser-Based Additive Manufacturing Using Powder Feedstock. *JOM* **2016**, *68*, 724–734. [CrossRef]
59. Liu, Z.Y.; Li, C.; Fang, X.Y.; Guo, Y.B. Energy Consumption in Additive Manufacturing of Metal Parts. *Procedia Manuf.* **2018**, *26*, 834–845. [CrossRef]
60. Lee, H.; Lim, C.H.J.; Low, M.J.; Tham, N.; Murukeshan, V.M.; Kim, Y.-J. Lasers in additive manufacturing: A review. *Int. J. Precis. Eng. Manuf.-Green Technol.* **2017**, *4*, 307–322. [CrossRef]
61. de Formanoir, C.; Paggi, U.; Colebrants, T.; Thijs, L.; Li, G.; Vanmeensel, K.; Van Hooreweder, B. Increasing the productivity of laser powder bed fusion: Influence of the hull-bulk strategy on part quality, microstructure and mechanical performance of Ti-6Al-4V. *Addit. Manuf.* **2020**, *33*, 101129. [CrossRef]
62. Cobbinah, P.V.; Nzeukou, R.A.; Onawale, O.T.; Matizamhuka, W.R. Laser Powder Bed Fusion of Potential Superalloys: A Review. *Metals* **2021**, *11*, 58. [CrossRef]
63. Polozov, I.; Gracheva, A.; Popovich, A. Processing, Microstructure, and Mechanical Properties of Laser Additive Manufactured Ti2AlNb-Based Alloy with Carbon, Boron, and Yttrium Microalloying. *Metals* **2022**, *12*, 1304. [CrossRef]
64. Aconity3D, Aconity Inline Process-Monitoring. Available online: https://aconity3d.com/products/aconity-midi/ (accessed on 10 March 2023).
65. Uçak, N.; Çiçek, A.; Aslantas, K. Machinability of 3D printed metallic materials fabricated by selective laser melting and electron beam melting: A review. *J. Manuf. Process.* **2022**, *80*, 414–457. [CrossRef]
66. *ISO/ASTM52900-15*; Standard Terminology for Additive Manufacturing—General Principles—Terminology. ASTM: West Conshohocken, PA, USA, 2015.
67. Metalnikov, P.; Ben-Hamu, G.; Eliezer, D. Corrosion behavior of AM-Ti-6Al-4V: A comparison between EBM and SLM. *Prog. Addit. Manuf.* **2022**, *7*, 509–520. [CrossRef]
68. Tamayo, J.A.; Riascos, M.; Vargas, C.A.; Baena, L.M. Additive manufacturing of Ti-6Al-4V alloy via electron beam melting for the development of implants for the biomedical industry. *Heliyon* **2021**, *7*, e06892. [CrossRef] [PubMed]
69. Gong, H.; Rafi, K.; Gu, H.; Starr, T.; Stucker, B. Analysis of defect generation in Ti-6Al-4V parts made using powder bed fusion additive manufacturing processes. *Addit. Manuf.* **2014**, *1*, 87–98. [CrossRef]
70. Kang, C.-W.; Fang, F.-Z. State of the art of bioimplants manufacturing: Part I. *Adv. Manuf.* **2018**, *6*, 20–40. [CrossRef]
71. Rehme, O.; Emmelmann, C. Rapid manufacturing of lattice structures with selective laser melting. In Proceedings of the Lasers and Applications in Science and Engineering, San Jose, CA, USA, 25–26 January 2006; pp. 192–203.
72. Mazur, M.; Selvakannan, P.R. Laser Powder Bed Fusion—Principles, Challenges, and Opportunities. In *Additive Manufacturing for Chemical Sciences and Engineering*; Bhargava, S.K., Ramakrishna, S., Brandt, M., Selvakannan, P.R., Eds.; Springer Nature: Singapore, 2022; pp. 77–108.
73. Shipley, H.; McDonnell, D.; Culleton, M.; Coull, R.; Lupoi, R.; O'Donnell, G.; Trimble, D. Optimisation of process parameters to address fundamental challenges during selective laser melting of Ti-6Al-4V: A review. *Int. J. Mach. Tools Manuf.* **2018**, *128*, 1–20. [CrossRef]
74. Jang, T.-S.; Kim, D.; Han, G.; Yoon, C.-B.; Jung, H.-D. Powder based additive manufacturing for biomedical application of titanium and its alloys: A review. *Biomed. Eng. Lett.* **2020**, *10*, 505–516. [CrossRef]
75. Svetlizky, D.; Zheng, B.; Vyatskikh, A.; Das, M.; Bose, S.; Bandyopadhyay, A.; Schoenung, J.M.; Lavernia, E.J.; Eliaz, N. Laser-based directed energy deposition (DED-LB) of advanced materials. *Mater. Sci. Eng. A* **2022**, *840*, 142967. [CrossRef]
76. Duda, T.; Raghavan, L.V. 3D metal printing technology: The need to re-invent design practice. *AI Soc.* **2018**, *33*, 241–252. [CrossRef]
77. Dutta, B.; Froes, F.H. The Additive Manufacturing (AM) of titanium alloys. *Metal Powder Rep.* **2017**, *72*, 96–106. [CrossRef]
78. Svetlizky, D.; Das, M.; Zheng, B.; Vyatskikh, A.L.; Bose, S.; Bandyopadhyay, A.; Schoenung, J.M.; Lavernia, E.J.; Eliaz, N. Directed energy deposition (DED) additive manufacturing: Physical characteristics, defects, challenges and applications. *Mater. Today* **2021**, *49*, 271–295. [CrossRef]
79. Liu, S.; Shin, Y.C. Additive manufacturing of Ti-6Al-4V alloy: A review. *Mater. Des.* **2019**, *164*, 107552. [CrossRef]
80. Wilson, J.M.; Piya, C.; Shin, Y.C.; Zhao, F.; Ramani, K. Remanufacturing of turbine blades by laser direct deposition with its energy and environmental impact analysis. *J. Clean. Prod.* **2014**, *80*, 170–178. [CrossRef]
81. Gasser, A.; Backes, G.; Kelbassa, I.; Weisheit, A.; Wissenbach, K. Laser additive manufacturing: Laser Metal Deposition (LMD) and Selective Laser Melting (SLM) in turbo-engine applications. *Laser Tech. J.* **2010**, *7*, 58–63. [CrossRef]
82. Thompson, S.M.; Bian, L.; Shamsaei, N.; Yadollahi, A. An overview of Direct Laser Deposition for additive manufacturing; Part I: Transport phenomena, modeling and diagnostics. *Addit. Manuf.* **2015**, *8*, 36–62. [CrossRef]

83. Ahn, D.-G. Directed Energy Deposition (DED) Process: State of the Art. *Int. J. Precis. Eng. Manuf.-Green Technol.* **2021**, *8*, 703–742. [CrossRef]

84. Dumas, M.; Cabanettes, F.; Kaminski, R.; Valiorgue, F.; Picot, E.; Lefebvre, F.; Grosjean, C.; Rech, J. Influence of the finish cutting operations on the fatigue performance of Ti-6Al-4V parts produced by Selective Laser Melting. *Procedia CIRP* **2018**, *71*, 429–434. [CrossRef]

85. Fortunato, A.; Lulaj, A.; Melkote, S.; Liverani, E.; Ascari, A.; Umbrello, D. Milling of maraging steel components produced by selective laser melting. *Int. J. Adv. Manuf. Technol.* **2018**, *94*, 1895–1902. [CrossRef]

86. Liu, Z.; Gao, C.; Liu, X.; Liu, R.; Xiao, Z. Improved surface integrity of Ti-6Al-4V fabricated by selective electron beam melting using ultrasonic surface rolling processing. *J. Mater. Process. Technol.* **2021**, *297*, 117264. [CrossRef]

87. Cerri, E.; Ghio, E.; Bolelli, G. Effect of surface roughness and industrial heat treatments on the microstructure and mechanical properties of Ti-6Al-4V alloy manufactured by laser powder bed fusion in different built orientations. *Mater. Sci. Eng. A* **2022**, *851*, 143635. [CrossRef]

88. Sun, Y.Y.; Gulizia, S.; Oh, C.H.; Fraser, D.; Leary, M.; Yang, Y.F.; Qian, M. The Influence of As-Built Surface Conditions on Mechanical Properties of Ti-6Al-4V Additively Manufactured by Selective Electron Beam Melting. *JOM* **2016**, *68*, 791–798. [CrossRef]

89. Gorsse, S.; Hutchinson, C.; Gouné, M.; Banerjee, R. Additive manufacturing of metals: A brief review of the characteristic microstructures and properties of steels, Ti-6Al-4V and high-entropy alloys. *Sci. Technol. Adv. Mater.* **2017**, *18*, 584–610. [CrossRef]

90. Kasperovich, G.; Hausmann, J. Improvement of fatigue resistance and ductility of TiAl6V4 processed by selective laser melting. *J. Mater. Process. Technol.* **2015**, *220*, 202–214. [CrossRef]

91. Ding, S.; Yang, D.C.H.; Han, Z. Boundary-conformed machining of turbine blades. *Proc. Inst. Mech. Eng. B J. Eng. Manuf.* **2005**, *219*, 255–263. [CrossRef]

92. Yang, D.C.H.; Chuang, J.J.; Han, Z.; Ding, S. Boundary-conformed toolpath generation for trimmed free-form surfaces via Coons reparametrization. *J. Mater. Process. Technol.* **2003**, *138*, 138–144. [CrossRef]

93. Pan, W.; Ding, S.; Mo, J. The prediction of cutting force in end milling titanium alloy (Ti-6Al-4V) with polycrystalline diamond tools. *Proc. Inst. Mech. Eng. B J. Eng. Manuf.* **2017**, *231*, 3–14. [CrossRef]

94. Sun, S.; Brandt, M.; Dargusch, M.S. Characteristics of cutting forces and chip formation in machining of titanium alloys. *Int. J. Mach. Tools Manuf.* **2009**, *49*, 561–568. [CrossRef]

95. Shunmugavel, M.; Polishetty, A.; Goldberg, M.; Singh, R.; Littlefair, G. A comparative study of mechanical properties and machinability of wrought and additive manufactured (selective laser melting) titanium alloy–Ti-6Al-4V. *Rapid Prototyp. J.* **2017**, *23*, 1051–1056. [CrossRef]

96. Polishetty, A.; Shunmugavel, M.; Goldberg, M.; Littlefair, G.; Singh, R.K. Cutting Force and Surface Finish Analysis of Machining Additive Manufactured Titanium Alloy Ti-6Al-4V. *Procedia Manuf.* **2017**, *7*, 284–289. [CrossRef]

97. Shunmugavel, M.; Polishetty, A.; Goldberg, M.; Singh, R.; Littlefair, G. Tool Wear and Surface Integrity Analysis of Machined Heat Treated Selective Laser Melted Ti-6Al-4V. *Int. J. Mater. Form. Mach. Process.* **2016**, *3*, 50–63. [CrossRef]

98. Ming, W.; Chen, J.; An, Q.; Chen, M. Dynamic mechanical properties and machinability characteristics of selective laser melted and forged Ti-6Al-4V. *J. Mater. Process. Technol.* **2019**, *271*, 284–292. [CrossRef]

99. Schur, R.; Ghods, S.; Wisdom, C.; Pahuja, R.; Montelione, A.; Arola, D.; Ramulu, M. Mechanical anisotropy and its evolution with powder reuse in Electron Beam Melting AM of Ti-6Al-4V. *Mater. Des.* **2021**, *200*, 109450. [CrossRef]

100. Gokcekaya, O.; Ishimoto, T.; Hibino, S.; Yasutomi, J.; Narushima, T.; Nakano, T. Unique crystallographic texture formation in Inconel 718 by laser powder bed fusion and its effect on mechanical anisotropy. *Acta Mater.* **2021**, *212*, 116876. [CrossRef]

101. Zheng, M.; Li, C.; Zhang, L.; Zhang, X.; Ye, Z.; Yang, X.; Gu, J. In-situ investigation of deformation behavior in additively manufactured FeCoCrNiMn high entropy alloy. *Mater. Sci. Eng. A* **2022**, *840*, 142933. [CrossRef]

102. Kallel, A.; Duchosal, A.; Hamdi, H.; Altmeyer, G.; Morandeau, A.; Méo, S. Analysis of the surface integrity induced by face milling of Laser Metal Deposited Ti-6Al-4V. *Procedia CIRP* **2020**, *87*, 345–350. [CrossRef]

103. Lizzul, L.; Bertolini, R.; Ghiotti, A.; Bruschi, S. Effect of AM-induced Anisotropy on the Surface Integrity of Laser Powder Bed Fused Ti-6Al-4V Machined Parts. *Procedia Manuf.* **2020**, *47*, 505–510. [CrossRef]

104. Sharma, H.; Parfitt, D.; Syed, A.K.; Wimpenny, D.; Muzangaza, E.; Baxter, G.; Chen, B. A critical evaluation of the microstructural gradient along the build direction in electron beam melted Ti-6Al-4V alloy. *Mater. Sci. Eng. A* **2019**, *744*, 182–194. [CrossRef]

105. Simonelli, M.; Tse, Y.Y.; Tuck, C. Effect of the build orientation on the mechanical properties and fracture modes of SLM Ti-6Al-4V. *Mater. Sci. Eng. A* **2014**, *616*, 1–11. [CrossRef]

106. Bertolini, R.; Lizzul, L.; Pezzato, L.; Ghiotti, A.; Bruschi, S. Improving surface integrity and corrosion resistance of additive manufactured Ti-6Al-4V alloy by cryogenic machining. *Int. J. Adv. Manuf. Technol.* **2019**, *104*, 2839–2850. [CrossRef]

107. Bordin, A.; Sartori, S.; Bruschi, S.; Ghiotti, A. Experimental investigation on the feasibility of dry and cryogenic machining as sustainable strategies when turning Ti-6Al-4V produced by Additive Manufacturing. *J. Clean. Prod.* **2017**, *142*, 4142–4151. [CrossRef]

108. Sartori, S.; Pezzato, L.; Dabalà, M.; Maurizi Enrici, T.; Mertens, A.; Ghiotti, A.; Bruschi, S. Surface Integrity Analysis of Ti-6Al-4V After Semi-finishing Turning Under Different Low-Temperature Cooling Strategies. *J. Mater. Eng. Perform.* **2018**, *27*, 4810–4818. [CrossRef]

109. Sartori, S.; Bordin, A.; Ghiotti, A.; Bruschi, S. Analysis of the Surface Integrity in Cryogenic Turning of Ti-6Al-4V Produced by Direct Melting Laser Sintering. *Procedia CIRP* **2016**, *45*, 123–126. [CrossRef]
110. Oyelola, O.; Crawforth, P.; M'Saoubi, R.; Clare, A.T. Machining of Additively Manufactured Parts: Implications for Surface Integrity. *Procedia CIRP* **2016**, *45*, 119–122. [CrossRef]
111. Arunachalam, R.M.; Mannan, M.A.; Spowage, A.C. Surface integrity when machining age hardened Inconel 718 with coated carbide cutting tools. *Int. J. Mach. Tools Manuf.* **2004**, *44*, 1481–1491. [CrossRef]
112. Oyelola, O.; Crawforth, P.; M'Saoubi, R.; Clare, A.T. On the machinability of directed energy deposited Ti-6Al-4V. *Addit. Manuf.* **2018**, *19*, 39–50. [CrossRef]
113. Jawaid, A.; Che-Haron, C.H.; Abdullah, A. Tool wear characteristics in turning of titanium alloy Ti-6246. *J. Mater. Process. Technol.* **1999**, *92*, 329–334. [CrossRef]
114. Jaffery, S.I.; Mativenga, P.T. Assessment of the machinability of Ti-6Al-4V alloy using the wear map approach. *Int. J. Adv. Manuf. Technol.* **2009**, *40*, 687–696. [CrossRef]
115. Zareena, A.R.; Veldhuis, S.C. Tool wear mechanisms and tool life enhancement in ultra-precision machining of titanium. *J. Mater. Process. Technol.* **2012**, *212*, 560–570. [CrossRef]
116. Pan, W.; Ding, S.; Mo, J. Thermal characteristics in milling Ti-6Al-4V with polycrystalline diamond tools. *Int. J. Adv. Manuf. Technol.* **2014**, *75*, 1077–1087. [CrossRef]
117. Pan, W.; Kamaruddin, A.; Ding, S.; Mo, J. Experimental investigation of end milling of titanium alloys with polycrystalline diamond tools. *Proc. Inst. Mech. Eng. B* **2014**, *228*, 832–844. [CrossRef]
118. Ayed, Y.; Germain, G.; Ammar, A.; Furet, B. Degradation modes and tool wear mechanisms in finish and rough machining of Ti17 Titanium alloy under high-pressure water jet assistance. *Wear* **2013**, *305*, 228–237. [CrossRef]
119. Sartori, S.; Moro, L.; Ghiotti, A.; Bruschi, S. On the tool wear mechanisms in dry and cryogenic turning Additive Manufactured titanium alloys. *Tribol. Int.* **2017**, *105*, 264–273. [CrossRef]
120. Su, Y.; Li, L.; Wang, G. Machinability performance and mechanism in milling of additive manufactured Ti-6Al-4V with polycrystalline diamond tool. *J. Manuf. Process.* **2022**, *75*, 1153–1161. [CrossRef]
121. Al-Rubaie, K.S.; Melotti, S.; Rabelo, A.; Paiva, J.M.; Elbestawi, M.A.; Veldhuis, S.C. Machinability of SLM-produced Ti-6Al-4V titanium alloy parts. *J. Manuf. Process.* **2020**, *57*, 768–786. [CrossRef]
122. Wu, G.; Li, G.; Pan, W.; Raja, I.; Wang, X.; Ding, S. Experimental investigation of eco-friendly cryogenic minimum quantity lubrication (CMQL) strategy in machining of Ti-6Al-4V thin-wall part. *J. Clean. Prod.* **2022**, *357*, 131993. [CrossRef]
123. Khaliq, W.; Zhang, C.; Jamil, M.; Khan, A.M. Tool wear, surface quality, and residual stresses analysis of micro-machined additive manufactured Ti-6Al-4V under dry and MQL conditions. *Tribol. Int.* **2020**, *151*, 106408. [CrossRef]
124. Dang, J.; Zhang, H.; Ming, W.; An, Q.; Chen, M. New observations on wear characteristics of solid Al2O3/Si3N4 ceramic tool in high speed milling of additive manufactured Ti-6Al-4V. *Ceram. Int.* **2020**, *46*, 5876–5886. [CrossRef]
125. Zhang, H.; Dang, J.; Ming, W.; Xu, X.; Chen, M.; An, Q. Cutting responses of additive manufactured Ti-6Al-4V with solid ceramic tool under dry high-speed milling processes. *Ceram. Int.* **2020**, *46*, 14536–14547. [CrossRef]
126. Schulz, H.; Moriwaki, T. High-speed Machining. *CIRP Ann.* **1992**, *41*, 637–643. [CrossRef]
127. Diniz, A.E.; Ferrer, J.A.G. A comparison between silicon nitride-based ceramic and coated carbide tools in the face milling of irregular surfaces. *J. Mater. Process. Technol.* **2008**, *206*, 294–304. [CrossRef]
128. TAN, G.; ZHANG, Y.; LI, G.; LIU, G.; RONG, Y. Performance of a coated cemented carbide tool in high speed milling of Ti-6Al-4V alloy. *J. Adv. Manuf. Syst.* **2013**, *12*, 131–146. [CrossRef]
129. Bordin, A.; Bruschi, S.; Ghiotti, A.; Bariani, P.F. Analysis of tool wear in cryogenic machining of additive manufactured Ti-6Al-4V alloy. *Wear* **2015**, *328*, 89–99. [CrossRef]
130. Bruschi, S.; Bertolini, R.; Bordin, A.; Medea, F.; Ghiotti, A. Influence of the machining parameters and cooling strategies on the wear behavior of wrought and additive manufactured Ti-6Al-4V for biomedical applications. *Tribol. Int.* **2016**, *102*, 133–142. [CrossRef]
131. Oyelola, O.; Crawforth, P.; M'Saoubi, R.; Clare, A.T. Machining of functionally graded Ti-6Al-4V/WC produced by directed energy deposition. *Addit. Manuf.* **2018**, *24*, 20–29. [CrossRef]
132. Di Ilio, A.; Paoletti, A. Machinability Aspects of Metal Matrix Composites. In *Machining of Metal Matrix Composites*; Davim, J.P., Ed.; Springer: London, UK, 2012; pp. 63–77.
133. Li, G.; Munir, K.; Wen, C.; Li, Y.; Ding, S. Machinablility of titanium matrix composites (TMC) reinforced with multi-walled carbon nanotubes. *J. Manuf. Process.* **2020**, *56*, 131–146. [CrossRef]
134. Li, G.; Li, N.; Wen, C.; Ding, S. Investigation and modeling of flank wear process of different PCD tools in cutting titanium alloy Ti-6Al-4V. *Int. J. Adv. Manuf. Technol.* **2018**, *95*, 719–733. [CrossRef]
135. Barry, J.; Byrne, G.; Lennon, D. Observations on chip formation and acoustic emission in machining Ti-6Al-4V alloy. *Int. J. Mach. Tools Manuf.* **2001**, *41*, 1055–1070. [CrossRef]
136. Komanduri, R.; Hou, Z.-B. On thermoplastic shear instability in the machining of a titanium alloy (Ti-6Al-4V). *Metall. Mater. Trans. A* **2002**, *33*, 2995–3010. [CrossRef]
137. Le Coz, G.; Fischer, M.; Piquard, R.; D'Acunto, A.; Laheurte, P.; Dudzinski, D. Micro Cutting of Ti-6Al-4V Parts Produced by SLM Process. *Procedia CIRP* **2017**, *58*, 228–232. [CrossRef]

138. Wu, G.; Li, G.; Pan, W.; Raja, I.; Wang, X.; Ding, S. A state-of-art review on chatter and geometric errors in thin-wall machining processes. *J. Manuf. Process.* **2021**, *68*, 454–480. [CrossRef]

139. Yue, C.; Gao, H.; Liu, X.; Liang, S.Y.; Wang, L. A review of chatter vibration research in milling. *Chin. J. Aeronaut.* **2019**, *32*, 215–242. [CrossRef]

140. Siddhpura, M.; Paurobally, R. A review of chatter vibration research in turning. *Int. J. Mach. Tools Manuf.* **2012**, *61*, 27–47. [CrossRef]

141. Raval, J.K.; Kazi, A.A.; Guo, X.; Zvanut, R.; Lee, C.; Tai, B.L. Preliminary Study on Machining of Additively Manufactured Ti-6Al-4V. *JOM* **2022**, *74*, 1120–1125. [CrossRef]

142. Umbrello, D.; Bordin, A.; Imbrogno, S.; Bruschi, S. 3D finite element modelling of surface modification in dry and cryogenic machining of EBM Ti-6Al-4V alloy. *CIRP J. Manuf. Sci. Technol.* **2017**, *18*, 92–100. [CrossRef]

143. Al-Bermani, S.S.; Blackmore, M.L.; Zhang, W.; Todd, I. The Origin of Microstructural Diversity, Texture, and Mechanical Properties in Electron Beam Melted Ti-6Al-4V. *Metall. Mater. Trans. A* **2010**, *41*, 3422–3434. [CrossRef]

144. Gockel, J.; Beuth, J.; Taminger, K. Integrated control of solidification microstructure and melt pool dimensions in electron beam wire feed additive manufacturing of Ti-6Al-4V. *Addit. Manuf.* **2014**, *1*, 119–126. [CrossRef]

145. Simonelli, M.; McCartney, D.G.; Barriobero-Vila, P.; Aboulkhair, N.T.; Tse, Y.Y.; Clare, A.; Hague, R. The Influence of Iron in Minimizing the Microstructural Anisotropy of Ti-6Al-4V Produced by Laser Powder-Bed Fusion. *Metall. Mater. Trans. A* **2020**, *51*, 2444–2459. [CrossRef]

146. Wu, M.-W.; Lai, P.-H.; Chen, J.-K. Anisotropy in the impact toughness of selective laser melted Ti-6Al-4V alloy. *Mater. Sci. Eng. A* **2016**, *650*, 295–299. [CrossRef]

147. Singla, A.K.; Banerjee, M.; Sharma, A.; Singh, J.; Bansal, A.; Gupta, M.K.; Khanna, N.; Shahi, A.S.; Goyal, D.K. Selective laser melting of Ti-6Al-4V alloy: Process parameters, defects and post-treatments. *J. Manuf. Process.* **2021**, *64*, 161–187. [CrossRef]

148. Kobryn, P.A.; Semiatin, S. Microstructure and texture evolution during solidification processing of Ti-6Al-4V. *J. Mater. Process. Technol.* **2003**, *135*, 330–339. [CrossRef]

149. Kobryn, P.; Semiatin, S.L. The laser additive manufacture of Ti-6Al-4V. *JOM* **2001**, *53*, 40–42. [CrossRef]

150. Carroll, B.E.; Palmer, T.A.; Beese, A.M. Anisotropic tensile behavior of Ti-6Al-4V components fabricated with directed energy deposition additive manufacturing. *Acta Mater.* **2015**, *87*, 309–320. [CrossRef]

151. Kobryn, P.; Moore, E.; Semiatin, S.L. The effect of laser power and traverse speed on microstructure, porosity, and build height in laser-deposited Ti-6Al-4V. *Scr. Mater.* **2000**, *43*, 299–305. [CrossRef]

152. Baufeld, B.; Van der Biest, O.; Dillien, S. Texture and Crystal Orientation in Ti-6Al-4V Builds Fabricated by Shaped Metal Deposition. *Metall. Mater. Trans. A* **2010**, *41*, 1917–1927. [CrossRef]

153. de Formanoir, C.; Michotte, S.; Rigo, O.; Germain, L.; Godet, S. Electron beam melted Ti-6Al-4V: Microstructure, texture and mechanical behavior of the as-built and heat-treated material. *Mater. Sci. Eng. A* **2016**, *652*, 105–119. [CrossRef]

154. Barba, D.; Alabort, C.; Tang, Y.; Viscasillas, M.; Reed, R.; Alabort, E. On the size and orientation effect in additive manufactured Ti-6Al-4V. *Mater. Des.* **2020**, *186*, 108235. [CrossRef]

155. Hrabe, N.; Quinn, T. Effects of processing on microstructure and mechanical properties of a titanium alloy (Ti-6Al-4V) fabricated using electron beam melting (EBM), Part 2: Energy input, orientation, and location. *Mater. Sci. Eng. A* **2013**, *573*, 271–277. [CrossRef]

156. Fernandez-Zelaia, P.; Nguyen, V.; Zhang, H.; Kumar, A.; Melkote, S.N. The effects of material anisotropy on secondary processing of additively manufactured CoCrMo. *Addit. Manuf.* **2019**, *29*, 100764. [CrossRef]

157. Guo, P.; Zou, B.; Huang, C.; Gao, H. Study on microstructure, mechanical properties and machinability of efficiently additive manufactured AISI 316L stainless steel by high-power direct laser deposition. *J. Mater. Process. Technol.* **2017**, *240*, 12–22. [CrossRef]

158. Lizzul, L.; Sorgato, M.; Bertolini, R.; Ghiotti, A.; Bruschi, S. Anisotropy effect of additively manufactured Ti-6Al-4V titanium alloy on surface quality after milling. *Precis. Eng.* **2021**, *67*, 301–310. [CrossRef]

159. Lizzul, L.; Sorgato, M.; Bertolini, R.; Ghiotti, A.; Bruschi, S. Influence of additive manufacturing-induced anisotropy on tool wear in end milling of Ti-6Al-4V. *Tribol. Int.* **2020**, *146*, 106200. [CrossRef]

160. Zhai, Y.; Galarraga, H.; Lados, D.A. Microstructure, static properties, and fatigue crack growth mechanisms in Ti-6Al-4V fabricated by additive manufacturing: LENS and EBM. *Eng. Fail. Anal.* **2016**, *69*, 3–14. [CrossRef]

161. Biswal, R.; Zhang, X.; Syed, A.K.; Awd, M.; Ding, J.; Walther, F.; Williams, S. Criticality of porosity defects on the fatigue performance of wire + arc additive manufactured titanium alloy. *Int. J. Fatigue* **2019**, *122*, 208–217. [CrossRef]

162. Aliprandi, P.; Giudice, F.; Guglielmino, E.; Sili, A. Tensile and Creep Properties Improvement of Ti-6Al-4V Alloy Specimens Produced by Electron Beam Powder Bed Fusion Additive Manufacturing. *Metals* **2019**, *9*, 1207. [CrossRef]

163. Wang, S.; Ning, J.; Zhu, L.; Yang, Z.; Yan, W.; Dun, Y.; Xue, P.; Xu, P.; Bose, S.; Bandyopadhyay, A. Role of porosity defects in metal 3D printing: Formation mechanisms, impacts on properties and mitigation strategies. *Mater. Today* **2022**, *59*, 133–160. [CrossRef]

164. Sanaei, N.; Fatemi, A. Defects in additive manufactured metals and their effect on fatigue performance: A state-of-the-art review. *Prog. Mater. Sci.* **2021**, *117*, 100724. [CrossRef]

165. Li, J.; Fang, Q.; Liu, B.; Liu, Y. The effects of pore and second-phase particle on the mechanical properties of machining copper matrix from molecular dynamic simulation. *Appl. Surf. Sci.* **2016**, *384*, 419–431. [CrossRef]

166. Ahmad, S.; Mujumdar, S.; Varghese, V. Role of porosity in machinability of additively manufactured Ti-6Al-4V. *Precis. Eng.* **2022**, *76*, 397–406. [CrossRef]

167. Varghese, V.; Mujumdar, S. Micromilling-induced Surface Integrity of Porous Additive Manufactured Ti-6Al-4V Alloy. *Procedia Manuf.* **2021**, *53*, 387–394. [CrossRef]

168. Vrancken, B.; Thijs, L.; Kruth, J.-P.; Van Humbeeck, J. Heat treatment of Ti-6Al-4V produced by Selective Laser Melting: Microstructure and mechanical properties. *J. Alloy. Compd.* **2012**, *541*, 177–185. [CrossRef]

169. Vilaro, T.; Colin, C.; Bartout, J.D. As-Fabricated and Heat-Treated Microstructures of the Ti-6Al-4V Alloy Processed by Selective Laser Melting. *Metall. Mater. Trans. A* **2011**, *42*, 3190–3199. [CrossRef]

170. Brandl, E.; Greitemeier, D. Microstructure of additive layer manufactured Ti-6Al-4V after exceptional post heat treatments. *Mater. Lett.* **2012**, *81*, 84–87. [CrossRef]

171. Leuders, S.; Thöne, M.; Riemer, A.; Niendorf, T.; Tröster, T.; Richard, H.A.; Maier, H.J. On the mechanical behaviour of titanium alloy TiAl6V4 manufactured by selective laser melting: Fatigue resistance and crack growth performance. *Int. J. Fatigue* **2013**, *48*, 300–307. [CrossRef]

172. Qiu, C.; Adkins, N.J.E.; Attallah, M.M. Microstructure and tensile properties of selectively laser-melted and of HIPed laser-melted Ti-6Al-4V. *Mater. Sci. Eng. A* **2013**, *578*, 230–239. [CrossRef]

173. Hrabe, N.; Gnäupel-Herold, T.; Quinn, T. Fatigue properties of a titanium alloy (Ti-6Al-4V) fabricated via electron beam melting (EBM): Effects of internal defects and residual stress. *Int. J. Fatigue* **2017**, *94*, 202–210. [CrossRef]

174. Herzog, D.; Seyda, V.; Wycisk, E.; Emmelmann, C. Additive manufacturing of metals. *Acta Mater.* **2016**, *117*, 371–392. [CrossRef]

175. Shui, X.; Yamanaka, K.; Mori, M.; Nagata, Y.; Kurita, K.; Chiba, A. Effects of post-processing on cyclic fatigue response of a titanium alloy additively manufactured by electron beam melting. *Mater. Sci. Eng. A* **2017**, *680*, 239–248. [CrossRef]

176. Brandl, E.; Leyens, C.; Palm, F. Mechanical Properties of Additive Manufactured Ti-6Al-4V Using Wire and Powder Based Processes. *IOP Conf. Ser. Mater. Sci. Eng.* **2011**, *26*, 012004. [CrossRef]

177. Bruschi, S.; Bertolini, R.; Ghiotti, A. Coupling machining and heat treatment to enhance the wear behaviour of an Additive Manufactured Ti-6Al-4V titanium alloy. *Tribol. Int.* **2017**, *116*, 58–68. [CrossRef]

178. Malekipour, E.; El-Mounayri, H. Common defects and contributing parameters in powder bed fusion AM process and their classification for online monitoring and control: A review. *Int. J. Adv. Manuf. Technol.* **2018**, *95*, 527–550. [CrossRef]

179. Mostafaei, A.; Zhao, C.; He, Y.; Reza Ghiaasiaan, S.; Shi, B.; Shao, S.; Shamsaei, N.; Wu, Z.; Kouraytem, N.; Sun, T.; et al. Defects and anomalies in powder bed fusion metal additive manufacturing. *Curr. Opin. Solid State Mater. Sci.* **2022**, *26*, 100974. [CrossRef]

180. Li, S.; Zhang, B.; Bai, Q. Effect of temperature buildup on milling forces in additive/subtractive hybrid manufacturing of Ti-6Al-4V. *Int. J. Adv. Manuf. Technol.* **2020**, *107*, 4191–4200. [CrossRef]

181. Du, W.; Bai, Q.; Zhang, B. Machining characteristics of 18Ni-300 steel in additive/subtractive hybrid manufacturing. *Int. J. Adv. Manuf. Technol.* **2018**, *95*, 2509–2519. [CrossRef]

182. Bai, Q.; Wu, B.; Qiu, X.; Zhang, B.; Chen, J. Experimental study on additive/subtractive hybrid manufacturing of 6511 steel: Process optimization and machining characteristics. *Int. J. Adv. Manuf. Technol.* **2020**, *108*, 1389–1398. [CrossRef]

183. Yang, Y.; Gong, Y.; Li, C.; Wen, X.; Sun, J. Mechanical performance of 316 L stainless steel by hybrid directed energy deposition and thermal milling process. *J. Mater. Process. Technol.* **2021**, *291*, 117023. [CrossRef]

184. Ye, Z.-P.; Zhang, Z.-J.; Jin, X.; Xiao, M.-Z.; Su, J.-Z. Study of hybrid additive manufacturing based on pulse laser wire depositing and milling. *Int. J. Adv. Manuf. Technol.* **2017**, *88*, 2237–2248. [CrossRef]

185. Jeng, J.-Y.; Lin, M.-C. Mold fabrication and modification using hybrid processes of selective laser cladding and milling. *J. Mater. Process. Technol.* **2001**, *110*, 98–103. [CrossRef]

186. Sun, S.; Brandt, M.; Dargusch, M.S. Thermally enhanced machining of hard-to-machine materials—A review. *Int. J. Mach. Tools Manuf.* **2010**, *50*, 663–680. [CrossRef]

187. Lauwers, B. Surface Integrity in Hybrid Machining Processes. *Procedia Eng.* **2011**, *19*, 241–251. [CrossRef]

188. Gao, Y.; Wang, G.; Bermingham, M.J.; Dargusch, M.S. Cutting force, chip formation, and tool wear during the laser-assisted machining a near-alpha titanium alloy BTi-6431S. *Int. J. Adv. Manuf. Technol.* **2015**, *79*, 1949–1960. [CrossRef]

189. Bermingham, M.J.; Kent, D.; Dargusch, M.S. A new understanding of the wear processes during laser assisted milling 17-4 precipitation hardened stainless steel. *Wear* **2015**, *328–329*, 518–530. [CrossRef]

190. Dargusch, M.S.; Sivarupan, T.; Bermingham, M.; Rashid, R.A.R.; Palanisamy, S.; Sun, S. Challenges in laser-assisted milling of titanium alloys. *Int. J. Extreme Manuf.* **2021**, *3*, 015001. [CrossRef]

191. Dargusch, M.S.; Sun, S.; Kim, J.W.; Li, T.; Trimby, P.; Cairney, J. Effect of tool wear evolution on chip formation during dry machining of Ti-6Al-4V alloy. *Int. J. Mach. Tools Manuf.* **2018**, *126*, 13–17. [CrossRef]

192. Garcí, V.; Arriola, I.; Gonzalo, O.; Leunda, J. Mechanisms involved in the improvement of Inconel 718 machinability by laser assisted machining (LAM). *Int. J. Mach. Tools Manuf.* **2013**, *74*, 19–28. [CrossRef]

193. Swarnakar, A.K.; Van der Biest, O.; Baufeld, B. Thermal expansion and lattice parameters of shaped metal deposited Ti-6Al-4V. *J. Alloys Compd.* **2011**, *509*, 2723–2728. [CrossRef]

194. Moritz, J.; Seidel, A.; Kopper, M.; Bretschneider, J.; Gumpinger, J.; Finaske, T.; Riede, M.; Schneeweiß, M.; López, E.; Brückner, F.; et al. Hybrid manufacturing of titanium Ti-6Al-4V combining laser metal deposition and cryogenic milling. *Int. J. Adv. Manuf. Technol.* **2020**, *107*, 2995–3009. [CrossRef]

195. Du, W.; Bai, Q.; Wang, Y.; Zhang, B. Eddy current detection of subsurface defects for additive/subtractive hybrid manufacturing. *Int. J. Adv. Manuf. Technol.* **2018**, *95*, 3185–3195. [CrossRef]
196. Hojati, F.; Daneshi, A.; Soltani, B.; Azarhoushang, B.; Biermann, D. Study on machinability of additively manufactured and conventional titanium alloys in micro-milling process. *Precis. Eng.* **2020**, *62*, 1–9. [CrossRef]
197. Oyelola, O.; Jackson-Crisp, A.; Crawforth, P.; Pieris, D.M.; Smith, R.J.; M'Saoubi, R.; Clare, A.T. Machining of directed energy deposited Ti-6Al-4V using adaptive control. *J. Manuf. Process.* **2020**, *54*, 240–250. [CrossRef]
198. Sartori, S.; Bordin, A.; Moro, L.; Ghiotti, A.; Bruschi, S. The Influence of Material Properties on the Tool Crater Wear When Machining Ti-6Al-4V Produced by Additive Manufacturing Technologies. *Procedia CIRP* **2016**, *46*, 587–590. [CrossRef]
199. de Oliveira Campos, F.; Araujo, A.C.; Jardini Munhoz, A.L.; Kapoor, S.G. The influence of additive manufacturing on the micromilling machinability of Ti-6Al-4V: A comparison of SLM and commercial workpieces. *J. Manuf. Process.* **2020**, *60*, 299–307. [CrossRef]
200. Gong, X.; Manogharan, G. Machining Behavior and Material Properties in Additive Manufacturing Ti-6Al-4V Parts. In Proceedings of the ASME 2020 15th International Manufacturing Science and Engineering Conference, Virtual, 3 September 2020.
201. Lizzul, L.; Sorgato, M.; Bertolini, R.; Ghiotti, A.; Bruschi, S. Ball end milling machinability of additively and conventionally manufactured Ti-6Al-4V tilted surfaces. *J. Manuf. Process.* **2021**, *72*, 350–360. [CrossRef]
202. Lizzul, L.; Bertolini, R.; Ghiotti, A.; Bruschi, S. Turning of Additively Manufactured Ti-6Al-4V: Effect of the Highly Oriented Microstructure on the Surface Integrity. *Materials* **2021**, *14*, 2842. [CrossRef] [PubMed]
203. Lizzul, L.; Sorgato, M.; Bertolini, R.; Ghiotti, A.; Bruschi, S. Surface finish of additively manufactured Ti-6Al-4V workpieces after ball end milling. *Procedia CIRP* **2021**, *102*, 228–233. [CrossRef]
204. Su, Y.; Li, L. Surface integrity of ultrasonic assisted dry milling of SLM Ti 6Al 4V using polycrystalline diamond tool. *Int. J. Adv. Manuf. Technol.* **2022**, *119*, 5947–5956. [CrossRef]
205. Airao, J.; Kishore, H.; Nirala, C.K. Measurement and analysis of tool wear and surface characteristics in micro turning of SLM Ti-6Al-4V and wrought Ti-6Al-4V. *Measurement* **2023**, *206*, 112281. [CrossRef]
206. Li, G.; Rahman Rashid, R.A.; Ding, S.; Sun, S.; Palanisamy, S. Machinability Analysis of Finish-Turning Operations for Ti-6Al-4V Tubes Fabricated by Selective Laser Melting. *Metals* **2022**, *12*, 806. [CrossRef]
207. Ni, C.; Wang, X.; Zhu, L.; Liu, D.; Wang, Y.; Zheng, Z.; Zhang, P. Machining performance and wear mechanism of PVD TiAlN/AlCrN coated carbide tool in precision machining of selective laser melted Ti-6Al-4V alloys under dry and MQL conditions. *J. Manuf. Process.* **2022**, *79*, 975–989. [CrossRef]
208. Zhang, B.; Wang, Z. Effects of Heat Treatment on Sliding Wear and Milling Properties of Ti-6Al-4V Prepared by Selective Laser Melting. *J. Tribol.* **2023**, *145*, 061701. [CrossRef]
209. Km, R.; Sahoo, A.K.; Routara, B.C.; Panda, A.; Kumar, R. Study on machinability characteristics of novel additive manufactured titanium alloy (Ti-6Al-4V) fabricated by direct metal laser sintering. *Proc. Inst. Mech. Eng. C-J. Mec.* **2022**, *237*, 865–885. [CrossRef]
210. Cai, C.; An, Q.; Ming, W.; Chen, M. Microstructure- and cooling/lubrication environment-dependent machining responses in side milling of direct metal laser-sintered and rolled Ti-6Al-4V alloys. *J. Mater. Process. Technol.* **2022**, *300*, 117418. [CrossRef]
211. Alves, U.C.; Hassui, A.; de Oliveira, M.F.; Neto, P.I.; Ventura, C.E.H. Microstructural and machinability aspects of electron beam melted Ti-6Al-4V with different building orientations. *Progr. Addit. Manuf.* **2022**, 1–11. [CrossRef]

Article

Analysis of the Effect of Machining of the Surfaces of WAAM 18Ni 250 Maraging Steel Specimens on Their Durability

Daren Peng [1,2], Andrew S. M. Ang [1], Alex Michelson [3], Victor Champagne [4], Aaron Birt [3] and Rhys Jones [1,2,*]

[1]   ARC Training Centre on Surface Engineering for Advanced Materials (SEAM), School of Engineering, Swinburne University of Technology, Hawthorn, VIC 3122, Australia

[2]   Centre of Expertise for Structural Mechanics, Department of Mechanical and Aerospace Engineering, Monash University, Clayton, VIC 3800, Australia

[3]   Solvus Global, 104 Prescott Street, Worcester, MA 01605, USA

[4]   US Army Research Laboratory, U.S. Army Combat Capabilities Development Command Weapons and Materials Research Directorate, Aberdeen Proving Ground, Aberdeen, MD 21005, USA

*   Correspondence: rhys.jones@monash.edu

Abstract: It is now well-known that the interaction between surface roughness and surface-breaking defects can significantly degrade the fatigue life of additively manufactured (AM) parts. This is also aptly illustrated in the author's recent study on the durability of wire and arc additively manufactured (WAAM) 18Ni 250 Maraging steel specimens, where it was reported that failure occurred due to fatigue crack growth that arose due to the interaction between the surface roughness and surface-breaking material defects. To improve the durability of an AM part, several papers have suggested the machining of rough surfaces. However, for complex geometries the fully machining of the entire rough surface is not always possible and the effect of the partial machining on durability is unknown. Therefore, this paper investigates if partial machining of WAAM 18Ni 250 Maraging steel surfaces will help to improve the durability of these specimens. Unfortunately, the result of this investigation has shown that partial machining may not significantly improve durability of WAAM 18Ni 250 Maraging steel specimens. Due to the order of surface roughness seen in WAAM 250 Maraging steel, the improvement to durability is only realized by full machining to completely remove the remnants of any print artefacts.

Keywords: additive manufacturing; rough surfaces; partial machining; WAAM 18Ni 250 Maraging steel; durability; crack growth

Citation: Peng, D.; Ang, A.S.M.; Michelson, A.; Champagne, V.; Birt, A.; Jones, R. Analysis of the Effect of Machining of the Surfaces of WAAM 18Ni 250 Maraging Steel Specimens on Their Durability. *Materials* **2022**, *15*, 8890. https://doi.org/10.3390/ma15248890

Academic Editors: Bartłomiej Wysocki, Joseph Buhagiar, Tomasz Durejko and Carlos Garcia-Mateo

Received: 27 October 2022
Accepted: 9 December 2022
Published: 13 December 2022

Publisher's Note: MDPI stays neutral with regard to jurisdictional claims in published maps and institutional affiliations.

1. Introduction

Practitioners in the field of additively manufactured structures have long been aware [1–19] that the interaction of surface-roughness with surface-breaking material discontinuities and surface-breaking porosity/lack of fusion can significantly degrade the durability and damage tolerance (DADT) of an additively manufactured part. Indeed, ref. [7] concluded that the fatigue life of AM Ti-6Al-4V parts built using either electron beam melt or direct metal laser sintering was determined by surface roughness effects (a picture illustrating how surface-breaking cracks can develop as a result of surface-breaking material discontinuities/defects is shown in Figure 1). The author's recent paper [1] on the durability of WAAM 18NI 250 Maraging steel specimens tested with their surfaces left in the "as built" condition aptly illustrated this phenomenon in that the specimens failed due to the interaction between the rough-surface and surface-breaking material discontinuities. Although not reported in [1], this test program also found that the presence of large, wholly contained, internal voids/porosity and large near-surface voids did not result in failure. Examples of this are shown in Figures 1 and 2. In each case, it was found that there was little crack growth from these internal and near-surface internal voids/porosity. Indeed, this observation is consistent with the seminal finding reported by Schijve [20], and subsequently confirmed in [21–24], that for conventionally built parts internal cracks grow much slower than the surface-breaking cracks.

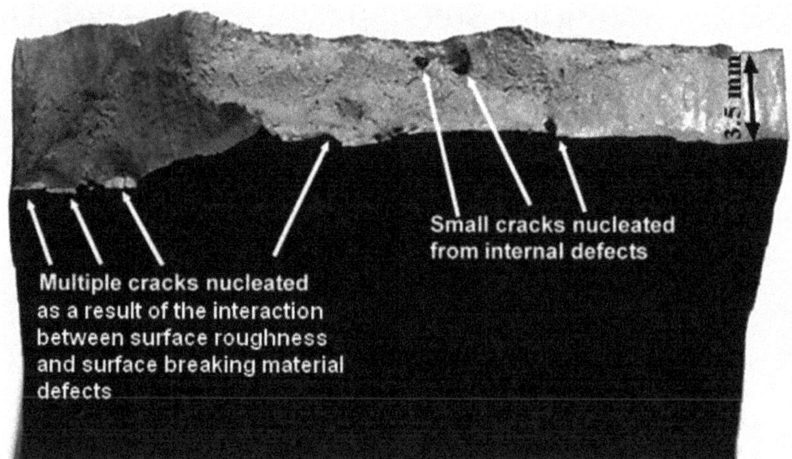

Figure 1. Picture of the failure surface of a WAAM 18Ni 250 Maraging steel test showing multiple cracks nucleating due to the interaction between surface roughness and surface-breaking material discontinuities, but little cracking associated with internal pores/voids.

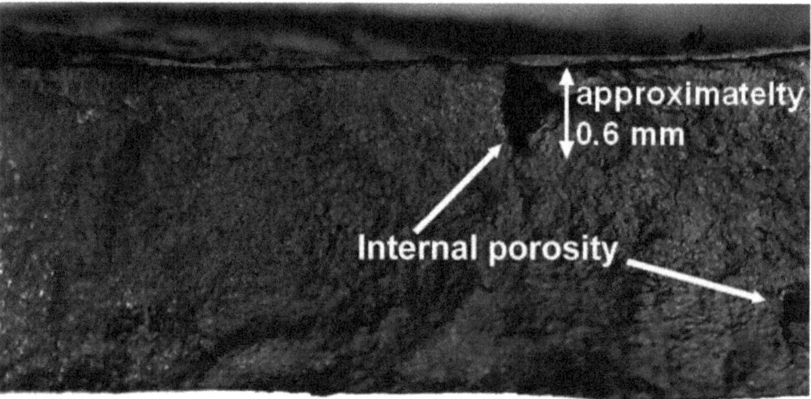

Figure 2. A close-up view of the failure surface shown in Figure 2 showing minimal crack growth associated with relatively large internal porosity/voids.

Whereas [17–19] suggested that machining the rough surfaces may improve the DADT of an AM part, the studies presented in [2,15] suggested that partial machining may be of little benefit. However, the analysis presented in [2] examined an idealized (rough) surface where the surface roughness had a wave-like (sinusoidal) pattern. Consequently, noting that:

i.   The United States Air Force (USAF) MIL-STD-1530D [25] states that the certification of a load-bearing part must be based on analysis, and that the role of testing is to validate the analysis.

ii.   Ref. [1] revealed that the durability and the associated crack growth histories of WAAM 18Ni 250 Maraging steel specimens with rough surfaces could be predicted in a fashion that was consistent with the linear elastic fracture mechanics approach mandated in the United States Air Force (USAF) MIL-STD-1530D.

The present paper addresses the question: Since failure of the WAAM 18Ni 250 Maraging steel specimens, which were built by Solvus Global in Worcester, in the United States of America (USA), in [1] was due to the interaction between surface-roughness and surface-breaking materials discontinuities, would partial machining of the surface help to improve the durability of these WAAM 18Ni 250 Maraging steel specimens?

Here it should be stressed that in contrast to the author's prior study [2], the analysis reported in the present paper is based on actual (measured) surface roughness and uses the same crack growth equation that has been shown [1] to predict crack growth in specimens with as-built surface roughness.

2. Materials and Methods

As stated in the introduction, this paper is motivated by the research gap that exists when attempting to certify WAAM parts in accordance with the requirements inherent with USAF MIL-STD-1530D and USAF Structures Bulletin EZ-19-01 and the linear elastic fracture mechanics-based building block approach mandated in USAF MIL-STD-1530D. Furthermore, as mandated in MIL-STD-1530D certification requires the use of linear elastic fracture mechanics to predict durability. This approach is mandated both for WAAM parts in the as-built state, and for WAAM built parts that have had their rough surfaces partially machined. It should also be noted that Section 5.3 of MIL-STD-1530D states that analysis is the key to certification, and that the role of testing is to validate/correct the analysis. However, to the best of the author's knowledge, other than the authors previous paper [1], there are no publicly available papers that have shown an ability to predict the durability, and the associated crack growth history, of WAAM steel specimens. It should be stressed that this is an essential step in the building block approach to certification, and is required before the analysis can assess the effect of partial machining. Consequently, this paper builds on the analysis methodology validated in [1] for predicting the durability of WAAM 18Ni 250 Maraging steel specimens to assess the effect of partial machining on WAAM 18Ni 250 Maraging steel specimen that was analyzed, in its "as-built" state, in [1].

For the sake of completion, it should also be noted that for the specimen with a rough surface analyzed in [1], the surface topography was first measured using an Artec3D Leo laser scanner (Santa Clara, CA, USA) that has a 3D point accuracy of approximately 0.1 mm. Consequently, as in [1], for the various durability analyses presented in this paper, the surface topography measurements obtained in [1] were first used to create a three-dimensional solid model. This CAD model was then auto-meshed to produce a three-dimensional finite element model of the particular specimen under consideration.

In the various crack growth analyses presented in this paper, the analyses began by assuming an initial crack size that was taken from the experimental measurements given in [1]. The stress intensity factor (K) solutions around this initial crack were determined, as per [1], using the multi-crack finite element analysis program developed as part of the US Federal Aviation Aging (FAA) Aircraft Program [26–28] and the stress field associated with the corresponding uncracked finite element model. The increment in the crack size (da/dN) around a given crack was then computed using the small crack growth equation for this material given in [1], viz:

$$da/dN = 2 \times 10^{-10} \left((\Delta K - 0.1)/\sqrt{(1 - K_{\max}/150)} \right)^{2.0} \tag{1}$$

here $\Delta K = K_{\max} - K_{\min}$, where $K_{\max}$ and $K_{\min}$ are the maximum and minimum values of K, in a given cycle (N), and "a" is the crack length. Armed with this knowledge, a new crack size and shape is then determined. This process was repeated until fatigue failure, i.e., until at some point around the crack front the value of $K_{\max}$ exceeded the cyclic fracture toughness value of the material, which has a value of 150 MPa $\sqrt{m}$. An alternative implementation of this approach, which uses standard finite element analysis rather than the alternating finite element approach to compute the stress intensity factors around the crack, is now available in the commercial finite element programs ABAQUS®, NASTRAN®, and ANSYS® via version 9 of the Zencrack® software module [29].

The twenty journal papers referenced in this paper are listed in either the World of Science and/or SCOPUS. The book chapter referenced is available on the Elsevier website. The three United States (US) Government references related to the Department of Defence certification requirement are publicly available, and their web addresses are given. Keywords used in the search for these references were: Additive manufacturing, surface roughness, crack growth, and DADT.

3. On the Effect of Rough Surfaces on Durability

Recently, a paper [1] focused on the ability to compute the effect of the rough surfaces on the durability of as-built WAAM 18Ni 250 Maraging steel. However, it did not address the reduction in performance (durability) due to the rough surfaces. To illustrate, this let us first consider a 4 mm thick dogbone specimen with the same plan view as the specimen analyzed in [1], but with a uniform thickness of 4 mm, see Figure 3. In other words, the specimen has a smooth surface. As in [1], the specimen geometry was auto-meshed to produce both a fine and a coarse (finite element) mesh. The fine mesh consisted of approximately 45,936 21-noded iso-parametric elements and 211,085 nodes. The coarse mesh consisted of approximately 29,568 21-noded iso-parametric elements and 136,189 nodes. The difference in stress field between these two finite element models, at an applied load of 29 kN, was less than 0.5%. The stress field associated with the fine mesh is shown in Figure 4.

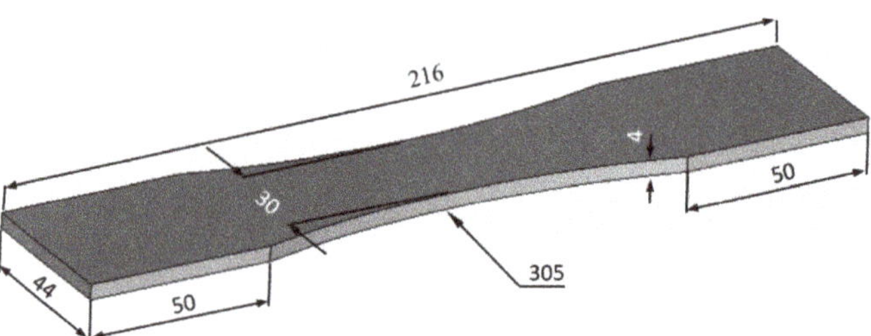

Figure 3. The geometry of the smooth surface dogbone specimen.

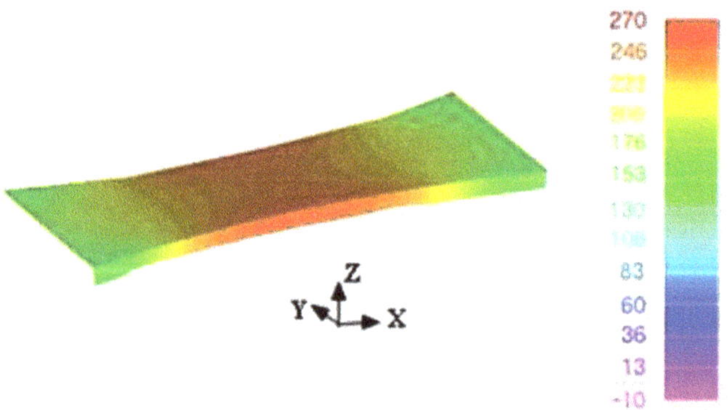

Figure 4. The computed stress field in the fine mesh finite element model at a remote load of 29 kN. The stress units in the picture are in MPa.

As in [1], the analysis assumed an initial surface-breaking semi-elliptical crack that was 0.228 mm deep and had a tip-to-tip surface length of 0.680 mm and was located in the same position reported in [1], i.e., it was near the center of the specimen. This crack size was taken from fractography measurements associated with the failed specimen analyzed in [1]. Noting that, as can also be seen in Figure 5, the measured and predicted crack growth histories reported in [1], for the specimen with a rough "as-built" surface, were in excellent agreement, we used the same analysis approach to examine the durability of this 4 mm thick dogbone specimen. Furthermore, this "smooth surface" specimen was assumed to be subjected to the same repeated load block as in [1], with each load block consisting of 1200 cycles at $R = 0.1$, and 8000 cycles at $R = 0.5$. As in [1], the maximum load in the load block was held constant at $P_{max} = 29$ kN (here the term R is the ratio of the minimum applied load divided by the maximum applied load). As previously mentioned, the durability analysis was performed using the small crack growth equation for this material given in [1], namely Equation (1). The resultant computed crack growth history is shown in Figure 5, together with that associated with the measured and computed histories given in [1] for the specimen with a rough surface (for the sake of, completion it should be noted that, for the specimen with a rough surface analyzed in [1], the surface topography was first measured using an Artec3D Leo laser scanner (Santa Clara, CA, USA) that has a 3D point accuracy of approximately 0.1 mm. The surface topography measurements were then used to create a three-dimensional solid model, which was then auto-meshed to produce a three-dimensional finite element model of the specimen, see Figure 6).

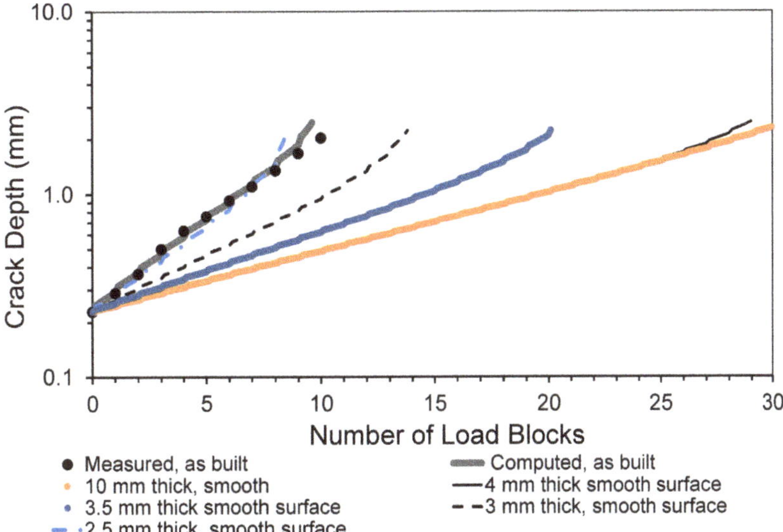

Figure 5. The measured and computed crack depth curves for a surface-breaking crack that is 0.228 mm deep and has a tip-to-tip length of 0.68 mm.

To study the effect of specimen thickness, the analysis was repeated for specimens with the same plan view and remote stress, but were either 2.5, 3, 3.5 or 10 mm thick, see Figure 5. Here we see that the effect of the surface roughness is to significantly increase the rate of crack growth in the specimen analyzed in [1], in comparison to the 4 mm thick specimen with a smooth surface. It is also seen that the crack growth rate associated with the specimen configuration tested in [1] is similar to that of a specimen that had a uniform thickness of 2.5 mm thick and had an initial crack that was 0.228 mm deep and had a tip-to-tip surface length of 0.680 mm.

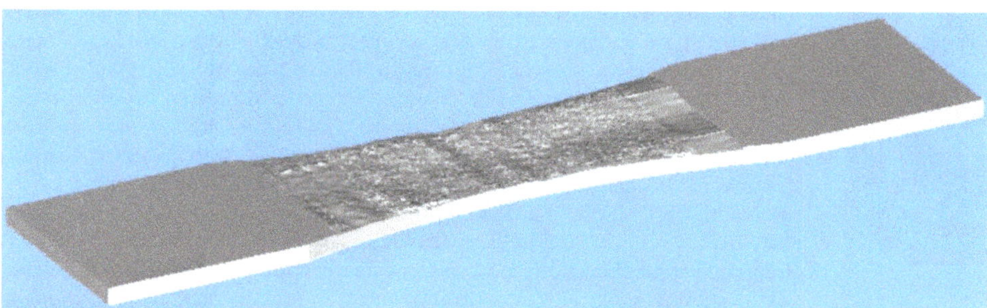

Figure 6. A typical CAD model with the surface roughness as measured in [1]. The in-plane dimensions are as shown in Figure 3.

To further highlight the effect of the rough surface on the life of the test specimen, Figure 7 presents a comparison of measured and computed crack growth histories given in [1] with the computed crack growth histories for a 4 mm specimen with a smooth surface and either a 0.228 mm radius or a 0.342 mm radius semi-circular surface-breaking crack.

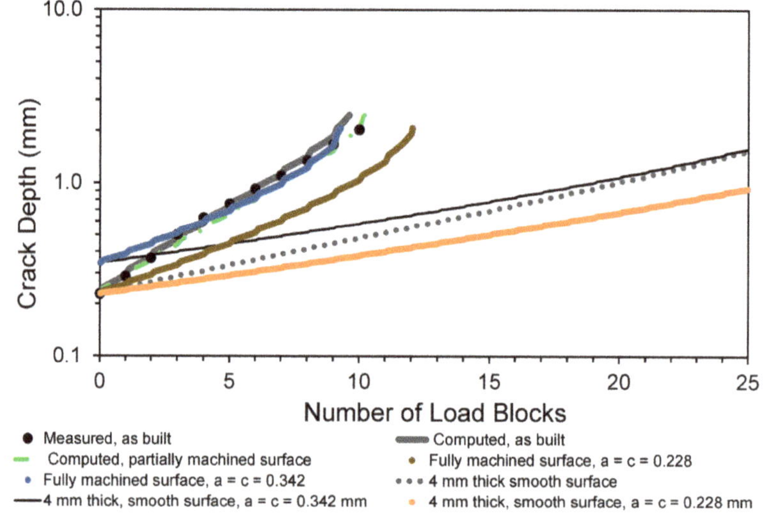

Figure 7. The measured and computed crack depth curves shown in Figure 5, together with the corresponding curve obtained for the partially machined surface and a specimen with a fully machined surface with a semi-circular surface-breaking crack.

4. Analysis of the Effect of Partial Machining on Durability

Having seen that the surface roughness would appear to significantly reduce the durability of the specimen, let us next address the question: Will partial machining of the surface help to improve the durability of WAAM built 18Ni 250 Maraging steel specimens?

To this end, the finite element model of the specimens developed [1] for the as-built WAAM 18Ni 250 Maraging steel specimens was modified such that 0.72 mm of the (upper) as-built surface was removed (this equates to roughly one-half of the maximum height of the as-built surface).

This "partially machined" model was assumed to be subjected to the same repeated load block spectrum, and a comparison between the stress states in the "as-built" and the "partially machined" states is given in Figure 8a. As previously, the initial crack size assumed in the analysis was 0.228 mm deep and had a tip-to-tip surface length of 0.680 mm. Since the modification of the CAD model to reflect partial machining of the surface did not result in the region where the surface-breaking crack was located being removed, the location of this surface-breaking crack remained as in [1], i.e., on the rough surface near the center of the specimen. Furthermore, as in [1], the crack growth analysis used Equation (1).

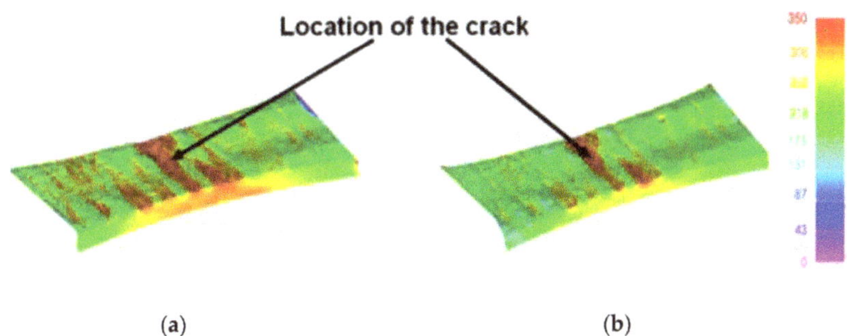

Figure 8. The stress fields in the (**a**) as-built, and (**b**) partially machined specimens.

This analysis gave a durability of approximately 10.0 load blocks. A comparison between the computed crack growth history for specimens in both the as-built and the partially machined state is also given in Figure 7. Here we see that for the measured initial crack size, there is little difference between the crack growth histories associated with the as-built and the partially machined specimens. The reason for this becomes clear upon inspecting Figure 8, where we see that partial machining did not significantly improve the stress field at the critical location.

4.1. Thickness Effects

The question now arises as to what would happen if the specimen was thicker? To investigate this question, the analysis was repeated with the thickness increased to 10 mm. To ensure that the remote stress was the same as for the 4 mm thick specimen, the load was increased to 72.5 kN. Analyses were again performed for 10 mm thick specimens with the same as-built and partially machined surface profiles, and for a 10 mm thick specimen with a smooth surface. In each case, to establish convergence, the specimen geometry was auto-meshed to produce both a fine and a coarse (finite element) mesh. For example, in the case of the partially machined specimen, the coarse mesh consisted of approximately 300,233 ten nodded iso-parametric tetrahedral elements and 437,380 nodes, see Figure 9. The fine mesh consisted of approximately 609,768 ten nodded iso-parametric tetrahedral elements and 876,866 nodes, see Figure 10. The difference in stress field between these two finite element models was less than 1.6%. In all cases, the durability analyses used the finer of the two meshes.

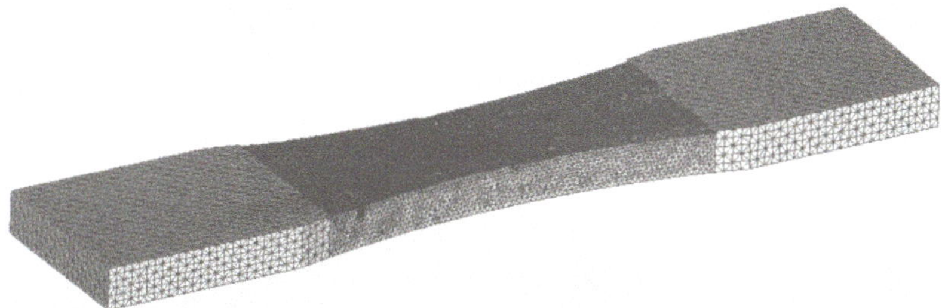

Figure 9. The coarse mesh, 300,233 ten nodded iso-parametric tetrahedral elements and 437,380 nodes.

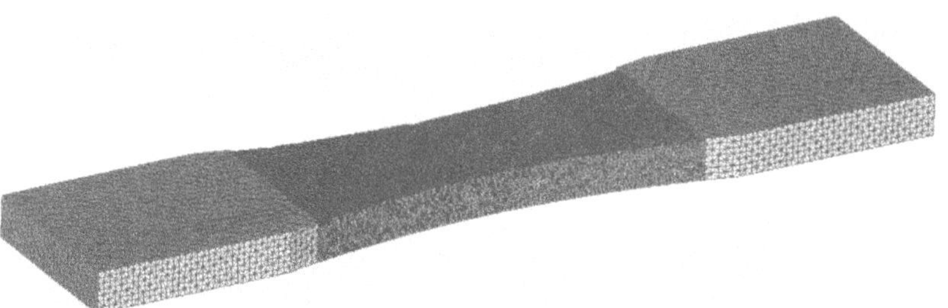

Figure 10. The fine mesh, 609,768 ten nodded iso-parametric tetrahedral elements and 876,866 nodes.

Figure 11 presents a comparison between the computed crack growth histories for 10 mm thick specimens in both the as-built and the partially machined state and for a 10 mm thick specimen with a smooth surface. Since MIL-STD-1530D [25] and the United States Joint Services Structural Guideline JSSG2006 [30] suggest that, for a conventionally manufactured part, the aspect ratio of the initial crack should be one, i.e., $c/a = 1$, analyses were performed with the initial crack size assumed being either:

(i)   A 0.228 mm deep semi-elliptical surface-breaking crack with a tip-to-tip length surface length of 0.680 mm.
(ii)   A 0.228 mm radius semi-circular surface-breaking crack.
(iii) A 0.342 mm radius semi-circular surface-breaking crack.

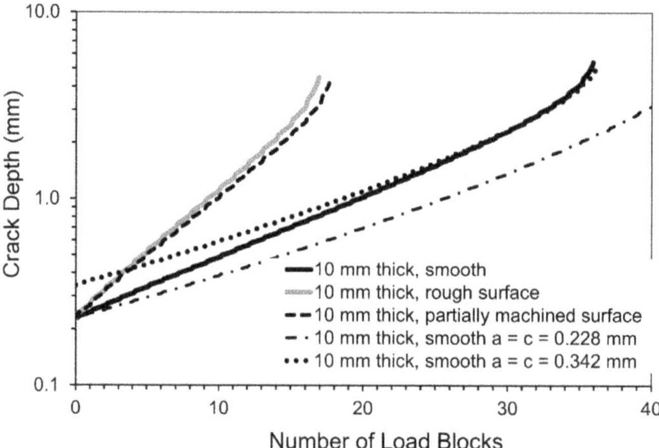

Figure 11. The computed crack depth curves for a 10 mm thick specimen.

Here we see that, despite the increased thickness, partial machining only resulted in a relatively small increase in durability. Furthermore, regardless of the assumptions used in the analysis of the durability of the smooth specimen, the durability of the partially machined specimen was significantly less than that of a specimen with a smooth surface. This finding, when taken in conjunction with the results presented in [2,15] and the above sections of this paper, adds support to the conclusion that partial machining may not be particularly effective in improving the durability of these WAAM-built steel specimens.

4.2. A Fully Machined Surface

The question now arises: What would happen if the rough surface was machined entirely flat? For the rough surface specimen discussed in [1] and Section 3, this would result in an approximately 2.6 mm thick specimen. The resultant crack growth history, for this "fully machined" specimen with the same size initial crack, is also shown in Figure 7. Unfortunately, machining the specimen flat increases the stress in the specimen. Consequently, as a result of the subsequent increase in the stress in the section, the resultant crack growth history is not significantly improved, see Figure 7.

The analysis of the fully machined specimen was repeated assuming an initial 0.228 mm radius and a 0.342 mm radius semi-circular surface-breaking crack. The results of this analysis are also shown in Figure 7. Here, we see that for a 0.228 mm radius semi-elliptical crack the life of the fully machined part, as computed using the aspect ratio $c/a = 1$ recommended in [25,30] for a conventionally manufactured specimen, is approximately 20% greater than that computed for the as-built part with the measured 0.228 mm deep and 0.680 mm tip-to-tip length crack. However, if the radius of the semi-elliptical crack is 0.342 mm then the life of the fully machined part is essentially the same as that computed for the as-built part.

Here it should be remembered that this specimen was only (nominally) four mm thick. Consequently, removing approximately 1.4 mm to ensure a smooth surface has a significant effect on the stress in the specimen. The increase in stress would be less for thicker specimens. As such, the benefit of completely machining away the surface roughness may need to be studied on a case-by-case basis.

5. Conclusions

The author's prior study into the durability of wire and arc additively manufactured (WAAM) 18Ni 250 Maraging steel specimens revealed that failure occurred due to the interaction between the surface roughness and surface-breaking material discontinuities. The present paper has revealed that in these tests, there was little crack growth from internal and near-surface, internal voids/porosity. This observation is consistent with that reported in tests with internal voids/porosity in conventionally manufactured parts. As a result, the present paper has addressed the questions:

(i) How severely does the rough "as-built" surface degrade the durability of the specimen?
(ii) Will partial machining of the surface help to improve the durability of WAAM built steel specimens?

Unfortunately, the results of this investigation suggest that:

(a) Surface roughness of the order of that seen in the WAAM 18Ni 250 Maraging steel specimens significantly degrades its durability.
(b) Partial machining of the rough surface may not significantly improve durability.
(c) The benefit of fully machining a rough surface is best suited to relatively thick specimens where the loss of material due to machining does not significantly increase the stress in the remaining material.

Author Contributions: Project direction—R.J.; surface topography measurements and CAD model development—D.P. and A.M.; durability analysis—D.P.; funding—V.C. and A.S.M.A.; conceptualization A.B. and A.M.; overview of the report and its relationship to durability—V.C. and R.J.; first draft of the paper —R.J. and A.B.; rewriting of the paper—R.J. and A.S.M.A. All authors have read and agreed to the published version of the manuscript.

Funding: Rhys Jones, Andrew Ang and Daren Peng would like to acknowledge funding provided by the US Army International Technology Center, Indo-Pacific (ITC-IPAC), Tokyo, Contract No. FA520921P0164.

Informed Consent Statement: Not applicable.

Data Availability Statement: The data are not yet publicly available due to the ongoing nature of this project. The data will be available on completion of the study.

Acknowledgments: The findings and conclusions or recommendations expressed in this paper are those of the authors and do not necessarily reflect the views of the ITC-IPAC.

Conflicts of Interest: The authors declare no conflict of interest.

References

1. Peng, D.; Jones, R.; Ang, A.S.M.; Michelson, A.; Champagne, V.; Birt, A.; Pinches, S.; Kundu, S.; Alankar, A.; Singh, R.R.K. Computing the durability of WAAM 18Ni 250 maraging steel specimens. *Fatigue Fract. Eng. Mater. Struct.* **2022**, *45*, 3535–3545. [CrossRef]
2. Peng, D.; Jones, R.; Ang, A.S.M.; Champagne, V.; Birt, A.; Michelson, A. A Numerical Study into the Effect of Machining on the Interaction between Surface Roughness and Surface Breaking Defects on the Durability of WAAM Ti-6Al-4V Parts. *Metals* **2022**, *12*, 1121. [CrossRef]
3. Cao, F.; Zhang, T.; Ryder, M.A.; Lados, D.A. A Review of the Fatigue Properties of Additively Manufactured Ti-6Al-4V. *JOM* **2018**, *70*, 349–357. [CrossRef]
4. Strano, G.; Hao, L.; Everson, R.M.; Evans, K.E. Surface roughness analysis, modelling and prediction in selective laser melting. *J. Mater. Process. Technol.* **2013**, *213*, 589–597. [CrossRef]
5. Molaei, R.; Fatemi, A.; Sanaei, N.; Pegues, J.; Shamsaei, N.; Shao, S.; Lie, P.; Warner, D.H.; Phan, N. Fatigue of additive manufactured Ti-6Al-4V, Part II: The relationship between microstructure, material cyclic properties, and component performance. *Int. J. Fatigue* **2020**, *132*, 105363. [CrossRef]
6. Yadollahi, A.; Shamsaei, N. Additive manufacturing of fatigue resistant materials: Challenges and opportunities. *Int. J. Fatigue* **2017**, *98*, 14–31. [CrossRef]
7. Greitemeier, D.; Donne, C.D.; Syassen, F.; Eufinger, J.; Melz, T. Effect of surface roughness on fatigue performance of additive manufactured Ti-6Al-4V. *Mater. Sci. Technol.* **2016**, *32*, 629–634. [CrossRef]
8. Fatemi, A.; Molaei, R.; Sharifimehr, S.; Phan, N.; Shamsaei, N. Multiaxial fatigue behavior of wrought and additive manufactured Ti-6Al-4V including surface finish effect. *Int. J. Fatigue* **2017**, *100*, 347–366. [CrossRef]
9. Lee, S.; Rasoolian, B.; Silva, D.F.; Pegues, J.W.; Shamsaei, N. Surface roughness parameter and modeling for fatigue behavior of additive manufactured parts: A non-destructive data-driven approach. *Addit. Manuf.* **2021**, *46*, 102094. [CrossRef]
10. Andrews, S.; Sehitoglu, H. A computer model for fatigue crack growth from rough surfaces. *Int. J. Fatigue* **2000**, *22*, 619–630. [CrossRef]
11. Sanaei, N.; Fatemi, A. Defect-based fatigue life prediction of L-PBF additive manufactured metals. *Eng. Fract. Mech.* **2021**, *244*, 107541. [CrossRef]
12. Sanaei, N.; Fatemi, A. Defects in additive manufactured metals and their effect on fatigue performance: A state-of-the-art review. *Prog. Mater. Sci.* **2020**, *117*, 100724. [CrossRef]
13. Shamir, M.; Zhang, X.; Syed, A.K. Characterising and representing small crack growth in an additive manufactured titanium alloy. *Eng. Fract. Mech.* **2021**, *253*, 108876. [CrossRef]
14. Structures Bulletin EZ-SB-19-01, Durability and Damage Tolerance Certification for Additive Manufacturing of Aircraft Structural Metallic Parts, Wright Patterson Air Force Base, OH, USA, 10 June 2019. Available online: https://daytonaero.com/usaf-structures-bulletins-library/ (accessed on 2 February 2020).
15. Raab, M.; Bambach, M. Fatigue properties of Scalmalloy®processed by laser powder bed fusion in as-built, chemically and conventionally machined surface condition. *J. Mat. Pro. Tech.* **2023**, *311*, 117811. [CrossRef]
16. Renzo, D.A.; Maletta, C.; Sgambitterra, E.; Furgiuele, F.; Berto, F. Surface roughness effect on multiaxial fatigue behavior of additively manufactured Ti6Al4V alloy. *Int. J. Fatigue* **2022**, *163*. [CrossRef]
17. Bagehorn, S.; Wehr, J.; Maier, H. Application of mechanical surface finishing processes for roughness reduction and fatigue improvement of additively manufactured Ti-6Al-4V parts. *Int. J. Fatigue* **2017**, *102*, 135–142. [CrossRef]
18. Li, P.; Warner, D.; Fatemi, A.; Phan, N. Critical assessment of the fatigue performance of additively manufactured Ti–6Al–4V and perspective for future research. *Int. J. Fatigue* **2016**, *85*, 130–143. [CrossRef]
19. Nezhadfar, P.; Shrestha, R.; Phan, N.; Shamsaei, N. Fatigue behavior of additively manufactured 17-4 PH stainless steel: Synergistic effects of surface roughness and heat treatment. *Int. J. Fatigue* **2019**, *124*, 188–204. [CrossRef]
20. Schijve, J. Internal fatigue cracks are growing in vacuum. *Eng. Fract. Mech.* **1978**, *10*, 359–370. [CrossRef]
21. Yoshinaka, F.; Nakamura, T.; Takeuchi, A.; Uesugi, M.; Uesugi, K. Initiation and growth behaviour of small internal fatigue cracks in Ti-6Al-4V via synchrotron radiation microcomputed tomography. *Fatigue Fract. Eng. Mater. Struct.* **2019**, *42*, 2093–2105. [CrossRef]
22. Schijve, J. Fatigue crack growth, physical understanding and practical application. *Fatigue Fract. Eng. Mater. Struct.* **2009**, *32*, 867–871. [CrossRef]
23. Jeddi, D.; Palin-Luc, T. A review about the effects of structural and operational factors on the gigacycle fatigue of steels. *Fatigue Fract. Eng. Mater. Struct.* **2018**, *41*, 969–990. [CrossRef]
24. Wu, S.C.; Hu, Y.N.; Song, Z.; Ding, S.S.; Fu, Y.N. Fatigue behaviors of laser hybrid welded AA7020 due to defects via synchrotron X-ray microtomography. *Fatigue Fract. Eng. Mater. Struct.* **2019**, *42*, 2232–2246. [CrossRef]
25. MIL-STD-1530D, Department Of Defense Standard Practice Aircraft Structural Integrity Program (ASIP), 13 October 2016. Available online: http://everyspec.com/MIL-STD/MIL-STD.../download.php?spec=MIL-STD-1530D (accessed on 2 February 2020).
26. Jones, R.; Atluri, S.N.; Hammond, P.S.; Williams, J.F. Developments in the analysis of interacting cracks. *Eng. Fail. Assess.* **1995**, *2*, 307–320. [CrossRef]

27. Pitt, S.; Jones, R.; Atluri, S.N. Further studies into interacting 3D cracks. *Comput. Struct.* **1999**, *70*, 583–597. [CrossRef]
28. Atluri, S.N.; Park, J.H.; Punch, E.F.; O'Donohue, P.E.; Jones, R. *Composite Repairs of Cracked Metallic Aircraft*; DOT/FAA/CT-92/32; Federal Aviation Administration: Washington, DC, USA, 1993.
29. Zencrack Fracture Mechanics Software. Available online: https://www.zentech.co.uk/zencrack_publications.htm (accessed on 12 December 2022).
30. Department of Defense Joint Service Specification Guide. Aircraft Structures, JSSG-2006, October 1998. Available online: http://everyspec.com/USAF/USAF-General/JSSG-2006_10206/ (accessed on 10 July 2020).

Article

Quality Quantification and Control via Novel Self-Growing Process-Quality Model of Parts Fabricated by LPBF Process

Xinyi Xiao [1], Beibei Chu [2] and Zhengyan Zhang [2,*]

[1]   Mechanical and Manufacturing Engineering Department, Miami University, Oxford, OH 45056, USA
[2]   School of Mechanical Engineering, Hebei University of Technology, Tianjin 300130, China
*   Correspondence: zzy@hebut.edu.cn

Citation: Xiao, X.; Chu, B.; Zhang, Z. Quality Quantification and Control via Novel Self-Growing Process-Quality Model of Parts Fabricated by LPBF Process. *Materials* **2022**, *15*, 8520. https://doi.org/10.3390/ma15238520

Academic Editors: Bartłomiej Wysocki, Joseph Buhagiar and Tomasz Durejko

Received: 1 September 2022
Accepted: 14 November 2022
Published: 29 November 2022

Publisher's Note: MDPI stays neutral with regard to jurisdictional claims in published maps and institutional affiliations.

Abstract: Laser Powder Bed Fusion (LPBF) presents a more extensive allowable design complexity and manufacturability compared with the traditional manufacturing processes by depositing materials in a layer-wised manner. However, the process variability in the LPBF process induces quality uncertainty and inconsistency. Specifically, the mechanical properties, e.g., tensile strength, are hard to be predicted and controlled in the LPBF process. Much research has recently been reported exploring the qualitative influence of single/two process parameters on tensile strength. In fact, mechanical properties are comprehensively affected by multiple correlated process parameters with unclear and complex interactions. Thus, the study on the quantitative process-quality model of the metal LPBF process is urgently needed to provide an enough-strength component via the metal LPBF process. Recent progress in artificial intelligence (AI) and machine learning (ML) provides new insight into quality prediction in terms of computational accuracy and speed. However, the predictive model quality through the traditional AL/ML is heavily determined by the training data size, and the experimental analysis can be expansive on LPBF. This paper explores the comprehensive effect of the tensile strength of 316L stainless-steel parts on LPBF and proposes a valid quantitative predictive model through a novel self-growing machine-learning framework. The self-growing framework can autonomously expand and classify the growing dataset to provide a high-accuracy prediction with fewer input data. To verify this predictive model of tensile strength, specimens manufactured by the LPBF process with different group process parameters (laser power, scanning speed, and hatch spacing) are collected. The experimental results validate the predicted tensile strengths within a less than 3% deviation.

Keywords: self-growing; process-quality model; machine learning; tensile strength; 316L stainless-steel; LPBF

1. Introduction

Laser Powder Bed Fusion (LPBF) is a material accumulation process by selectively melting and solidifying metal powders to form three-dimensional objects in a layer-wised manner, contrasting with the traditional subtractive manufacturing processes. This process is capable of handling a large variety of materials and with a much larger design space compared with subtractive manufacturing. However, this process involves repeated rapid heating and cooling during the melting and solidifying of the metal powders processes. Thus, this process comes with high process variability and low repeatability compared with traditional manufacturing processes [1]. Therefore, it is always necessary to conduct destructive/non-destructive qualification methods to measure the qualities of LPBF-ed parts, such as porosity, tensile strength, and fatigue. In addition, post-processing techniques, for example, heat treatment and machining techniques, are always subsequent to the LPBF process to further improve the part finish and enhance the as-built mechanical properties.

Figure 1 indicates the workflow from a digital 3D model to an end-use part produced through the LPBF manufacturing process. This process can be separated into prediction and qualification sections. The process quality prediction involves the consideration of all potential influential parameters' effects on the as-built qualities. And the quality quantification process heavily relies on the measurement systems, through either monitoring or post-measurements. The process involves extensive human-to-machine interactions and high process uncertainties. Thus, repeatability and reproducibility cannot be ensured.

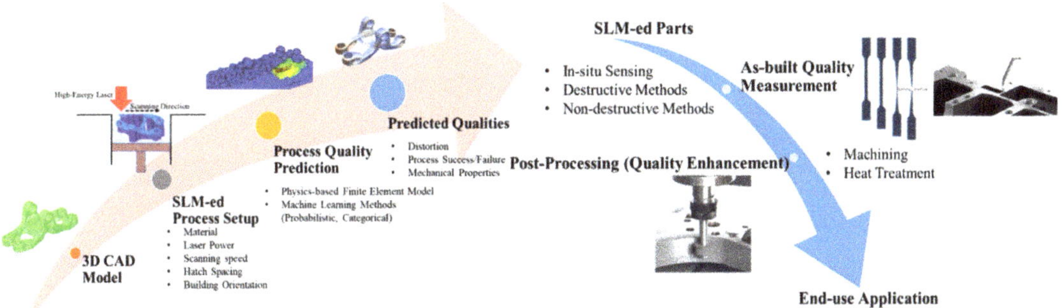

Figure 1. LPBF-ed Part Quality Prediction-Measurement-Enhancement Flow.

The typical workflow to predict, measure, and enhance the LPBF-ed components can be seen in Figure 1. To fulfill the end-use requirements, every LPBF-ed part must be verified and validated in its as-built quality. If the parts fail to pass the as-desired requirement, the part needs to be re-printed or processed through certain quality enhancement post-processing techniques. Such an iterative process increases the production cost and lowers the producibility.

Past research mainly focuses on process-quality prediction, which lies in thermal distortion [2] and mechanical properties, such as tensile strength [3] and porosity [4]. However, these studies have the following limitations:

- Physical-based finite element models for distortion prediction vary from software to software and always require high computational time;
- Most probabilistic and categorical prediction models require heavy amounts of training data to increase the accuracy of the model;
- The location of the differently labeled training data, i.e., the distance from the labeled data to the decision boundary, would make it quite difficult for neural networks to make correct classification or prediction.

The scope of the previous research [1–5] is limited to exploring the mechanical properties affected by single/two specific process parameters, such as laser energy and layer thickness. However, mechanical properties are comprehensive performance from the LPBF process and are heavily affected by multiple process parameters but with unclear and complex interactions with each other. Thus, a comprehensive quantitative predictive model for analyzing the mechanical properties of the part before the real printing process is urgently needed to ensure component functionality. For example, suppose the predictive model of mechanical properties affected by multiple process parameters comprehensively can be established. In that case, the mechanical properties of the LPBF parts can be predicted and ensured, and the optimal process parameters to fabricate parts with higher performance can be obtained based on this predictive model. The proposed process-quality framework overall workflow is shown in Figure 2.

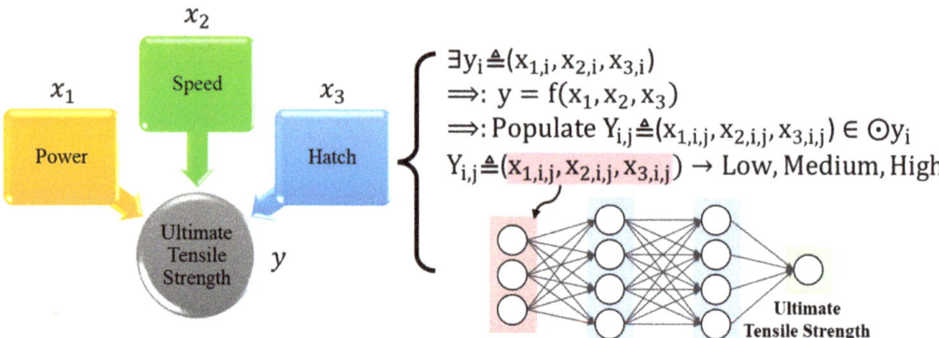

Figure 2. Process-Quality Framework Overall Workflow.

Experimental data are collected to determine the correct relationship between the process and the ultimate property. However, if the training dataset is small, the machine learning prediction model cannot provide accurate insight to analyze and control the process repeatability quantitatively. Therefore, the self-growing algorithm (Section 4) will populate and self-classify these data. Then these data will be used for developing the predictive model.

In conclusion, this paper presents a comprehensive study on the as-built ultimate tensile strength of 316L stainless steel caused and controlled by the LPBF process. The predictive model contains three key parameters (laser power, scanning speed, and hatch spacing) with intercorrelated effects on tensile strength. In addition, experimental data are collected to verify the effectiveness and accuracy of the established self-growing predictive model. The proposed approach is also applicable to establish predictive models for other as-built properties that are correlated with the multiple process parameters.

2. Literature Review

LPBF process is capable of fabricating a large variety of materials, such as steel 300 [5], Ti-6Al-4V [6–8], Al-Si10-Mg [9], CoCr alloy [10], Inconel 718 [11], 304 stainless steel [12], and 316L stainless steel [13] is one of the common-use materials for discussing its as-built properties. In this section, we conduct a thorough literature review on the existing AM-ed components' quality analysis based on experimental and computational analysis.

Numerous studies focus on the SS316L as-built properties in the LPBF process [13–18]. Specifically, the surface finish has been mostly discussed [15]. For example, Wang used the experimental method to analyze the surface roughness [16], and Yusuf analyzed the porosity-microhardness relationship to study the microhardness [17]. In addition, Lin analyzed the effects of plasma co-alloying treatments to conduct surface modification [18] on LPBF-ed parts. Besides the surface character, overall thermal distortion has been widely studied and commonly used before the experiment to prevent build failure. Commercial software [19–21] uses voxels as the simulation base, and other research works focus on developing accurate laser models and improving computational accuracy and speed [22–24]. However, the mechanical properties are the crucial factors to ensure the functionality of the build besides the shape of the products. Recently, many researchers have studied mechanical properties [25,26] and the optimization/control from the process parameters, such as process time interval and heat treatment [27], manufacturing parameters [28–30], the particle size distribution effects [31], layer thickness [32], energy input [33], laser parameters [34–38], and environment variables [39]. Specifically, the qualitative relationship between one/two process parameters with one mechanical property has been widely reported. For example, Leicht et al. [40] concluded that higher energy would result in a higher density component. Wang et al. [41] reported a proportional linear relationship between the grain diameter in the direction of applied tensile load and the yield strength.

Cherry et al. [42] investigated the inverse proportional relationship between the as-built porosity and the hardness level. In addition, Liverani et al. [43] provided an optimal processing zone with laser power and hatch spacing to guarantee a density greater than 98%. Meier et al. [44] found a dependent relationship between the density and the tensile strength of the SS-316L through experiments.

Besides the abovementioned experimental analysis for developing a qualitative relationship between the process and the desired quality, other researchers used the data analytical methods [45], and machine learning model [46–48] to provide a probabilistic predictive model for providing guidance in assuring the as-built quality in the pre-processing stage. However, few researchers developed a novel multi-dimensional process-quality framework [49–53] that can provide a quantitative relation between the multiple process parameter and the build quality and provide a certain printable zone to control the quality as desired.

3. Experimental Setup

The 316L stainless-steel powder was selected as the experimental material purchased from the Germany TLS company. The chemical composition is shown in Table 1, and the average particle size was 30 μm.

Table 1. Chemical composition of the 316L stainless steel.

Chemical Composition	Ni	Cr	Mn	Si	Mo	C	P	S	Cu
Percentage (%)	10.76	16.78	1.23	0.58	2.42	0.018	0.011	0.007	0.08

The BLT S200 is adopted in this experiment with a maximum building space of 105 mm × 105 mm × 200 mm, and a 500-W fiber laser was used for vibrating-mirror laser scanning with a wavelength of 1070 mm. The forming substrate was 304 stainless steel. Before the experiment, the surface of the substrate was pretreated with industrial alcohol to ensure that the forming process was not affected by surface oil or dust. Before processing, the substrate was preheated to 80 °C. Argon was used as a protective gas in the molding process, and the oxygen content was kept below 800 ppm. Tensile test pieces were designed according to GB/T 228.1–2010, as shown in Figure 3. Three key process parameters (laser power-P, scanning speed-V, and hatch scan-D) were selected to explore the comprehensive predictive model of tensile strength. The process parameters experimental setup was designed for analyzing the correlated effects towards tensile strength, as shown in Table 2. The tensile test procedure follows the ASTM E8 specification [54]. All specimens are pulled at a strain rate of 0.005in/in/min.

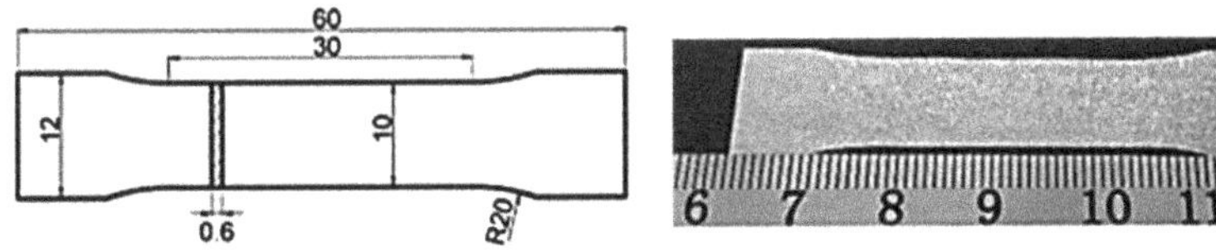

Figure 3. (**left**) Tensile Test Specimen Dimensions, (**right**) the as-built part.

Table 2. The Experimental Setup.

Process Parameters	Experimental Range
P (W)	[160, 280]
V (mm/s)	[800, 1200]
D (mm)	[0.06, 0.22]

4. Proposed Methodology

This research proposes a novel self-growing dynamic neural network framework to establish the comprehensive process-quality predictive model of the LPBF process. One of the main limitations of applying traditional neural networks to additive manufacturing data is the shortage of the dataset. The other drawback of neural networks is over-parameterized neurons and synapses. To overcome such limitations, a point-wise population and self-growing dynamic interference training techniques are developed in this paper without sacrificing the desired task performance (e.g., classification accuracy or prediction quality).

The point-wise population is developed to overcome the limitations of the shortage of the dataset. The surface response models based on the collected data are first established. In addition, point-wise population data volumes are created through the constructed response models. These data volumes are designed to adapt to the significance level of the parameter. And the density of the data in the volume is varied based on the level of the tensile strength. The self-growing dynamic interference training schemes are based on easy and hard data. The easy data refers to the data that is located far from the hyperplane, in contrast to the hard data, which sits close to the decision boundary. To leverage the existence of the easy and hard data to accelerate the tensile strength classification within the process-quality model, the highly correlated synapses for the easy data are collected (marked in bold in Figure 4). These synapses are emphasized while processing hard data through the front-end part of the neural network. Similarly, the less-relevant synapses from the easy data are gathered (marked in dashed in Figure 3). These synapses are ignored when processing hard data.

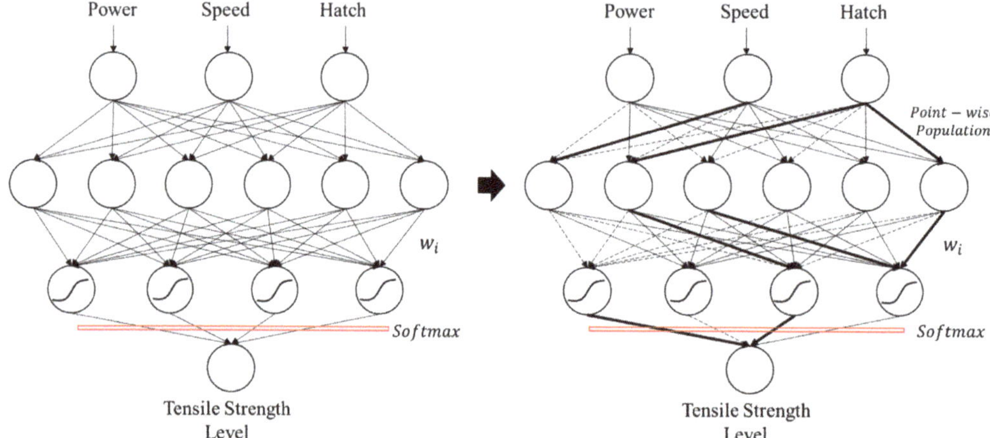

Figure 4. Self-growing Dynamic Neural Network.

Figure 4 presents the self-growing dynamic neural network scheme, which first takes the input parameters to the hidden sigmoid layers, then processes with SoftMax output neurons to predict the level of tensile strength. The key idea of the self-growing training scheme is based on the phenomenon of easy and hard data (as shown in Figure 5). Each data has different locations to the hyperplane/decision boundary in the entire dataset and the constructed surface response model. The closer ones are called hard data, since it is quite hard for the neural network to classify and distinguish them. In contrast, the easy data are far from the decision boundary, making the network easier to classify and recognize. The self-growing scheme heavily depends on the easy/hard data phenomena in training/validating datasets. Through the proposed algorithm, the populated data on the easy ones will be denser and with a larger tolerance zone. In comparison, the self-generated data on the hard ones will be sparser and with a smaller defined zone of collection.

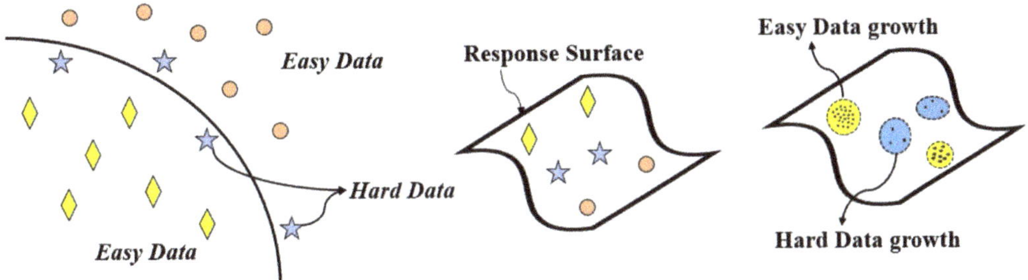

Figure 5. Point-wise population based on the easy/hard data.

Aiming at the three factors, laser power P, scanning speed V, and hatch scan D, the response model can be formulated as:

$$y = b_0 + b_1 P + b_2 V + b_3 D + b_{11} P^2 + b_{22} V^2 + b_{33} D^2 + b_{12} PV + b_{13} PD + b_{23} VD + \varepsilon \quad (1)$$

The total effect of the fitting is as follows: R-Sq is close to R-Sq (adj), the fitting effect of the model is better, and R-Sq (pred) is close to R-Sq, which shows that the prediction of the model is reliable. The ANOVA table is presented in Table 3.

Table 3. The ANOVA table that displays the results of a statistical analysis of variance.

	Freedom	Seq SS	Adj SS	Adj MS	F	W
Regression	9	989.22	989.22	109.914	11.04	0.000
Linear	3	209.47	482.54	160.847	16.16	0.000
P	1	0.24	106.93	106.926	10.74	0.008
V	1	33.33	204.84	204.842	20.58	0.001
D	1	175.90	470.83	470.828	47.31	0.000
Square	3	611.10	346.07	115.356	11.59	0.001
P*P	1	366.22	326.59	326.594	32.81	0.000
V*V	1	182.25	132.32	132.319	13.29	0.004
D*D	1	62.63	24.40	24.403	2.45	0.148
Interaction	3	168.65	168.65	56.216	5.65	0.016
P*V	1	69.23	25.03	25.026	2.51	0.144
P*D	1	8.77	56.62	56.616	5.69	0.038
V*D	1	90.65	90.65	90.652	9.11	0.013
Residual Error	10	99.53	99.53	9.953		
Misuse	7	85.42	85.42	12.203	2.60	0.233
Pure error	3	14.11	14.11	4.702		
Total	19	1088.75				

S = 3.15478 PRESS = 499.659 R-Sq = 90.86% R-Sq (pred) = 78.11% R-Sq (adj) = 82.63%

The fitted model has the following: R-Sq = 90.86%, R-Sq (pred) = 78.11%, and R-Sq (adj) = 82.63%, with the probability values of P, V, D, P*P, V*V, P*D, and V*D, which are less than the significance level of 0.05, indicating that these effects are all significant. In contrast, the corresponding probability values (W) of the interaction effects of D*D and P*V are 0.148 and 0.144, respectively, greater than the significance level, and the effect is insignificant.

Based on the fitted model, the self-growing zone on the P–V, P–H, and V–H response model based on the classified easy/hard data is indicated in Figure 6. The size and the gradient of the populated data volumes are varied based on the significance of the process parameter and the deviation between the fitted model and the input data.

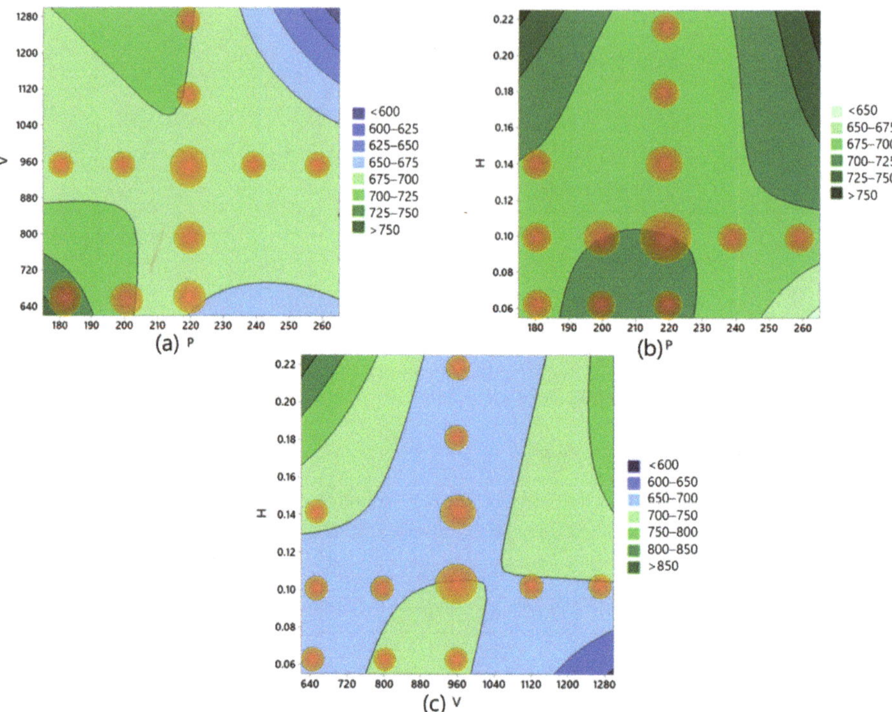

Figure 6. (**a**) Self-growing populated data volume in the P–V–TS model, (**b**) self-growing populated data volume in the P–H–TS model, and (**c**) self-growing populated data volume in the V–H–TS model.

After obtaining the self-populated data, 80% of the data are used for training and 20% for testing in the following analysis. The accuracy and validation of the proposed machine-learning network will be presented in Section 5.

5. Results and Discussions

The self-growing machine learning model performance is shown in Figure 7, comparing the model without the self-growing feature.

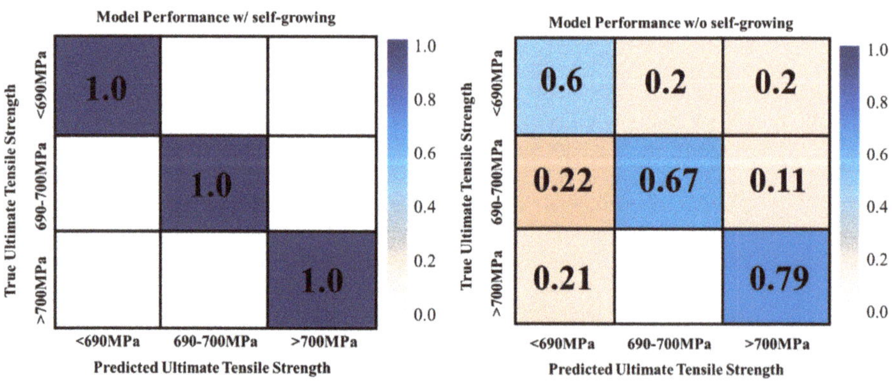

Figure 7. Machine Learning Models Performance.

The tensile strength has been categorized as low, medium, and high, and the diagonal cells in the matrixes represent the percentage of the correct response level prediction. The higher the number represents that the model is accurate. Furthermore, the cells that are not lying on the diagonal in the matrixes show incorrect classification. Figure 6 shows that the proposed self-growing model presents a 100% accurate prediction for levels 0, 1, and 2. In contrast, the traditional machine learning model presents the highest accuracy of 80%. There is a significant increase in the prediction by adding the self-growing feature to the model.

The convergence of the neural network is the process by which the network gradually adjusts its weights and biases so that the output of the network is closer and closer to the desired target values. This is conducted through a process of iteration, in which the network is repeatedly exposed to new training data, and the weights and biases are adjusted accordingly. The convergence of the neural network is a key factor in determining the accuracy of the network's predictions. The convergence of the one with and without the proposed self-growing neural network is presented in Figure 8.

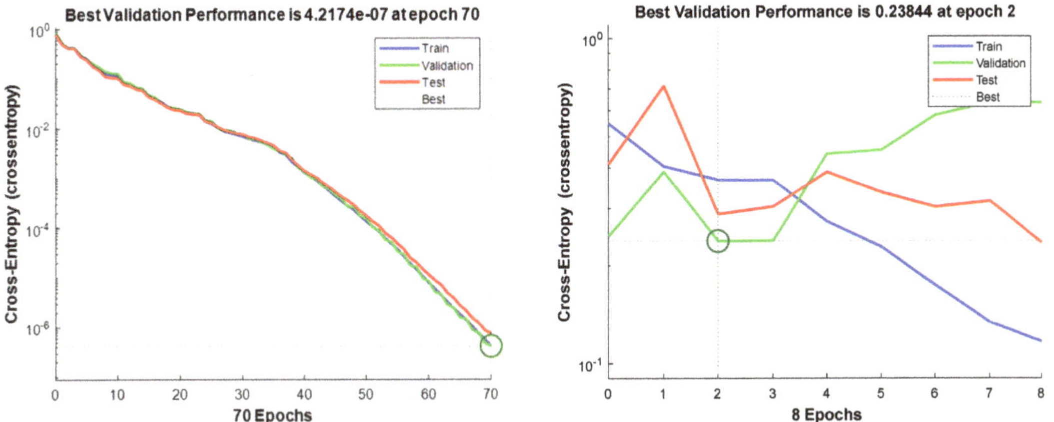

Figure 8. Neural network convergence for Tensile Strength (**left**) with self-growing and (**right**) without self-growing.

Cross-entropy is a measure of how close a neural network is to convergence. It is calculated by taking the sum of the products of the weights and the errors for each neuron in the network. This measure determines when a neural network has reached a stable state. For example, from the neural network models' convergence plots in Figure 8, the model with the self-growing function can reach a lower cross-entropy, which means that the network can learn the tensile strength more accurately. This is achieved by ensuring that the network is properly configured and has enough training data provided autonomously through the self-growing function.

Five groups of specimens that varied three analyzed parameters (P, V, D) were fabricated and tested further to verify the correctness of the proposed machine learning model. The test value and calculated values based on the comprehensive predictive model using the self-growing feature are shown in Table 4. In addition, the prediction levels of the tensile without using the self-growing technique are also presented to be used as the comparison group. The categorical level predictions are also presented in the table which are shown in the bracket () manner.

Table 4. The Validation Experimental Setup.

Experimental Data (P, V, H)	Predict Level (Without Self-Growing)	Predicted TS (MPa) (With Self-Growing)	Measured TS (MPa)
(220,960,0.10)	2	700.284 (2)	705.076 (2)
(220,900,0.10)	1	700.398 (2)	701.782 (2)
(220,960,0.15)	0	689.849 (1)	693.550 (1)
(240,1120,0.10)	1	695.245 (1)	697.371 (1)
(240,960,0.06)	0	712.138 (2)	713.792 (2)

The tensile strength prediction through the proposed machine learning model matches the experimental measures, which indicates the credibility of the prediction model. In addition, from Table 3, the ultimate tensile strength prediction by comparing the two machine learning models has a significant difference, specifically for predicting the high tensile strength. Therefore, the presented self-growing technique can significantly improve prediction accuracy.

The topography of the fracture analysis was measured over the entire area of the test specimen with an optical three-dimensional surface measurement system. Three representative groups of specimens are selected from Table 2 with the representative tensile strength level (low, medium, high), as shown in Table 5.

Table 5. Selected groups of specimens for fracture analysis.

Group Num.	P (W)	V (mm/s)	D (mm)	Measured UTS (MPa)	Level
8	220	1120	0.1	693.691	Medium
14	180	640	0.06	682.111	Low
20	220	960	0.1	705.076	High

Figures 9–11 present the fracture morphology of the selected three groups of tensile specimens, indicating that the tensile specimens are all ductile fractures with numerous dimples. As shown in Figure 9, it is observed that the dimples are non-uniform, and their size and depth are bigger than the ones that are shown in Figure 9, but smaller than that in Figure 11. In addition, some defects exist, such as the micro-holes (as shown by the red arrows in Figure 10c) intermingled with small dimples and showing holes in some local regions. These fractures with obvious fluctuation, pure shear stress, and crack propagation along the direction of principal stress are characterized by serpentine glide, a typical ductile fracture of dimple micro-groove. Figure 9 shows that it is fractured along the grain boundary, and the dimples with uniform shape and size are smaller than those in Figures 9 and 11. The dimples are shallow in-depth, and small edges with different orientations can be seen on the fracture surface, which is composed of a large number of dimples, and there will be a particle at the bottom of the dimples. Energy-dispersive X-ray spectroscopy (EDS) is performed to analyze the chemical characterization/elemental analysis of the particles indicated in Figure 10c. The obtained results are shown in Figure 12. The outlined areas indicate the un-melted SS316L particles. The un-melted SS316L will reduce the tensile strength. This further verifies the tensile strength test results.

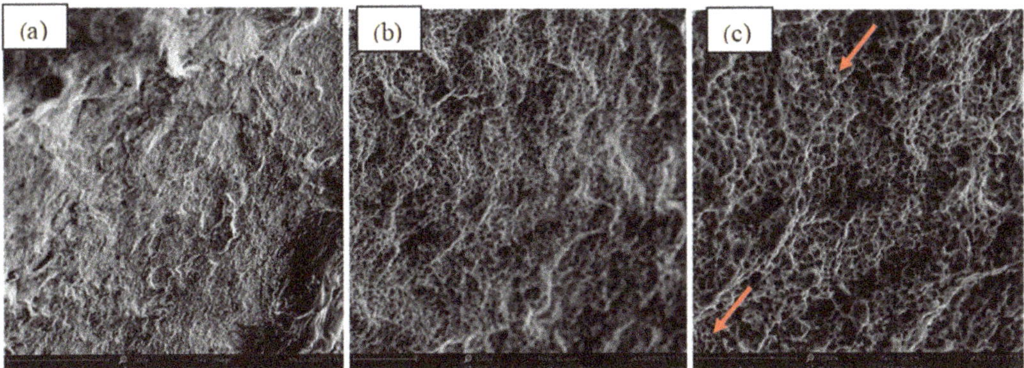

Figure 9. The fracture morphology of group 8—Medium Tensile Strength in Table 4, (**a**) ×1000, (**b**) ×5000, (**c**) ×10,000 (red arrow represents the micro-holes).

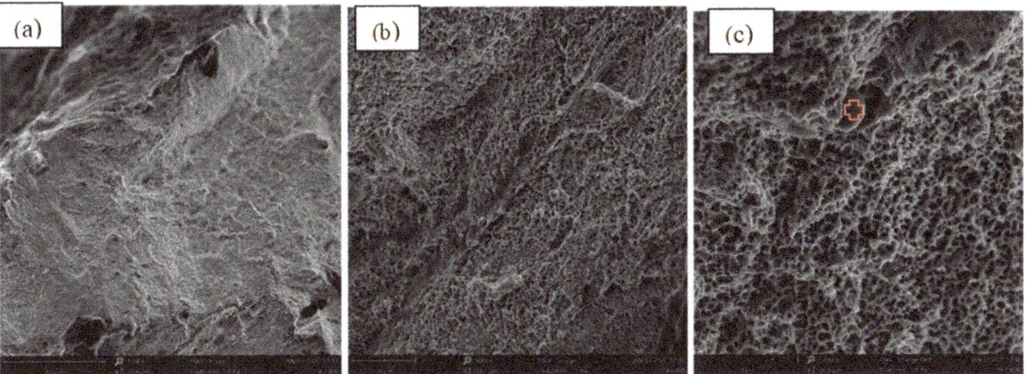

Figure 10. The fracture morphology of group 14—Low Tensile Strength in Table 4, (**a**) ×1000, (**b**) ×5000, (**c**) ×10,000 (circled area represents the EDS performed zone).

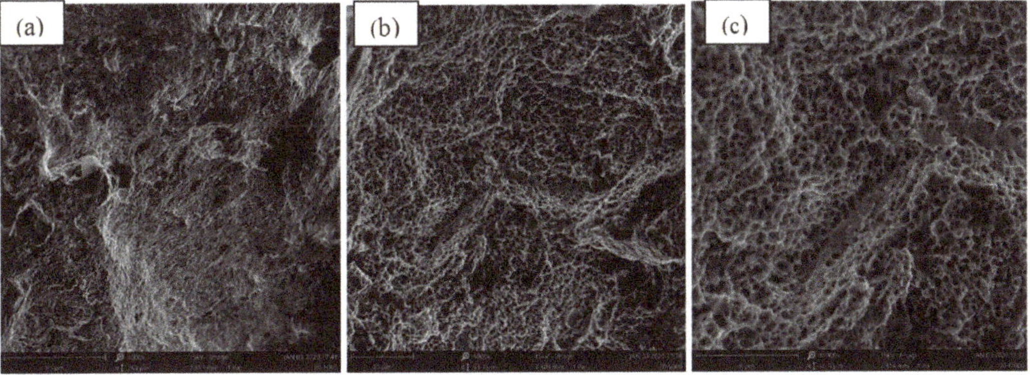

Figure 11. The fracture morphology of group 20—High Tensile Strength in Table 4, (**a**) ×1000, (**b**) ×5000, (**c**) ×10,000.

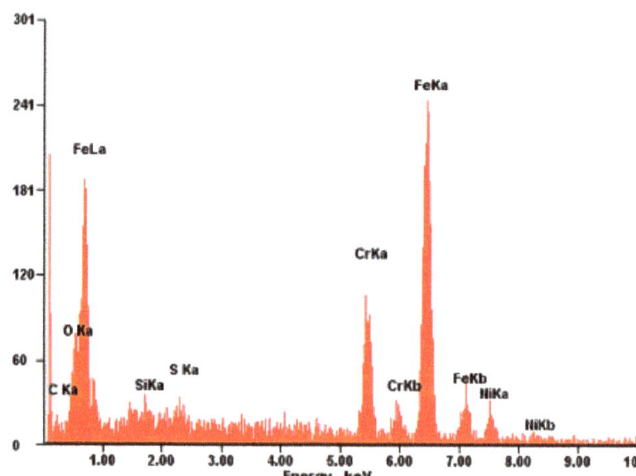

Figure 12. The EDS analysis result for the area that is indicated in Figure 10c.

Furthermore, the stress state formed by equiaxial dimples is a uniform strain. From Figure 11, the fracture behavior is a ductile fracture, which illustrates the phenomenon of corner fractures compared to the specimens shown in Figures 8 and 9. This indicates that this group requires a higher strength to break, confirming the data obtained in Table 4. A radial herringbone ridge pattern also appears when the scale is larger. The dimple is elongated parallel to the fracture direction, which generates a ductile sliding stress state.

The fracture morphology and the EDS analysis match the measurement and prediction results. This further reinforces the accuracy of the process-quality prediction model.

6. Conclusions

Tensile strength is one of the most important mechanical properties for ensuring additive manufacturing parts' functionality. However, it is comprehensively affected by multiple process parameters rather than single process parameters. Therefore, selecting the optimal processing condition to fabricate the parts with the desired high performance is difficult since the comprehensive effect is very complicated. Typically, the process parameters behave highly correlated but with unclear and complex interactions with each other. Thus, an accurate and computationally efficient tensile strength predictive model by multiple process parameters is desired and significant to assist when selecting the optimal processing condition. The following qualitative conclusions can be obtained from the proposed research to achieve a higher tensile strength component:

- An increase in power should be coupled with a higher scanning velocity;
- A smaller hatch spacing would be set with a higher laser power;
- Velocity and the hatch spacing should be inverse proportional relation.

The research of this paper focuses on comprehensively analyzing the tensile strength 316L stainless-steel parts and establishing a self-growing machine learning-based predictive model with three key parameters (laser power, scanning speed, and hatch scan). The experimental tensile test results and the fracture morphology both indicate that this proposed predictive model and its corresponding establishment approach can explicitly predict the ultimate tensile strength and establish an optimal printable zone that ensures the high-end components of the LPBF process. This research paves the path for building a digital twin on the metal AM process from design, material, process, and quality in an effective manner.

Author Contributions: Conceptualization, X.X. and Z.Z.; methodology, X.X.; software, X.X.; validation, X.X., B.C., and Z.Z.; formal analysis, X.X.; investigation, X.X.; resources, X.X. and Z.Z.; data curation, X.X. and B.C.; writing—original draft preparation, X.X.; writing—review and editing,

X.X. and Z.Z.; visualization, X.X.; supervision, X.X. and Z.Z.; project administration, Z.Z.; funding acquisition, Z.Z. All authors have read and agreed to the published version of the manuscript.

Funding: This paper is sponsored by the special project for the Central government to guide local scientific and technological development of Hebei Province (226Z1903G), National Natural Science Foundation of China (61802108).

Informed Consent Statement: Not applicable.

Conflicts of Interest: The authors declare no conflict of interest.

References

1. Xiao, X.; Joshi, S. Process planning for five-axis support free additive manufacturing. *Addit. Manuf.* **2020**, *36*, 101569. [CrossRef]
2. Zou, S.; Pang, L.; Xu, C.; Xiao, X. Effect of Process Parameters on Distortions Based on the Quantitative Model in the LPBF Process. *Appl. Sci.* **2022**, *12*, 1567. [CrossRef]
3. Zou, S.; Xiao, X.; Li, Z.; Liu, M.; Zhu, C.; Zhu, Z.; Chen, C.; Zhu, F. Comprehensive investigation of residual stress in selective laser melting based on cohesive zone model. *Mater. Today Commun.* **2022**, *31*, 103283. [CrossRef]
4. Roh, B.-M.; Kumara, S.R.T.; Witherell, P.; Simpson, T.W. Ontology-based Process Map for Metal Additive Manufacturing. *J. Mater. Eng. Perform.* **2021**, *30*, 8784–8797. [CrossRef]
5. Bai, Y.; Yang, Y.; Wang, D.; Zhang, M. Influence mechanism of parameters process and mechanical properties evolution mechanism of maraging steel 300 by selective laser melting. *Mater. Sci. Eng. A* **2017**, *703*, 116–123. [CrossRef]
6. Simonelli, M.; Tse, Y.Y.; Tuck, C. Effect of the build orientation on the mechanical properties and fracture modes of LPBF Ti-6Al-4V. *Mater. Sci. Eng. A* **2014**, *616*, 1–11. [CrossRef]
7. Traini, T.; Mangano, C.; Sammons, R.L.; Mangano, F.; Macchi, A.; Piattelli, A. Direct laser metal sintering as a new approach to fabrication of an isoelastic functionally graded material for manufacture of porous titanium dental implants. *Dent. Mater.* **2008**, *24*, 1525–1533. [CrossRef] [PubMed]
8. Wu, Q.; Lu, J.; Liu, C.; Fan, H.; Shi, X.; Fu, J.; Ma, S. Effect of Molten Pool Size on Microstructure and Tensile Properties of Wire Arc Additive Manufacturing of Ti-6Al-4V Alloy. *Materials* **2017**, *10*, 749. [CrossRef]
9. Taheri Andani, M.; Dehghani, R.; Karamooz-Ravari, M.R.; Mirzaeifar, R.; Ni, J. Spatter formation in selective laser melting process using multi-laser technology. *Mater. Des.* **2017**, *131*, 460–469. [CrossRef]
10. Wang, D.; Wu, S.; Fu, F.; Mai, S.; Yang, Y.; Liu, Y.; Song, C. Mechanisms and characteristics of spatter generation in LPBF processing and its effect on the properties. *Mater. Des.* **2017**, *117*, 121–130. [CrossRef]
11. Clare, A.T.; Chalker, P.R.; Davies, S.; Sutcliffe, C.J.; Tsopanos, S. Selective laser melting of high aspect ratio 3D nickel–titanium structures two way trained for MEMS applications. *Int. J. Mech. Mater. Des.* **2007**, *4*, 181–187. [CrossRef]
12. Guan, K.; Wang, Z.; Gao, M.; Li, X.; Zeng, X. Effects of processing parameters on tensile properties of selective laser melted 304 stainless steel. *Mater. Des.* **2013**, *50*, 581–586. [CrossRef]
13. Leicht, A.; Rashidi, M.; Klement, U.; Hryha, E. Effect of process parameters on the microstructure, tensile strength and productivity of 316L parts produced by laser powder bed fusion. *Mater. Charact.* **2020**, *159*, 110016. [CrossRef]
14. Ahmadi, A.; Mirzaeifar, R.; Moghaddam, N.S.; Turabi, A.S.; Karaca, H.E.; Elahinia, M. Effect of manufacturing parameters on mechanical properties of 316L stainless steel parts fabricated by selective laser melting: A computational frame-work. *Mater. Des.* **2016**, *112*, 328–338. [CrossRef]
15. Spierings, A.B.; Bourell, D.; Herres, N.; Levy, G. Influence of the particle size distribution on surface quality and mechanical properties in AM steel parts. *Rapid Prototyp. J.* **2011**, *17*, 195–202. [CrossRef]
16. Wang, D.; Liu, Y.; Yang, Y. Theoretical and experimental study on surface roughness of 316L stainless steel metal parts obtained through selective laser melting. *Rapid Prototyp. J.* **2016**, *22*, 706–716. [CrossRef]
17. Yusuf, S.M.; Chen, Y.; Boardman, R.; Yang, S.; Gao, N. Investigation on Porosity and Microhardness of 316L Stainless Steel Fabricated by Selective Laser Melting. *Metals* **2017**, *7*, 64. [CrossRef]
18. Lin, K.; Li, X.; Dong, H.; Du, S.; Lu, Y.; Ji, X.; Gu, D. Surface modification of 316 stainless steel with platinum for the application of bipolar plates in high performance proton exchange membrane fuel cells. *Int. J. Hydrog. Energy* **2017**, *42*, 2338–2348. [CrossRef]
19. Ansys Additive Manufacturing Solutions. Available online: https://www.ansys.com/products/additive (accessed on 13 November 2022).
20. Additive Manufacturing. Available online: https://www.plm.automation.siemens.com/global/en/products/manufacturing-planning/additive-manufacturing.html (accessed on 13 November 2022).
21. Fusion 360 with Netfabb Features. Available online: https://www.autodesk.com/products/netfabb/features (accessed on 13 November 2022).
22. Chen, Q.; Taylor, H.; Takezawa, A.; Liang, X.; Jimenez, X.; Wicker, R.; To, A.C. Island scanning pattern optimization for residual deformation mitigation in laser powder bed fusion via sequential inherent strain method and sensitivity analysis. *Addit. Manuf.* **2021**, *46*, 102116. [CrossRef]
23. Cheng, L.; To, A. Part-scale build orientation optimization for minimizing residual stress and support volume for metal additive manufacturing: Theory and experimental validation. *Comput. Des.* **2019**, *113*, 1–23. [CrossRef]

24. Ren, K.; Chew, Y.; Zhang, Y.F.; Bi, G.J.; Fuh, J.Y.H. Thermal analyses for optimal scanning pattern evaluation in laser aided additive manufacturing. *J. Mater. Process. Technol.* **2019**, *271*, 178–188. [CrossRef]

25. Suryawanshi, J.; Prashanth, K.G.; Ramamurty, U. Mechanical behavior of selective laser melted 316L stainless steel. *Mater. Sci. Eng. A* **2017**, *696*, 113–121. [CrossRef]

26. Miranda, G.; Faria, S.; Bartolomeu, F.; Pinto, E.; Madeira, S.; Mateus, A.; Carreira, P.; Alves, N.; Silva, F.S.; Carvalho, O. Predictive models for physical and mechanical properties of 316L stainless steel produced by selective laser melting. *Mater. Sci. Eng. A* **2016**, *657*, 43–56. [CrossRef]

27. Yadollahi, A.; Shamsaei, N.; Thompson, S.M.; Seely, D.W. Effects of process time interval and heat treatment on the mechanical and microstructural properties of direct laser deposited 316L stainless steel. *Mater. Sci. Eng. A* **2015**, *644*, 171–183. [CrossRef]

28. Liverani, E.; Toschi, S.; Ceschini, L.; Fortunato, A. Effect of selective laser melting (LPBF) process parameters on microstructure and mechanical properties of 316L austenitic stainless steel. *J. Mater. Process. Technol.* **2017**, *249*, 255–263. [CrossRef]

29. Roh, B.M.; Kumara, S.R.; Yang, H.; Simpson, T.W.; Witherell, P.; Lu, Y. In-Situ Observation Selection for Quality Management in Metal Additive Manufacturing. In *International Design Engineering Technical Conferences and Computers and Information in Engineering Conference*; American Society of Mechanical Engineers: New York, NY, USA, 2021; Volume 85376, p. V002T02A069.

30. Xiao, X.; Joshi, S. Automatic toolpath generation for heterogeneous objects manufactured by directed energy deposition additive manufacturing process. *J. Manuf. Sci. Eng.* **2018**, *140*, 071005. [CrossRef]

31. Calignano, F.; Manfredi, D.; Ambrosio, E.P.; Iuliano, L.; Fino, P. Influence of process parameters on surface roughness of aluminum parts produced by DMLS. *Int. J. Adv. Manuf. Technol.* **2013**, *67*, 2743–2751. [CrossRef]

32. Wang, S.; Liu, Y.; Shi, W.; Qi, B.; Yang, J.; Zhang, F.; Han, D.; Ma, Y. Research on High Layer Thickness Fabricated of 316L by Selective Laser Melting. *Materials* **2017**, *10*, 1055. [CrossRef]

33. Di, W.; Yongqiang, Y.; Xubin, S.; Yonghua, C. Study on energy input and its influences on single-track, multi-track, and multi-layer in LPBF. *Int. J. Adv. Manuf. Technol.* **2011**, *58*, 1189–1199. [CrossRef]

34. Gu, D.; Hagedorn, Y.-C.; Meiners, W.; Meng, G.; Batista, R.J.o.S.; Wissenbach, K.; Poprawe, R. Densification behavior, microstructure evolution, and wear performance of selective laser melting processed commercially pure titanium. *Acta Mater.* **2012**, *60*, 3849–3860. [CrossRef]

35. Prashanth, K.G.; Scudino, S.; Klauss, H.J.; Surreddi, K.B.; Löber, L.; Wang, Z.; Chaubey, A.K.; Kühn, U.; Eckert, J. Microstructure and mechanical properties of Al–12Si produced by selective laser melting: Effect of heat treatment. *Mater. Sci. Eng. A* **2014**, *590*, 153–160. [CrossRef]

36. Thijs, L.; Verhaeghe, F.; Craeghs, T.; Humbeeck, J.V.; Kruth, J.-P. A study of the microstructural evolution during selective laser melting of Ti–6Al–4V. *Acta Mater.* **2010**, *58*, 3303–3312. [CrossRef]

37. Wang, D.; Song, C.; Yang, Y.; Bai, Y. Investigation of crystal growth mechanism during selective laser melting and mechanical property characterization of 316L stainless steel parts. *Mater. Des.* **2016**, *100*, 291–299. [CrossRef]

38. Zhang, B.; Bi, G.; Nai, S.; Sun, C.-N.; Wei, J. Microhardness and microstructure evolution of TiB2 reinforced Inconel 625/TiB2 composite produced by selective laser melting. *Opt. Laser Technol.* **2016**, *80*, 186–195. [CrossRef]

39. Zhang, B.; Dembinski, L.; Coddet, C. The study of the laser parameters and environment variables effect on mechanical properties of high compact parts elaborated by selective laser melting 316L powder. *Mater. Sci. Eng. A* **2013**, *584*, 21–31. [CrossRef]

40. Leicht, A.; Yu, C.H.; Luzin, V.; Klement, U.; Hryha, E. Effect of scan rotation on the microstructure development and mechanical properties of 316L parts produced by laser powder bed fusion. *Mater. Charact.* **2020**, *163*, 110309. [CrossRef]

41. Wang, Z.; Palmer, T.A.; Beese, A.M. Effect of processing parameters on microstructure and tensile properties of austenitic stainless steel 304L made by directed energy deposition additive manufacturing. *Acta Mater.* **2016**, *110*, 226–235. [CrossRef]

42. Cherry, J.A.; Davies, H.M.; Mehmood, S.; Lavery, N.P.; Brown, S.G.R.; Sienz, J. Investigation into the effect of process parameters on microstructural and physical properties of 316L stainless steel parts by selective laser melting. *Int. J. Adv. Manuf. Technol.* **2015**, *76*, 869–879. [CrossRef]

43. Cai, C.; Wu, X.; Liu, W.; Zhu, W.; Chen, H.; Qiu JC, D.; Shi, Y. Selective laser melting of near-α titanium alloy Ti-6Al-2Zr-1Mo-1V: Parameter optimization, heat treatment and mechanical performance. *J. Mater. Sci. Technol.* **2020**, *57*, 51–64. [CrossRef]

44. Meier, H.; Haberland, C. Experimental studies on selective laser melting of metallic parts. *Mater. Werkst.* **2008**, *39*, 665–670. [CrossRef]

45. Chen, R.; Lu, Y.; Witherell, P.; Simpson, T.W.; Kumara, S.; Yang, H. Ontology-Driven Learning of Bayesian Network for Causal Inference and Quality Assurance in Additive Manufacturing. *IEEE Robot. Autom. Lett.* **2021**, *6*, 6032–6038. [CrossRef]

46. Chen, R.; Rao, P.; Lu, Y.; Reutzel, E.W.; Yang, H. Recurrence network analysis of design-quality interactions in additive manufacturing. *Addit. Manuf.* **2021**, *39*, 101861. [CrossRef] [PubMed]

47. Liu, X.; Kan, C.; Ye, Z.; Liu, B. In-process multiscale performance evaluation of FDM-based honeycomb structures with geometric defects. In Proceedings of the AIAA SCITECH Forum, San Diego, CA, USA, 3–7 January 2022; p. 1425.

48. Liu, X.; Kan, C.; Ye, Z. Real-time multiscale prediction of structural performance in material extrusion additive manufacturing. *Addit. Manuf.* **2021**, *49*, 102503. [CrossRef]

49. Xiao, X.; Roh, B.M.; Hamilton, C. Porosity management and control in powder bed fusion process through pro-cess-quality interactions. *CIRP J. Manuf. Sci. Technol.* **2022**, *38*, 120–128. [CrossRef]

50. Xiao, X.; Waddell, C.; Hamilton, C.; Xiao, H. Quality Prediction and Control in Wire Arc Additive Manufacturing via Novel Machine Learning Framework. *Micromachines* **2022**, *13*, 137. [CrossRef] [PubMed]

51. Roh, B.-M.; Kumara, S.R.T.; Yang, H.; Simpson, T.W.; Witherell, P.; Jones, A.T.; Lu, Y. Ontology Network-Based In-Situ Sensor Selection for Quality Management in Metal Additive Manufacturing. *J. Comput. Inf. Sci. Eng.* **2022**, *22*, 060905. [CrossRef]
52. Zhang, Z.; Wang, S.; Liu, H.; Wang, L.; Xiao, X. Effects of Hatch Distance on the Microstructure and Mechanical Anisotropy of 316 L Stainless Steel Fabricated by Laser Powder Bed Fusion. *J. Mater. Eng. Perform.* **2022**, 1–11. [CrossRef]
53. Roh, B.M.; Yang, H.; Simpson, T.W.; Jones, A.T.; Witherell, P. A Hierarchical V-Network Framework for Part Qualification in Metal Additive Manufacturing. In Proceedings of the 2022 International Solid Freeform Fabrication Symposium, University of Texas at Austin, Austin, TX, USA, 25–27 July 2022.
54. Standard Test Methods for Tension Testing of Metallic Materials. ASTM International—Standards Worldwide. (n.d.). Available online: https://www.astm.org/e0008_e0008m-22.html (accessed on 13 October 2022).

Article

Surface Morphology, Compressive Behavior, and Energy Absorption of Graded Triply Periodic Minimal Surface 316L Steel Cellular Structures Fabricated by Laser Powder Bed Fusion

Bharath Bhushan Ravichander *,†, Shweta Hanmant Jagdale † and Golden Kumar

Mechanical Engineering, University of Texas at Dallas, Richardson, TX 75080, USA
* Correspondence: bharathbhushan.ravichander@utdallas.edu
† These authors contributed equally to this work.

Abstract: Laser powder bed fusion (LPBF) is an emerging technique for the fabrication of triply periodic minimal surface (TPMS) structures in metals. In this work, different TPMS structures such as Diamond, Gyroid, Primitive, Neovius, and Fisher–Koch S with graded relative densities are fabricated from 316L steel using LPBF. The graded TPMS samples are subjected to sandblasting to improve the surface finish before mechanical testing. Quasi-static compression tests are performed to study the deformation behavior and energy absorption capacity of TPMS structures. The results reveal superior stiffness and energy absorption capabilities for the graded TPMS samples compared to the uniform TPMS structures. The Fisher–Koch S and Primitive samples show higher strength whereas the Fisher–Koch S and Neovius samples exhibit higher elastic modulus. The Neovius type structure shows the highest energy absorption up to 50% strain among all the TPMS structures. The Gibson–Ashby coefficients are calculated for the TPMS structures, and it is found that the C_2 values are in the range suggested by Gibson and Ashby while C_1 values differ from the proposed range.

Keywords: laser powder bed fusion; 316L steel; porous structure; triply periodic minimal surface; deformation behavior; energy absorption; surface morphology

Citation: Ravichander, B.B.; Jagdale, S.H.; Kumar, G. Surface Morphology, Compressive Behavior, and Energy Absorption of Graded Triply Periodic Minimal Surface 316L Steel Cellular Structures Fabricated by Laser Powder Bed Fusion. *Materials* **2022**, *15*, 8294. https://doi.org/10.3390/ma15238294

Academic Editors: Bartłomiej Wysocki, Joseph Buhagiar and Tomasz Durejko

Received: 25 September 2022
Accepted: 20 November 2022
Published: 22 November 2022

Publisher's Note: MDPI stays neutral with regard to jurisdictional claims in published maps and institutional affiliations.

1. Introduction

The advances in computer-aided design (CAD) and additive manufacturing (AM), have led to the realization of complex geometries in metallic materials [1,2]. The new design concepts, such as topology optimization, can enable the reduction in weight without compromising the performance of a metal part. The combination of AM techniques and topology optimization has resulted in fabrication of light weight triply periodic minimal surface (TPMS) structures with high energy absorption efficiency, better vibration reduction, and good biocompatibility [3,4]. The TPMS structures are used as shock absorbers in spacecrafts and high-speed trains due to their high energy absorption capabilities [5,6]. LPBF is one of the most widely used metal AM techniques to fabricate complex metal TPMS structures with high accuracy [7–11]. The LPBF is a layer-by-layer manufacturing process in which the laser beam selectively melts and joins the metal powder layers until the desired part is fabricated [11–13].

The TPMS structures are classified as bending- or stretching-dominated. The bending-dominated TPMS structures like Tetrakaidecahedron and BCC display longer stress plateau regions, whereas the stretching-dominated TPMS structures like octet-truss soften after yielding [14,15]. The desired yield strength and Young's modulus can be obtained by optimizing the size and the geometry of the unit cells in metal TPMS structures. The periodic unit cells in TPMS structures are typically defined by using mathematical equations. These equations allow controlling the pore size, the zero-mean curvature, and the surface area in TPMS unit cells. Diamond, Gyroid, and Schwartz are the commonly used unit cell types in metal TPMS structures. Numerous studies have been conducted to understand the correlation between the unit cell characteristics and the mechanical properties of metal TPMS structures [16–18]. Gibson et al. established a model to determine the effect of relative density on the mechanical properties of the TPMS structures [14]. Al-Saedi et al. conducted a similar study and determined the Gibson–Ashby co-efficients for functionally graded F2BCC lattice structure with the help of experiments and calculations [19]. The TPMS structures with smooth and continuous curves are known for their high-strength and energy absorption capabilities [20,21]. Likewise, Al-Ketan et al. concluded that the sheet-based TPMS structures tend to show a higher energy absorption and better load bearing capacity compared to solid TPMS structures [22].

316L steel has been extensively studied using LPBF because of its widespread use in structural and functional applications [23]. Bonatti et al. compared the shell-based TPMS structures with that of truss lattices in 316L steel and found that the TPMS structures showcased superior mechanical properties compared to the truss lattices with the same density [24]. Yang et al. compared the compressive behavior of SLM-manufactured 316L steel uniform and graded Gyroid lattices. It was established that the graded lattices had improved mechanical properties compared to the uniform Gyroid lattices [25]. In a similar study conducted by Yang et al., the effect of sandblasting on the compressive and fatigue behavior of uniform 316L steel Gyroid lattice, it was found that the overall mechanical behavior of the Gyroid sample improved after sandblasting [26]. The surface morphology, mechanical response, and microstructure of uniform TPMS structures have been extensively studied [27–29].

The relationship between mechanical properties and grading approaches for functionally graded TPMS samples have been studied [25,30–33]. The mechanical properties of functionally graded Diamond sample was investigated by Han et al. [33]. They found that the functionally graded nature of the diamond sample led to layer-by-layer failure mechanism. This is different from the uniform Diamond structure where the sample failed due to the formation of diagonal shear bands. In a study conducted by Liu et al. [31], the effect of changing cell type, cell size, and relative density was investigated. Acceleration of biodegradation of functionally graded TPMS structures were reported by Li et al. [32]. They found that the topology of TPMS lattices improved the degradation of metals. However, the functionally graded TPMS structures are underexplored especially in structures like Fisher–Koch S and Neovius compared to Gyroid, Diamond, and Primitive structures.

The aim of this study is to analyze the surface morphology, the mechanical properties, and the energy absorption of functionally graded TPMS structures such as Gyroid, Diamond, Primitive, Neovius and Fisher–Koch S fabricated from 316L steel using LPBF. The relative density of the structures was designed to vary from 30% to 70%. The samples were sandblasted to improve the surface finish and compressive tests were performed to measure the mechanical response and energy absorption capability. Further, Gibson–Ashby model was implemented to determine the pre-factor coefficients C_1 and C_2.

2. Fabrication and Experimental Procedure

2.1. Design of TPMS Structures

As shown in Figure 1, five TPMS structures (i.e., Diamond, Gyroid, Primitive, Neovius, and Fisher–Koch S) were designed by using an open source TPMS generator MS Lattice, [34]. The size of the samples was maintained at 8 mm × 8 mm × 8 mm with each unit cell being 2 mm × 2 mm × 2 mm. The relative density of the samples varied from 30% to 70% along the x-direction as shown in Figure 1. The TPMS designs were converted into 3D stereo-lithography (STL) files, which were imported into Solidworks. Two end plates, each with the thickness of 1 mm, were added at the top and the bottom of the structures to support the compressive loading. The files were then imported to Materialise Magics (© Copyright Materialise 2021, Leuven, Belgium) to assign the supports and the corresponding process parameters for 3D printing. Equations (1)–(5) represent the nodal approximations of the TPMS structures, and their porosity is governed by the constant K as defined by Al-Ketan et al. [34]:

$$\text{Diamond: } f\,(x,y,z) = sin(x) \times sin(y) \times sin(z) + sin(x) \times cos(y) \times cos(z) + cos(x) \times sin(y) \times cos(z) \\ + cos(x) \times cos(y) \times sin(z) - K \tag{1}$$

$$\text{Gyroid: } f\,(x,y,z) = sin(x) \times cos(y) + sin(y) \times cos(z) + sin(z) \times cos(x) - K \tag{2}$$

$$\text{Primitive: } f\,(x,y,z) = cos(x) + cos(y) + cos(z) - K \tag{3}$$

$$\text{Neovius: } f\,(x,y,z) = 3 \times (cos(x) + cos(y) + cos(z)) + 4 \times cos(x) \times cos(y) \times cos(z) - K \tag{4}$$

$$\text{Fisher-Koch S: } f\,(x,y,z) = cos(2x) \times sin(y) \times cos(z) + cos(2y) \times sin(z) \times cos(x) + cos(2z) \times sin(x) \times cos(y) - K \tag{5}$$

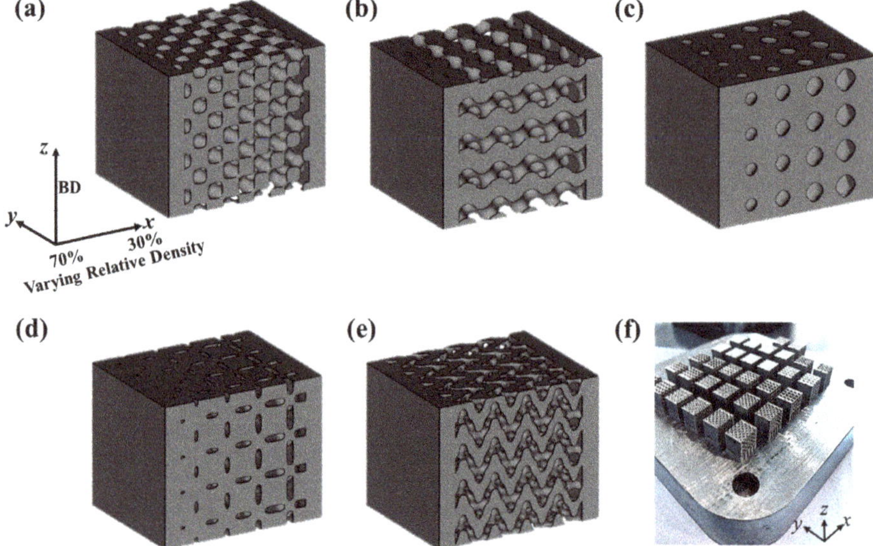

Figure 1. Designed TPMS structures with relative density varying from 30% to 70% and build direction (BD) indicated by the arrow and types: (**a**) Diamond; (**b**) Gyroid; (**c**) Primitive; (**d**) Neovius; (**e**) Fisher–Koch; (**f**) As-built samples on the build plate.

2.2. Powder Preparation and Fabrication

The TPMS structures shown in Figure 1 were fabricated from 316L steel using LPBF printer SLM 125 HL (SLM Solutions Group AG, Lübeck, Germany). The printer is equipped with a 400 W Ytterbium fiber laser and has a building volume of 125 mm × 125 mm × 125 mm. A total of three samples for each TPMS design were fabricated as shown in Figure 1f. The scanning electron microscopy (SEM) micrograph of the gas-atomized 316L steel powder obtained from SLM solutions AG is shown in Figure 2a. Image J [35] software was used to determine the particle size of the fresh powder and a histogram of the particle size distribution was obtained as seen in Figure 2b. The average powder particle size was found to be 30 μm. The composition of the fresh 316L steel powder is listed in Table 1.

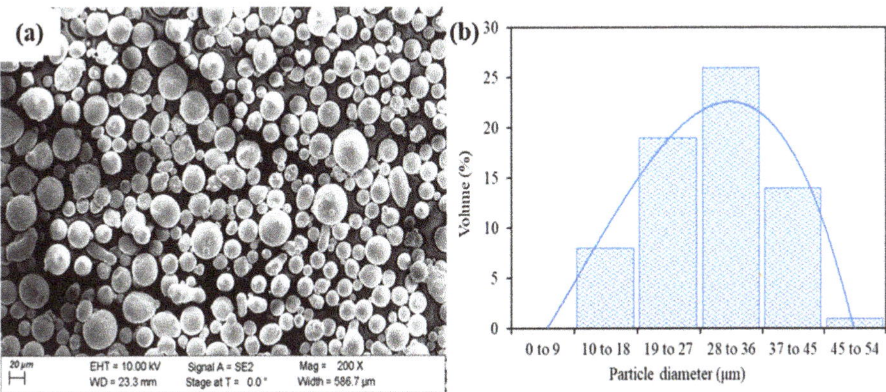

Figure 2. (**a**) SEM micrograph of powder; (**b**) particle size distribution of commercial SLM solutions 316L powder.

Table 1. The chemical composition of 316L steel powder obtained from SLM solutions.

Element	Fe	Cr	Ni	Mo	Nb + Ta	Mn	Si	P	S	C	N	O
Mass fraction (%)	Bal	16–18	10–14	2–3	-	2	1	0.045	0.03	0.03	0.1	-

The laser power (LP = 200 W), scan speed (SS = 800 mm/s), hatch spacing (HS = 120 μm), layer thickness (LT = 30 μm), and the stripes scan strategy were used for the fabrication of 316L steel TPMS samples. The above-mentioned process parameters were suggested by the manufacturer SLM solutions Group AG. The energy density (E_v) used during the fabrication process is 69.45 J/mm^3, which is calculated as [10,13,36]:

$$E_v = \frac{LP}{SS \times HS \times LT} \qquad (6)$$

2.3. Experimental Procedure

A wire electrical discharge machine was used to remove the as-built TPMS samples from the steel substrate. The samples were ultrasonically cleaned in iso-propyl alcohol. Subsequently, the samples were sandblasted using a Shop Fox M1114 benchtop Sandblaster and 100 grit aluminum oxide sand with a pressure of 60 psi for 30 s on each side to remove the unmelted powder particles uniformly from all sides. Surface morphology was analyzed using a Keyence VHX-970FN digital microscope.

Density measurements were performed using the Archimedes method as described by Ma et al. [27]. A Veritas Precision Balance with a sensitivity of 0.0001 g was used for the density measurements. The room temperature quasi-static compression tests were conducted using an Instron 5969 universal testing machine equipped with a 50 kN load cell. A constant strain rate of 10^{-3} sec^{-1} was applied along the x-direction (30% dense side) for a maximum displacement maintained at 50% for each sample. Using the load vs. extension data from the compression test, the stress vs. strain curves were obtained. The stress values were calculated by dividing the force by the cross-sectional area of the sample and the strain was determined by dividing the displacement by the sample height. A total of 3 samples were tested for each TPMS design and the average values for yield strength and Young's modulus are reported.

3. Results and Discussion

3.1. Surface Morphology and Ensity Analysis

The surface finish and the density of the TPMS samples significantly affect their mechanical properties [37]. Thus, it is important to ensure that the samples are fabricated without any large defects. The optical micrographs of the as-built TPMS samples are shown in Figure 3a–e. In general, no macroscale defects were observed in all the samples. However, rough surfaces were found along the struts due to the presence of unmelted powder particles on the side and the overhang surfaces. The unmelted powder particles form during the laser scanning which leads to the partial melting of adjacent powder particles on the side and the overhang surfaces along the designed path (highlighted with red arrows) as shown in Figure 3a–e.

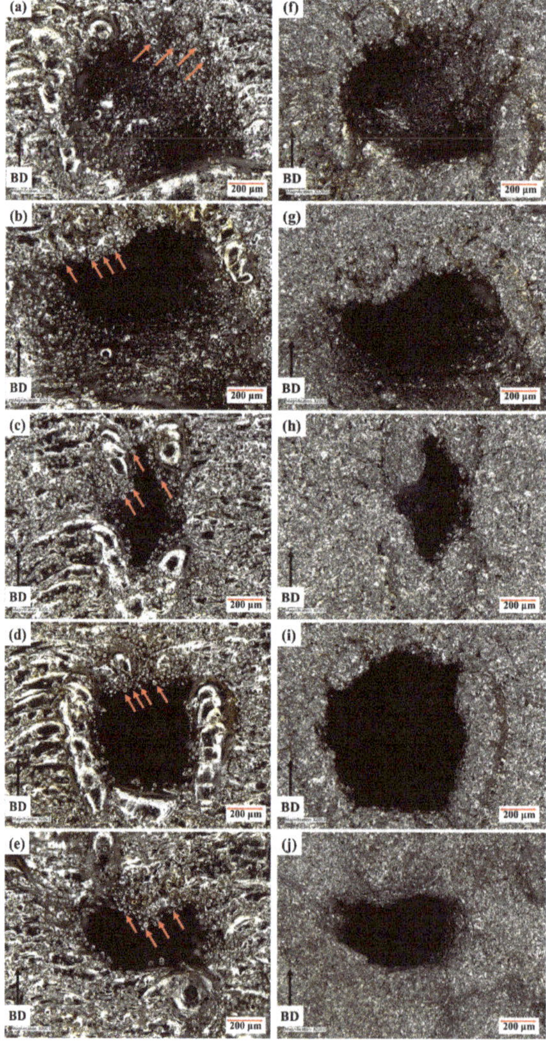

Figure 3. Optical micrographs along the build direction showing the unmelted powder particles highlighted with arrows attached to the surface for the as-built samples (**a**) Diamond, (**b**) Gyroid,

(c) Neovius, (**d**) Primitive, and (**e**) Fisher–Koch S and the corresponding surfaces after sandblasting for (**f**) Diamond, (**g**) Gyroid, (**h**) Neovius, (**i**) Primitive, and (**j**) Fisher–Koch S samples.

The samples were sandblasted to remove the unmelted powder particles and improve the surface finish of the TPMS samples as the presence of unmelted powder particles on LPBF samples can be detrimental to the mechanical behavior [26]. From Figure 3f–j, we can observe that the sandblasting effectively removed the unmelted powder particles. The surfaces of sandblasted TPMS samples at different regions are shown in Figure 4. The optical micrographs confirm that there are no macroscale defects such as cracks, pores, and deformation, thus indicating the good manufacturability of the TPMS structures using LPBF technique. One can also notice a gradual change in the dimensions of the struts as the part density changes from 70% porous at the top to 30% porous at the bottom.

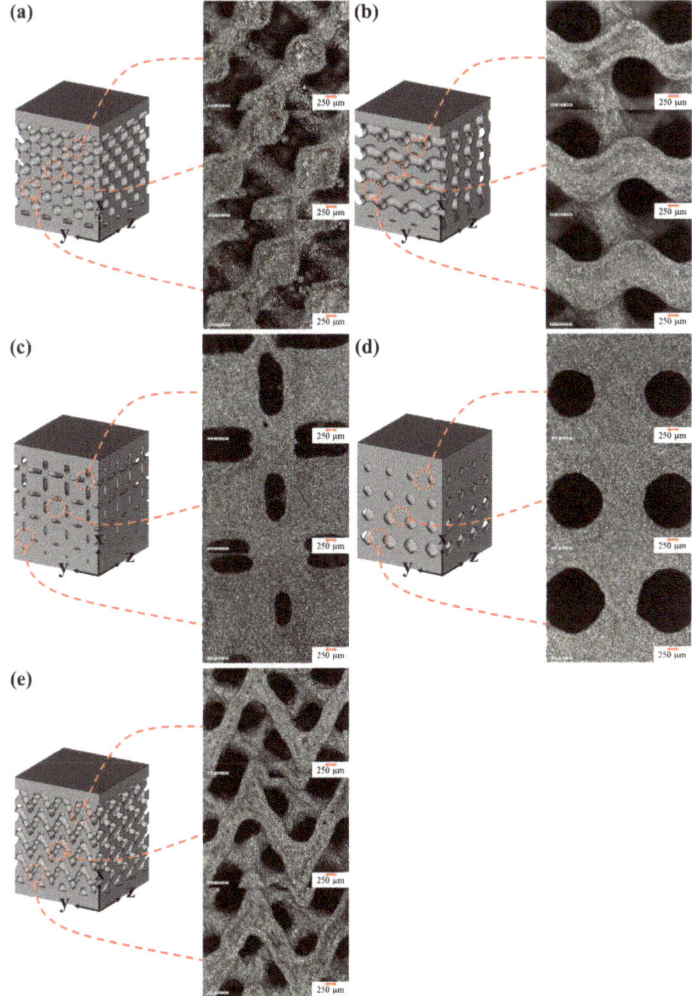

Figure 4. Surface morphologies of LPBF (**a**) Diamond, (**b**) Gyroid, (**c**) Neovius, (**d**) Primitive, and (**e**) Fisher–Koch S type TPMS samples after sandblasting.

The TPMS samples were cleaned In an ultrasonic bath after sandblasting and a precision balance was used to measure the mass of the samples in air and water. The part densities of the samples were calculated which are presented in Table 2. The achievement of near fully dense samples is usually expected in solid parts fabricated by the LPBF process due to layer-by-layer melting. Due to the functionally graded nature of the TPMS specimens and the small unit cell size of 2 mm, it was noticed that the part density values were lower, which is consistent with the literature [38]. The lower part density values can be due to shrinkage, lack of fusion between powder particles, and unmelted powder particles in the more porous (70% porous side) region consisting of larger overhang sections and thermal stresses as it was fabricated without support structures [16,39]. We observed that at more porous regions, there are some discontinuities in the struts of the samples leading to weak intersections as the features are closer to the threshold of the laser spot size. The melt pool size in LPBF technique is larger than the spot size of the laser. Thus, the scan contour tracks compensate for this by shifting inwards as reported by [40]. The contour tracks also partially melt the powder particles adjacent to the walls and therefore cause more powder particles to adhere to the surface. These process-related factors lead to marginal increase or decrease in the thickness of the struts which further justifies the small deviations in the part density of the samples as also noticed by Al-Ketan et al. [41].

Table 2. The part densities of the sandblasted TPMS samples.

Sample	Diamond	Gyroid	Neovius	Primitive	Fisher–Koch S
Part Density (%)	98.87 ± 0.12	99.71 ± 0.23	98.82 ± 0.13	98.99 ± 0.13	98.03 ± 0.12

3.2. Quasi-Static Compression Analysis

The compressive stress–strain curves of the TPMS samples are shown in Figure 5a (up to 50% strain) and Figure 5b (up to 10% strain). The tests were performed with the loading direction perpendicular to the build direction (see Figure 1). The Young's modulus was obtained by calculating the slope of the linear elastic region and the compressive yield strength was determined by using 0.2% offset approach [39]. The stress–strain curves indicate that after initial elastic region, the structures deform plastically and continue to absorb energy and it is followed by the densification stage, where the samples behave like a solid material as there is a large self-contact area. A similar deformation behavior was observed by Li et al. for 316L steel Gyroid structures [16]. Figure 6 compares the Young's modulus and yield strength values for the 5 TPMS structures. The Fisher–Koch S specimen shows the highest Young's modulus value of 6.96 GPa followed by the Neovius (6.74 GPa) and Gyroid (4.46 GPa) samples. The Diamond and the Primitive samples exhibit lower Young's modulus of 3.36 GPa and 2.28 GPa, respectively. The Fisher–Koch S sample has the highest yield strength of 129 MPa. The Primitive sample yields at 95.9 MPa. The Diamond and Neovius type samples have similar yield strength values in the range of 81–83 MPa and the Gyroid sample exhibits the lowest yield strength of 71.3 MPa.

The section views of the designed unit cells along with the loading direction are shown in Figure 5c. The stress–strain curve in Figure 5b indicated that the Gyroid sample has the lowest yield strength compared to the other samples and this can be attributed to the lack of structural continuity to the upward facing surface (highlighted in red). As the load is applied, the highlighted surface is not able to transfer the load to the layers below it and thus yields at a strain value of 1.7%. A similar trend can be observed for Diamond (2.6% strain) and the Neovius (1.4% strain) samples, where there is a lack of continuity between the struts and the main portion of the respective designs. The Primitive and the Fisher–Koch S samples have better structural continuity between the upward-facing surfaces and the rest of the main portion of the TPMS design. Thus, the Fisher–Koch S and Primitive TPMS designs have higher yield strength values as the load is transferred across the samples more uniformly compared to the other TPMS samples. The Young's modulus of functionally graded materials with loading applied parallel to the grading direction can be predicted by applying the rule of mixtures such that [41]:

$$\frac{1}{E_{graded}} = \frac{1}{m} \sum_{i=1}^{m} \frac{1}{E_i} \tag{7}$$

Here, the cross-sectional area is assumed to be constant and unchanged throughout the loading direction, and the thickness of each layer is L/m where L is the total length of the specimen and m is the number of layers. This has been discussed in detail by Maskery et al. [42].

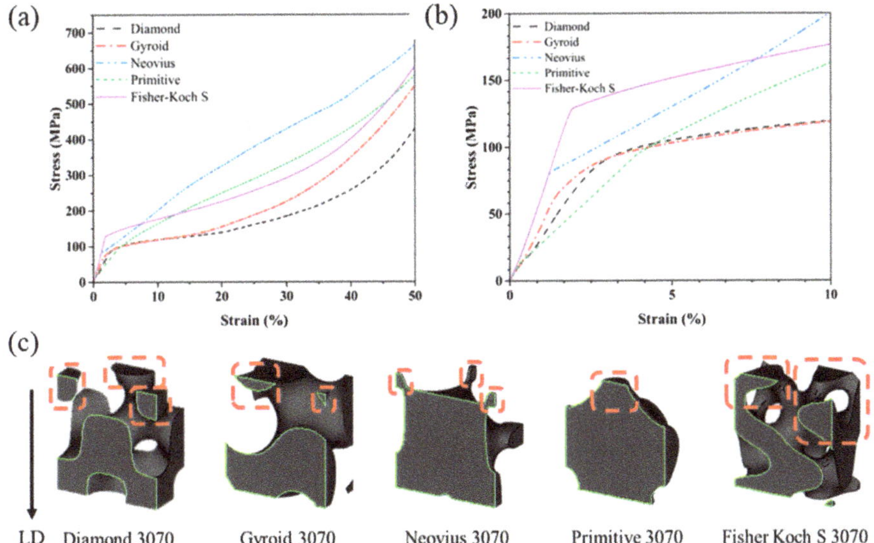

Figure 5. (**a**) Stress–strain curves of the different functionally graded lattice specimens from 0 to 50% strain. (**b**) Stress–strain curve of the fabricated samples with 0 to 10% strain. (**c**) Section views of the unit cells with loading direction (LD).

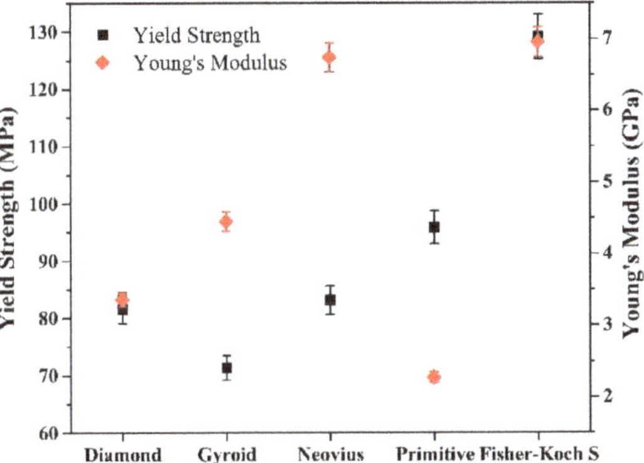

Figure 6. Young's Modulus and yield strength of the different functionally graded lattice samples.

The deformation of samples at different strains (ε) was captured by recording the images as shown in Figure 7. During the first 15% strain, the samples expand laterally by forming a radius along the edge of the TPMS cubes as shown in Figure 7 (highlighted with dotted lines for the 15% strain images). The plastic deformation starts at about 1.1% and 1.2% strain for the Gyroid and Neovius samples, respectively. Whereas the plastic region for the Diamond and the Fisher–Koch S samples begins at 1.3% and 2% strains. The Primitive sample has the largest elastic deformation range up to 4% strain. Until 50% strain, no sudden reduction in stress is observed indicating there is no brittle fracture during compression. The samples did not deform with a diagonal shear, but rather deformed layer-by-layer as observed in Figure 7. Similar findings were reported by Yang et al. and Han et al. [25,33].

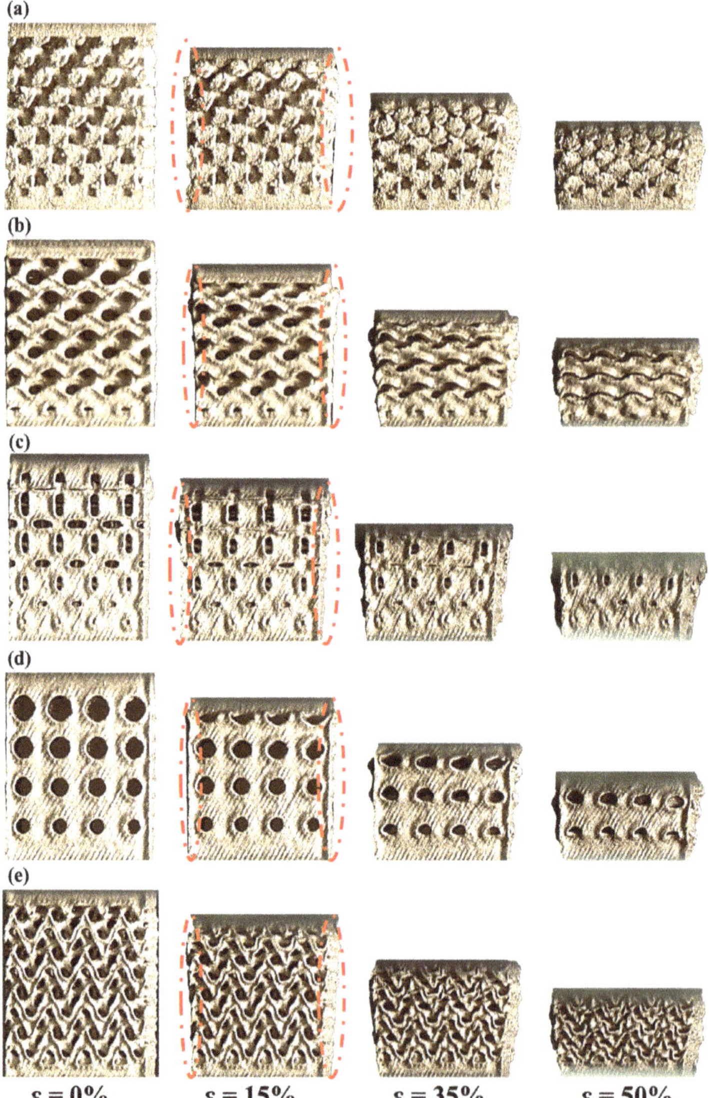

Figure 7. Images of (**a**) Diamond, (**b**) Gyroid, (**c**) Neovius, (**d**) Primitive, and (**e**) Fisher–Koch S TPMS structure recorded at different strains.

The Young's modulus and yield strength for different relative densities of metallic foams can be predicted using the Gibson–Ashby model [14]. The formulae of the Gibson–Ashby model are as follows:

$$\frac{E}{E_o} = C_1 \left(\frac{\rho}{\rho_o} \right)^2 \tag{8}$$

$$\frac{\sigma}{\sigma_o} = C_2 \left(\frac{\rho}{\rho_o} \right)^{1.5} \tag{9}$$

where E, ρ, and σ are the apparent compressive modulus, the density, and the compressive yield strength of open-cellular structures, respectively. E_0, ρ_0, and σ_0 are the respective values for the fully dense materials and C_1 and C_2 are Gibson–Ashby coefficients. Bulk compressive yield strength and elastic modulus of 316L steel alloy are taken to be 205 MPa and 193 GPa, respectively [2]. The E, ρ, and σ for the TPMS structures were determined from the experimental stress–strain measurements. The determined values for C_1 and C_2 for every sample are listed in Table 3. The obtained C_1 and C_2 values can be used in further studies for the TPMS structures with porosity levels varying from 30% to 70% without mechanical testing. The values for C_1 and C_2 depend on the material and the lattice type. From Table 3, we can observe that the C_1 values are out of the range given by Gibson and Ashby. This can be attributed to the differences in the parameters of the LPBF process as also noted by Zhao et al. and Yan et al. [21,43]. It is well known that the process parameters of the LPBF process play a significant role in determining the properties of the as-built specimens as discussed extensively by Ravichander et al. [8,12]. Additionally, the range of values defined by Gibson and Ashby was established for the mechanical properties of metallic foams. The C_2 values agree with the range given by Gibson and Ashby.

Table 3. Gibson–Ashby coefficient values for different TPMS lattices.

Coefficients	Given Range of Gibson–Ashby Coefficients	Diamond	Gyroid	Primitive	Neovius	Fisher–Koch S
C_1	0.1–4	0.0396	0.0567	0.0285	0.0753	0.0717
C_2	0.1–1	0.7369	0.6830	0.9053	0.7210	1.0554

3.3. Energy Absorption

The lattice structures are known for their low density and high energy absorption capabilities. Thus, they can be used in protective devices, implants, and in aerospace components. The cumulative energy absorption per unit volume (W_v) is widely utilized to determine the energy absorption capability of the lattice structures. The energy absorption per unit volume up to 50% strain was calculated using the following equation:

$$W_v = \int_0^{0.5} \sigma(\varepsilon)d\varepsilon \tag{10}$$

where ε is the strain, and $\sigma(\varepsilon)$ is the stress related to ε during the compression test. According to the ISO13314 standard [44], compressive stress–strain curves were integrated for the energy absorption characterization. Origin was used to integrate the area under the stress–strain curves. The cumulative energy absorption–strain curves are shown in Figure 8a, and the total energy absorbed values are presented in Table 4. It can be observed from Figure 8a that the cumulative energy absorption increases steadily, which is then followed by an exponential increase due to the increase of density as well as the structural stiffness after the layers collapse. From Table 4, we can note that the Neovius (479.62 MJ/m^3) sample absorbed the most energy followed by the Primitive (384.95 MJ/m^3) and Fisher–Koch S (373.08 MJ/m^3) samples. The Gyroid (296.67 MJ/m^3) and the Diamond (242.90 MJ/m^3) samples absorbed the second lowest and lowest energy per unit volume, respectively. From the sectional view of the Neovius design in Figure 4c, we can see that the highlighted portion of the struts collapse leading to the densification of the sample, and it is further supported from Figure 5a, where we can notice that the Neovius sample has a small plateau stress region and larger densification region compared to the other samples. A similar trend can be observed for the Fisher–Koch S and Primitive samples. On the contrary, the Diamond and the Gyroid TPMS samples have a larger plateau stress region and start densifying around the 15% strain mark. The energy absorption capabilities depend significantly on the TPMS structure and its deformation behavior than the density [4]. This can be observed in the initial energy absorption–strain curves presented in Figure 8b as the Fisher–Koch S sample absorbs more energy at the beginning and then the Neovius sample absorbs more energy from the 12.5% strain value. From Figure 8a, it can be observed that the W_v values gradually increase for all TPMS samples during compression, and it is due to the layer-by-layer deformation behavior with improved load-bearing capabilities to absorb more energy during the compression process.

Table 4. Energy absorbed values for different TPMS lattices.

	Diamond	Gyroid	Primitive	Neovius	Fisher–Koch S
Energy Absorbed (MJ/m^3)	242.90 ± 6.1	296.67 ± 3.9	384.95 ± 7.1	479.62 ± 9.8	373.08 ± 6.4

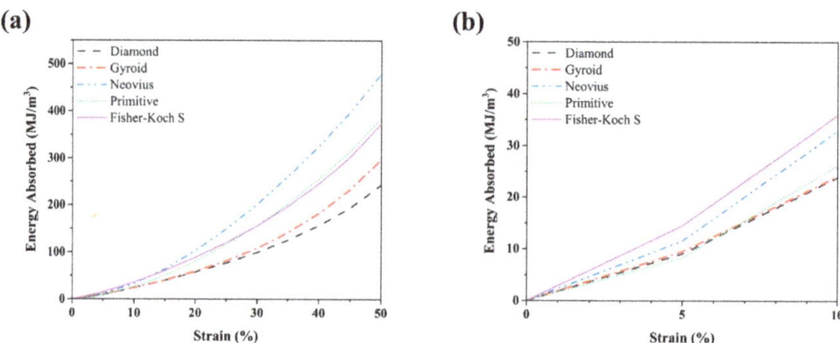

Figure 8. (**a**) Energy absorption–strain curves for the different functionally graded lattice specimens from 0 to 50% strain. (**b**) Energy absorption–strain curves of the fabricated samples with 0 to 10% strain.

The exponential increase in the energy absorption for functionally graded samples differs from the linear increase of the energy absorption values observed for uniformly graded samples as reported by Sun et al. [4]. This change is due to the functionally graded nature of the samples. The functionally graded samples thus provide a more predictable energy absorption profile leading to more applications where undesirable failure modes in lattices structures can be eliminated and the benefit of the high specific energy absorption can be retained.

4. Conclusions

In this work, different functionally graded 316L steel lattice structures were designed and fabricated using LPBF process. The fabricated samples were then sandblasted to improve the surface finish of the samples. Surface morphology, quasi-static compression, and energy absorption capabilities were studied. The key findings of this study are:

- Sandblasting was found to be a viable post processing technique to remove the adhered powder particles and thus improve the surface finish.

- Quasi-static compression analysis reveals that the all TPMS samples fail layer-by-layer deformation. The samples expanded laterally by forming a radius along the edges.
- The Fisher–Koch S sample showed the highest yield strength and Young's modulus. The Gyroid and the Primitive structures resulted in the lowest value for yield strength and Young's modulus.
- The Gibson–Ashby model was determined for all the functionally graded TPMS structures based on the experimental results.
- The Neovius structure was found to absorb the most energy (479.62 MJ/m^3) compared to the other TPMS structures up to 50% strain values.

The compressive and energy absorption behaviors will be beneficial for engineers to design and fabricate functionally graded 316L steel TPMS structures. The results indicate that the underexplored functionally graded Neovius and Fisher–Koch S offer greater energy absorption and yield strength values for impact and light weighting applications.

Author Contributions: G.K.: Conceptualization, Supervision, Writing—review and editing; B.B.R. and S.H.J.: Methodology, Analysis, Writing—original draft. All authors have read and agreed to the published version of the manuscript.

Funding: This research was funded by a University of Texas System STARs award.

Conflicts of Interest: The authors declare no conflict of interest.

References

1. Srihari, S.; Bharath, B.R.; Narges Shayesteh, M.; Nahid, S.; Amirhesam, A. Investigation of the strength of different porous lattice structures manufactured using selective laser melting. In Proceedings of the SPIE Smart Structures + Nondestructive Evaluation, Long Beach, CA, USA, 22 April 2020.
2. Maconachie, T.; Leary, M.; Lozanovski, B.; Zhang, X.; Qian, M.; Faruque, O.; Brandt, M. SLM lattice structures: Properties, performance, applications and challenges. *Mater. Des.* **2019**, *183*, 108137. [CrossRef]
3. Wang, Y.; Ramirez, B.; Carpenter, K.; Naify, C.; Hofmann, D.C.; Daraio, C. Architected lattices with adaptive energy absorption. *Extreme Mech. Lett.* **2019**, *33*, 100557. [CrossRef]
4. Sun, Q.; Sun, J.; Guo, K.; Wang, L. Compressive mechanical properties and energy absorption characteristics of SLM fabricated Ti6Al4V triply periodic minimal surface cellular structures. *Mech. Mater.* **2022**, *166*, 104241. [CrossRef]
5. Sychov, M.; Lebedev, L.; Dyachenko, S.; Nefedova, L. Mechanical properties of energy-absorbing structures with triply periodic minimal surface topology. *Acta Astronaut.* **2018**, *150*, 81–84. [CrossRef]
6. Marsolek, J.; Reimerdes, H.-G. Energy absorption of metallic cylindrical shells with induced non-axisymmetric folding patterns. *Int. J. Impact Eng.* **2004**, *30*, 1209–1223. [CrossRef]
7. Farhang, B.; Ravichander, B.B.; Venturi, F.; Amerinatanzi, A.; Moghaddam, N.S. Study on variations of microstructure and metallurgical properties in various heat-affected zones of SLM fabricated Nickel–Titanium alloy. *Mater. Sci. Eng. A* **2020**, *774*, 138919. [CrossRef]
8. Ravichander, B.B.; Amerinatanzi, A.; Moghaddam, N.S. Study on the Effect of Powder-Bed Fusion Process Parameters on the Quality of as-Built IN718 Parts Using Response Surface Methodology. *Metals* **2020**, *10*, 1180. [CrossRef]
9. Farhang, B.; Ravichander, B.B.; Ma, J.; Amerinatanzi, A.; Moghaddam, N.S. The evolution of microstructure and composition homogeneity induced by borders in laser powder bed fused Inconel 718 parts. *J. Alloys Compd.* **2022**, *898*, 162787. [CrossRef]
10. Ravichander, B.B.; Mamidi, K.; Rajendran, V.; Farhang, B.; Ganesh-Ram, A.; Hanumantha, M.; Moghaddam, N.S.; Amerinatanzi, A. Experimental investigation of laser scan strategy on the microstructure and properties of Inconel 718 parts fabricated by laser powder bed fusion. *Mater. Charact.* **2022**, *186*, 111765. [CrossRef]
11. Ravichander, B.B.; Thakare, S.; Ganesh-Ram, A.; Farhang, B.; Hanumantha, M.; Yang, Y.; Moghaddam, N.S.; Amerinatanzi, A. Cost-Aware Design and Fabrication of New Support Structures in Laser Powder Bed Fusion: Microstructure and Metallurgical Properties. *Appl. Sci.* **2021**, *11*, 10127. [CrossRef]
12. Bharath Bhushan, R.; Behzad, F.; Nahid, S.; Amirhesam, A.; Narges Shayesteh, M. Analysis of the deviation in properties of selective laser melted samples fabricated by varying process parameters. In Proceedings of the SPIE Smart Structures + Nondestructive Evaluation, Long Beach, CA, USA, 22 April 2020.
13. Ravichander, B.B.; Rahimzadeh, A.; Farhang, B.; Moghaddam, N.S.; Amerinatanzi, A.; Mehrpouya, M. A Prediction Model for Additive Manufacturing of Inconel 718 Superalloy. *Appl. Sci.* **2021**, *11*, 8010. [CrossRef]
14. Ashby, M.F.; Gibson, L.J. *Cellular Solids: Structure and Properties*; Press Syndicate of the University of Cambridge: Cambridge, UK, 1997; pp. 175–231.
15. Ashby, M.F. The properties of foams and lattices. *Philos. Trans. R. Soc. A Math. Phys. Eng. Sci.* **2006**, *364*, 15–30. [CrossRef] [PubMed]

16. Li, X.; Xiao, L.; Song, W. Compressive behavior of selective laser melting printed Gyroid structures under dynamic loading. *Addit. Manuf.* **2021**, *46*, 102054. [CrossRef]
17. Li, P.; Ma, Y.E.; Sun, W.; Qian, X.; Zhang, W.; Wang, Z. Fracture and failure behavior of additive manufactured Ti6Al4V lattice structures under compressive load. *Eng. Fract. Mech.* **2021**, *244*, 107537. [CrossRef]
18. Yu, S.; Sun, J.; Bai, J. Investigation of functionally graded TPMS structures fabricated by additive manufacturing. *Mater. Des.* **2019**, *182*, 108021. [CrossRef]
19. Al-Saedi, D.S.J.; Masood, S.H.; Faizan-Ur-Rab, M.; Alomarah, A.; Ponnusamy, P. Mechanical properties and energy absorption capability of functionally graded F2BCC lattice fabricated by SLM. *Mater. Des.* **2018**, *144*, 32–44. [CrossRef]
20. Han, L.; Che, S. An Overview of Materials with Triply Periodic Minimal Surfaces and Related Geometry: From Biological Structures to Self-Assembled Systems. *Adv. Mater.* **2018**, *30*, e1705708. [CrossRef]
21. Zhao, M.; Liu, F.; Fu, G.; Zhang, D.Z.; Zhang, T.; Zhou, H. Improved Mechanical Properties and Energy Absorption of BCC Lattice Structures with Triply Periodic Minimal Surfaces Fabricated by SLM. *Materials* **2018**, *11*, 2411. [CrossRef]
22. Al-Ketan, O.; Lee, D.-W.; Abu Al-Rub, R.K. Mechanical properties of additively-manufactured sheet-based gyroidal stochastic cellular materials. *Addit. Manuf.* **2021**, *48*, 102418. [CrossRef]
23. Yusuf, S.M.; Chen, Y.; Yang, S.; Gao, N. Microstructural evolution and strengthening of selective laser melted 316L stainless steel processed by high-pressure torsion. *Mater. Charact.* **2020**, *159*, 110012. [CrossRef]
24. Bonatti, C.; Mohr, D. Mechanical performance of additively-manufactured anisotropic and isotropic smooth shell-lattice materials: Simulations & experiments. *J. Mech. Phys. Solids* **2019**, *122*, 1–26. [CrossRef]
25. Yang, L.; Mertens, R.; Ferrucci, M.; Yan, C.; Shi, Y.; Yang, S. Continuous graded Gyroid cellular structures fabricated by selective laser melting: Design, manufacturing and mechanical properties. *Mater. Des.* **2019**, *162*, 394–404. [CrossRef]
26. Yang, L.; Yan, C.; Cao, W.; Liu, Z.; Song, B.; Wen, S.; Zhang, C.; Shi, Y.; Yang, S. Compression–compression fatigue behaviour of gyroid-type triply periodic minimal surface porous structures fabricated by selective laser melting. *Acta Mater.* **2019**, *181*, 49–66. [CrossRef]
27. Ma, S.; Tang, Q.; Han, X.; Feng, Q.; Song, J.; Setchi, R.; Liu, Y.; Liu, Y.; Goulas, A.; Engstrøm, D.S.; et al. Manufacturability, Mechanical Properties, Mass-Transport Properties and Biocompatibility of Triply Periodic Minimal Surface (TPMS) Porous Scaffolds Fabricated by Selective Laser Melting. *Mater. Des.* **2020**, *195*, 109034. [CrossRef]
28. Sander, G.; Thomas, S.; Cruz, V.; Jurg, M.; Birbilis, N.; Gao, X.; Brameld, M.; Hutchinson, C.R. On The Corrosion and Metastable Pitting Characteristics of 316L Stainless Steel Produced by Selective Laser Melting. *J. Electrochem. Soc.* **2017**, *164*, C250–C257. [CrossRef]
29. Tancogne-Dejean, T.; Spierings, A.B.; Mohr, D. Additively-manufactured metallic micro-lattice materials for high specific energy absorption under static and dynamic loading. *Acta Mater.* **2016**, *116*, 14–28. [CrossRef]
30. Leong, K.F.; Chua, C.K.; Sudarmadji, N.; Yeong, W.Y. Engineering functionally graded tissue engineering scaffolds. *J. Mech. Behav. Biomed. Mater.* **2008**, *1*, 140–152. [CrossRef]
31. Liu, F.; Mao, Z.; Zhang, P.; Zhang, D.Z.; Jiang, J.; Ma, Z. Functionally graded porous scaffolds in multiple patterns: New design method, physical and mechanical properties. *Mater. Des.* **2018**, *160*, 849–860. [CrossRef]
32. Li, Y.; Jahr, H.; Pavanram, P.; Bobbert, F.; Puggi, U.; Zhang, X.-Y.; Pouran, B.; Leeflang, M.; Weinans, H.; Zhou, J.; et al. Additively manufactured functionally graded biodegradable porous iron. *Acta Biomater.* **2019**, *96*, 646–661. [CrossRef]
33. Han, C.; Li, Y.; Wang, Q.; Wen, S.; Wei, Q.; Yan, C.; Hao, L.; Liu, J.; Shi, Y. Continuous functionally graded porous titanium scaffolds manufactured by selective laser melting for bone implants. *J. Mech. Behav. Biomed. Mater.* **2018**, *80*, 119–127. [CrossRef]
34. Al-Ketan, O.; Abu Al-Rub, R.K. MSLattice: A free software for generating uniform and graded lattices based on triply periodic minimal surfaces. *Mater. Des. Process. Commun.* **2021**, *3*, e205. [CrossRef]
35. Schneider, C.A.; Rasband, W.S.; Eliceiri, K.W. NIH Image to ImageJ: 25 Years of image analysis. *Nat. Methods* **2012**, *9*, 671–675. [CrossRef] [PubMed]
36. Ravichander, B.B.; Favela, C.; Amerinatanzi, A.; Moghaddam, N.S. *A Framework for the Optimization of Powder-Bed Fusion Process*; SPIE: Long Beach, CA, USA, 2021; Volume 11589.
37. Heinl, P.; Müller, L.; Körner, C.; Singer, R.F.; Müller, F.A. Cellular Ti–6Al–4V structures with interconnected macro porosity for bone implants fabricated by selective electron beam melting. *Acta Biomater.* **2008**, *4*, 1536–1544. [CrossRef]
38. Rajagopalan, S.; Robb, R. Schwarz meets Schwann: Design and fabrication of biomorphic and durataxic tissue engineering scaffolds. *Med. Image Anal.* **2006**, *10*, 693–712. [CrossRef]
39. Yánez, A.; Herrera, A.; Martel, O.; Monopoli, D.; Afonso, H. Compressive behaviour of gyroid lattice structures for human cancellous bone implant applications. *Mater. Sci. Eng. C* **2016**, *68*, 445–448. [CrossRef]
40. Van Bael, S.; Kerckhofs, G.; Moesen, M.; Pyka, G.; Schrooten, J.; Kruth, J. Micro-CT-based improvement of geometrical and mechanical controllability of selective laser melted Ti6Al4V porous structures. *Mater. Sci. Eng. A* **2011**, *528*, 7423–7431. [CrossRef]
41. Al-Ketan, O.; Lee, D.-W.; Rowshan, R.; Abu Al-Rub, R.K. Functionally graded and multi-morphology sheet TPMS lattices: Design, manufacturing, and mechanical properties. *J. Mech. Behav. Biomed. Mater.* **2020**, *102*, 103520. [CrossRef]
42. Maskery, I.; Aremu, A.; Parry, L.; Wildman, R.; Tuck, C.; Ashcroft, I. Effective design and simulation of surface-based lattice structures featuring volume fraction and cell type grading. *Mater. Des.* **2018**, *155*, 220–232. [CrossRef]

43. Yan, C.; Hao, L.; Yang, L.; Hussein, A.Y.; Young, P.G.; Li, Z.; Li, Y. (Eds.) Chapter 7-Functionally graded TPMS. In *Triply Periodic Minimal Surface Lattices Additively Manufactured by Selective Laser Melting*; Academic Press: Cambridge, MA, USA, 2021; pp. 219–281.
44. Zhang, M.; Yang, Y.; Wang, D.; Song, C.; Chen, J. Microstructure and mechanical properties of CuSn/18Ni300 bimetallic porous structures manufactured by selective laser melting. *Mater. Des.* **2019**, *165*, 107583. [CrossRef]

materials

Article

Design and Fabrication of an Additively Manufactured Aluminum Mirror with Compound Surfaces

Jizhen Zhang [1,2,3], Chao Wang [1,3,*], Hemeng Qu [1,3], Haijun Guan [1,3], Ha Wang [3], Xin Zhang [1,3], Xiaolin Xie [1,3], He Wang [3], Kai Zhang [1,2] and Lijun Li [3]

[1]   Changchun Institute of Optics, Fine Mechanics and Physics, Chinese Academy of Sciences, Changchun 130033, China
[2]   University of Chinese Academy of Sciences, Beijing 100049, China
[3]   Smart Optics Co., Ltd., Changchun 130102, China
*   Correspondence: wangchao05@ciomp.ac.cn; Tel.: +86-189-4675-7085

Abstract: Microsatellites have a great attraction to researchers due to their high reliability, resource utilization, low cost, and compact size. As the core component of the optical payload, the mirror directly affects the system package size. Therefore, the structural design of mirrors is critical in the compact internal space of microsatellites. This study proposes a closed-back mirror with composite surfaces based on additive manufacturing (AM). Compared with the open-back mirror, it provides excellent optomechanical performance. In addition, AM significantly reduces the intricate mechanical parts' manufacturing difficulty. Finally, the roughness was better than 2 nm. The surface shape of the AM aluminum mirror reached RMS $1/10\lambda$ (λ = 632.8 nm) with the aid of ultra-precision machining technologies such as single-point diamond turning (SPDT), surface modification, and polishing, and the maximum deviation of the surface shape was about RMS $1/42\lambda$ (λ = 632.8 nm) after the thermal cycle test, which verified the optical grade application of AM.

Keywords: additive manufacturing (AM); 3D printing; microsatellites; lightweight mirror; metal mirror

Citation: Zhang, J.; Wang, C.; Qu, H.; Guan, H.; Wang, H.; Zhang, X.; Xie, X.; Wang, H.; Zhang, K.; Li, L. Design and Fabrication of an Additively Manufactured Aluminum Mirror with Compound Surfaces. *Materials* **2022**, *15*, 7050. https://doi.org/10.3390/ma15207050

Academic Editor: Antonio Santagata

Received: 31 August 2022
Accepted: 21 September 2022
Published: 11 October 2022

Publisher's Note: MDPI stays neutral with regard to jurisdictional claims in published maps and institutional affiliations.

1. Introduction

The design concept for integrated and lightweight microsatellites was proposed to improve traditional satellites' reliability and resource utilization. According to the international general classification principle, microsatellites refer to satellites with a weight in the range of 10–100 kg [1]. Turan et al. reviewed the numerous advantages of microsatellites, such as short production cycles, low development, launch costs, high design function density, and mobility [2]., Moreover, optical microsatellites have been widely used in military and space science. Compared with the large aperture space optics, these integrated space-borne payloads have a greater application prospect in comprehensive low-cost earth observation and commercial development [3,4]. However, the internal space of the microsatellites is limited. It is necessary to carry out effective configuration and layout for utilizing the space fully, resulting in high integration for each piece of equipment. At the same time, the package size of the whole configuration is as tiny as possible to reduce the transportation cost, which generally requires the optical system to be compact.

Mirrors are the most critical elements in the optical system. Their aperture sizes and mounting methods directly affect the package size of the optical payload, and their optical and structural characteristics play a critical role in the performance of the entire system [5]. The excellent processing characteristics of metal materials make it easy to integrate optical and mechanical elements. Therefore, the advantages of metal mirrors can meet the requirements of microsatellites perfectly, and relying on modern ultra-precision machining technology, metal mirrors can be processed to ideal surface quality [6,7]. In the previous work by the authors, an off-axis aluminum (Al) mirror for visible-light imaging was designed and manufactured [8]. Actual optomechanical integrated manufacturing can be realized by optimizing the whole process chain of design, manufacturing, testing, and assembly. Xie et al. fabricated an all-Al imaging telescope, verifying that the free-form Al mirror benefited from a compact package size and a wide field of view for optical systems [9]. Additionally, the optical system could achieve the optimal athermal effect when the mirror and the mounting structures were made of the same materials. Liu et al. optimized and manufactured an Al mirror with an aperture of 160 mm [10]. According to the test results, the error budget showed that the mirror kept a fine surface shape at low temperatures. Therefore, Al mirrors have good prospects for low-temperature applications. According to Vukobratovich's calculation, the manufacturing cost of metal materials for the mirror is only about 50%, compared with ceramics, glass, and other materials [11]. Thus, the all-Al configuration is also a cost-effective product, with the advantages of athermalization and short lead time.

However, more and more deficiencies of the conventional lightweight mirrors have been highlighted in many cases to dates, such as the limitation of flexible structures, quilting deflection, and even the stress-induced surface deformation of the coating [12,13]. Especially in the compact optical systems of microsatellites, it is challenging for conventional mirrors to achieve the trade-off between lightweight and mechanical properties in such a limited space. The closed-back mirror can improve mechanical performance while maintaining the lightweight rate, but it is challenging to fabricate through conventional processing methods [14]. Fortunately, with the continuous advancement of metal additive manufacturing (AM) technology, the form of layer-by-layer printing dramatically reduces the difficulty of manufacturing intricate mechanical parts. Furthermore, AM metal parts have good fatigue properties [15]. Zhou et al. explored the fatigue properties of as-built $AlSi_{10}Mg$ parts with curved surfaces, which provides a good reference for the optical spheric and aspheric mirrors [16]. AM is a promising solution to optimize mass and dimensional stability to a higher level than conventional machining.

Nevertheless, it is also precisely because of the layer-by-layer printing that the as-built AM parts are arduous for obtaining optical surfaces. Extensive research has been carried out to develop the new-generation metal mirror based on AM. Sweeney et al. provided case studies utilizing multiple metal AM technologies for mirror designs and discussed the findings, such as stress relieving and isotropy of mechanical properties [17]. Hartung et al. designed and manufactured the Voronoi AM mirror for visible-light level application [18]. Robert et al. presented and incorporated the circular lattice into the mirror for a CubeSat telescope [19]. Yan et al. designed an assembly-level topology optimization of Al mirrors based on AM, which improved the efficiency of assembly and integration [20]. Moreover, the modeling rules are also essential for the manufacturing accuracy of precision optical elements. Sara et al. proposed design rules for improving the accuracy of selective laser melting (SLM) manufacturing using benchmark parts [21]. Table 1 gives a comparison study with the mentioned literature. While other work has shown successes in AM mirror preforms that were post-polished, discussions on the design and manufacturing methodology leveraging AM have not been discussed in detail.

Table 1. Comparison with the other literature.

Authors	Aperture (mm)	Material	Highlight
Sweeney et al. [17]	75–150	$AlSi_{10}Mg$ etc.	Process exploration and validation
Hartung et al. [18]	76	$AlSi_{40}$	Novel lightweight structure design
Robert et al. [4]	72	$AlSi_{10}Mg$	Lattice mirror design for CubeSat
Yan et al. [21]	58	$AlSi_{10}Mg$	Assembly-level topology optimization and integration design
This study	175	$AlSi_{10}Mg$	Large aperture and integration design with compound surfaces

Different from other research, this paper aims to balance the lightweight, optical, and structural performances of the large aperture mirror in the limited space for microsatellites. An Al mirror with composite surfaces based on AM is proposed for low material cost and compact design. Compared with the open-back scheme manufactured by conventional machining, the closed-back mirror provides better optomechanical properties through finite element analysis (FEA). Then, a complete manufacturing process of AM metal mirror, such as AM for mirror blank, single point diamond turning (SPDT), surface modification, and polishing, is described in detail. Then, the paper examines the surface-shape accuracy and the infrared imaging requirements. Finally, a discussion of achievements and future work closes this paper.

2. Optical System and Structural Design Input

Due to the constraints of microsatellite platform resources, reducing the weight and volume has become the focus of optical payload design and optimization. A compact coaxial reflective optical system with four mirrors is presented in this paper, as shown in Figure 1a. It is an F/10 optical system with a focal length of 1750 mm working in the infrared band. All four mirrors are aspheric surfaces. After reflection by the primary mirror (PM) and the secondary mirror (SM), the light passes through the central hole of the quaternary mirror (QM) to arrive at the tertiary mirror (TM). The central hole is the field stop of this optical system. Finally, the light forms a perfect image on the detector plane after four times of reflection, and the modulation transfer function (MTF) image is given in Figure 1b.

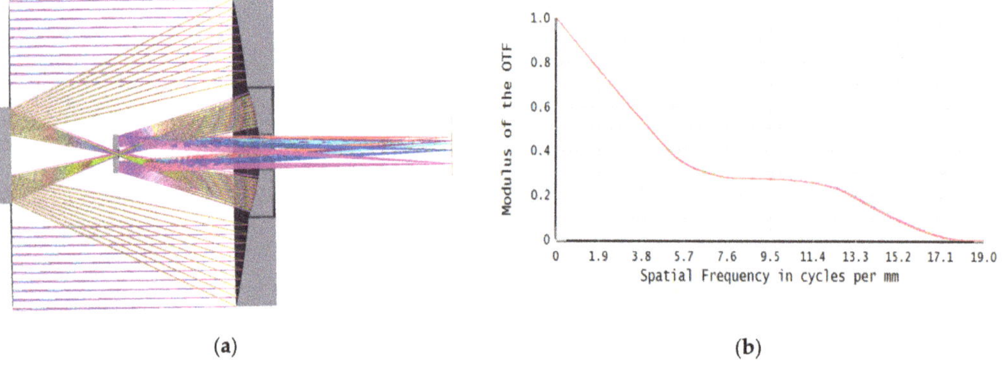

Figure 1. (**a**) Light traces of the optical system; (**b**) MTF image.

Figure 2 shows the relative position of the PM and the TM. The effective aperture of the PM and the TM is 175 mm and 70 mm, respectively. The size of the PM center hole is 75 mm. There is a 30 mm × 16 mm rectangular hole in the center of the TM resulting from the field stop. In the initial design, the distance difference between the PM's center hole and the TM's outer circle was 0.2 mm in the optical axis direction. Considering the environmental conditions in which the optical system is located, the structural elements should be as light as possible to ensure appropriate structural performance. Therefore, the integrated scheme of the PM and TM was adopted with a lightweight rate of 70%. Based on providing an effective aperture, the clear aperture of the TM was increased appropriately, and the center hole of the PM was reduced. Optical design optimization allows the PM and TM edges to be fused into a continuous sectional curve. This integrated scheme minimizes the difficulty of the system assembly and wavefront aberration correction. The parameters and specifications for the mirror with compound surfaces are given in Table 2.

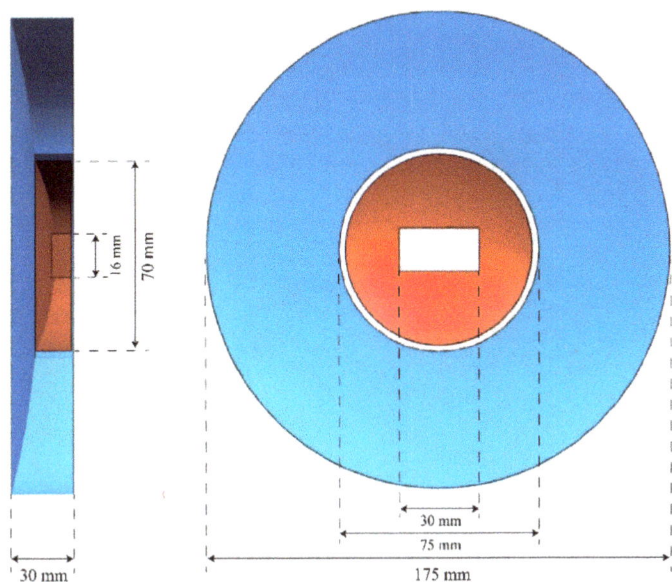

Figure 2. Schematic diagram of the relative position for the PM and the TM.

Table 2. Parameters and specifications for the mirror with compound surfaces.

Parameter	Specification (PM)	Specification (TM)
Optical prescription	Aspheric surface	Aspheric surface
Radius of curvature	396 mm	98 mm
Effective Aperture	175 mm	70 mm
Form error	Better than RMS 1/10λ (λ = 632.8 nm)	
Roughness	Better than 2 nm	
First-order modal frequency	2000 Hz	
Lightweight rate	70%	

3. Structural Design and Finite Element Analysis

3.1. Structural Design for Mirrors with Compound Surfaces

The PM and the TM were integrated into a whole during structural design based on their position relationship. PM is the component with the largest package size in the radial direction to reduce the system volume. In addition, three through-holes are evenly distributed on the mirror to pass through the support structures (Figure 3a). As mentioned, this structure is more compact, which is conducive to reducing the weight and volume of the system. The manufacturing process guarantees the relative position of the PM, the TM, and the rectangular hole, which simplifies of the optical system assembly. It can also reduce manufacturing time and cost. Conventional glass and SiC mirrors are made of hard and brittle materials, so it is difficult to realize the processing of intricate mechanical parts. Furthermore, because of the optical processing technology of grinding and polishing, the relative position relationship between the two mirrors cannot be accurately guaranteed [22]. Therefore, an all-Al alloy design was adopted, and the PM and the TM share one mirror substrate.

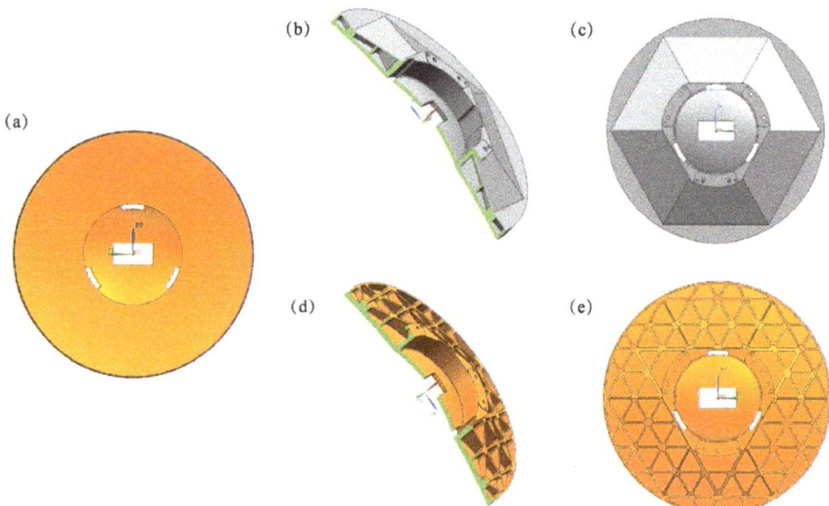

Figure 3. Schematic diagram of mirror models: (**a**) front view of the mirror; (**b**) half-section of the closed-back mirror; (**c**) back view of the closed-back mirror; (**d**) half-section of the open-back mirror; and (**e**) back view of the open-back mirror.

The open-back mirror and the closed-back mirror were designed in this paper. The closed-back structure has one more back plate of 2 mm than the open-back structure. Therefore, it is also referred to as the sandwich mirror. Because of the AM, the manufacturing difficulty of the complex sandwich lightweight structure is significantly reduced. Daniel Vukobratovich found that the sandwich mirror has the best mirror efficiency, defined as the total mirror height divided by the mechanical deflection [23]. The sandwich mirror places most mass in the face and back plates, with as little mass as possible in the shear core. It provides high stiffness and efficient bending when the mass of a structure is distributed as far as possible from the neutral or bending axis. The mirrors inside are supported by triangular geometry, which can trade off overall stiffness and lightweight rate. The average thickness of the ribs is about 1.5 mm, and the mirror surface thickness is 4 mm. The half-sectional and rear views of the two mirrors are shown in Figure 3. Next, FEA (modal analysis and gravity deformation analysis) is carried out to compare the two structural schemes' comprehensive optomechanical performance and verify whether they can be applied in engineering.

3.2. Finite Element Analysis

3.2.1. Modal Analysis

The modal frequency and mass are negatively correlated as follows:

$$\omega = \sqrt{\frac{k}{m}},$$

(1)

where ω is the modal angular frequency, k is the rigidity coefficient, and m is the mass. Thus, the masses of the two mirrors were adjusted to be equal to compare the mechanical properties accurately [24]. Since the closed-back scheme has a back plate, the mass is heavier under the same conditions. The surface thickness of the closed-back scheme was reduced (~1 mm) until the equal group. Then, the modal analysis of the mirrors was carried out by fixing the six mounting surfaces.

The curves of the first 10-order modal frequencies for the two structural schemes are shown in Figure 4. At the same time, the total deformation of the first three-order for the closed-back scheme is also given. It can be seen intuitively that all modes of the closed-back mirror are higher than those of the open-back mirror. Table 3 shows the specific data of each order mode for the two schemes. The first-order modal frequency of the closed-back scheme is 2149 Hz, and that of the open-back scheme is 1703 Hz; the 10-order modal frequencies of the two schemes are 6823 Hz and 4876 Hz, respectively. Therefore, the overall stiffness of the closed-back scheme is excellent.

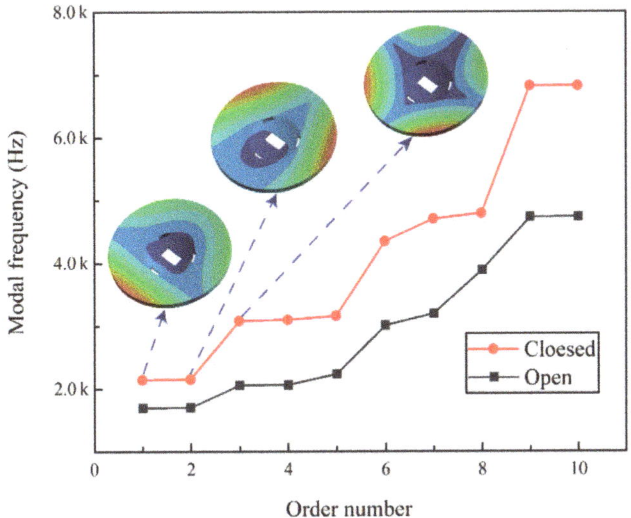

Figure 4. Modal analysis results.

Table 3. Modal analysis data.

Order Number	Open-Frequency (Hz)	Closed-Frequency (Hz)
1	1703.4	2149.6
2	1705.3	2157.2
3	2064.1	3082.5
4	2068.1	3099.7
5	2236.1	3160.1
6	3011.3	4351.0
7	3195.2	4705.6
8	3891.6	4795.5
9	4733.5	6821.2
10	4736.1	6823.9

3.2.2. Gravity Deformation Analysis

Both mirrors are rotationally symmetric, and the gravity deformation analysis was performed under the axial and radial conditions (Figure 5) [25]. Surface-shape-fitting nephograms after gravity deformation analysis are given in Figure 5. It can be concluded from the nephograms that there are no significant differences in the radial deformation of the two schemes. However, the difference in axial deformation is noticeable. As mentioned, for beams distributed in the sandwich mirror, most of their mass is far from the bending axis. It is an efficient structure in bending in which relatively little mass is placed in the web of the beams. Thus, the surface deformation caused by the axial shear force can be reduced. The fitting data corresponding to Figure 6 is shown in Table 4. The radial deformation value of the closed-back scheme is about 20% smaller than that of the open-back scheme. The axial deformation value of the closed-back scheme is about 15–25%, significantly less than that of the open-back scheme. Consequently, the closed-back mirror has a pronounced effect on improving the mirror's gravitational deformation.

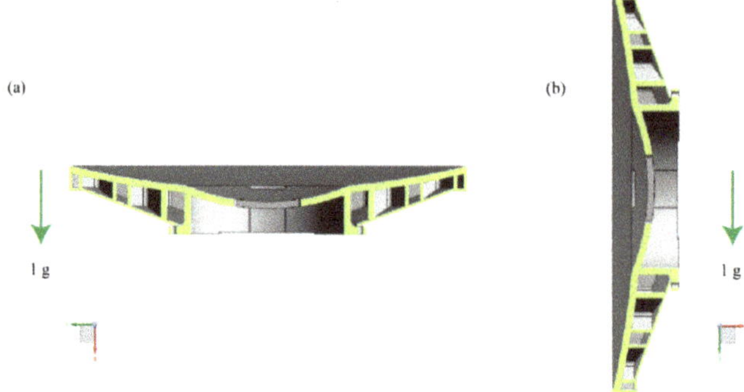

Figure 5. Directions of gravity deformation analysis: (**a**) axial gravity; (**b**) radial gravity.

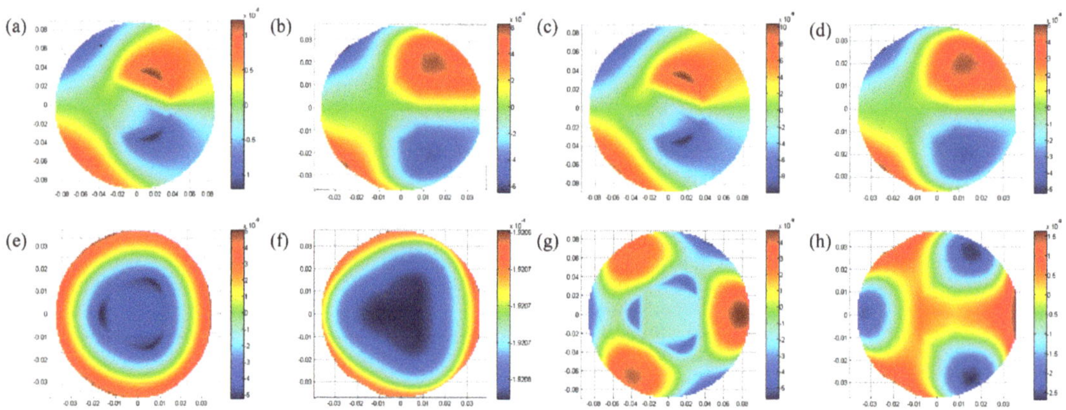

Figure 6. Surface-shape-fitting nephograms after gravity deformation analysis: (**a**) open-PM-radial; (**b**) open-TM-radial; (**c**) closed-PM-radial; (**d**) closed-TM-radial; (**e**) open-PM-axial; (**f**) open-TM-axial; (**g**) closed-PM-axial; (**h**) closed-TM-axial.

Table 4. Surface-shape-fitting results after gravity deformation analysis.

	Open-PM	Open-TM	Closed-PM	Closed-TM
Radial (P.V. nm)	22.8	12.9	20.6	10.5
Radial (RMS nm)	5.4	3.3	4.6	2.6
Axial (P.V. nm)	58.3	18.6	10.5	4.4
Axial (RMS nm)	16.7	4.0	2.5	1.0

4. Mirror Fabrication

As shown in Figure 7, the processing chain for AM metal mirrors was divided into three stages. The first stage is the mirror blank prefabrication stage. After the structural design, the mirror blank is printed via metal AM technology. Stress relieving through densification and heat treatment prepares for the next step. Optical manufacturing is to obtain a high-quality optical surface through ultra-precision processing technologies. Finally, the surface quality and thermal performance of mirrors were tested.

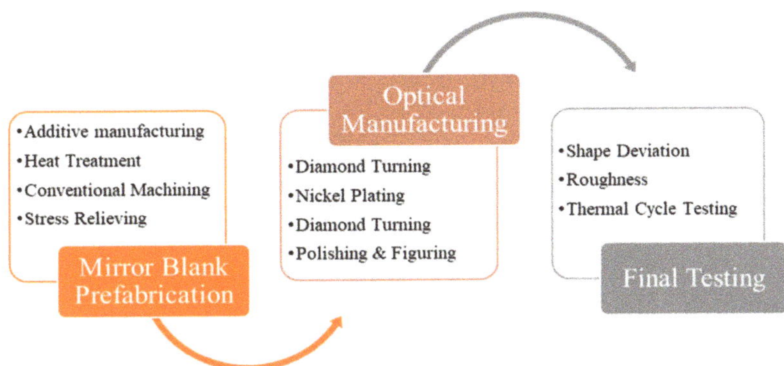

Figure 7. Processing chain for an AM metal mirror.

4.1. Mirror Blank Prefabrication

4.1.1. AM for the Mirror Blank

Laser powder bed fused (LPBF) is a specific kind of metal AM method that generates parts out of powder on a layer-by-layer principle. It converts the laser energy into heat energy to make parts form [26]. The laser focuses the spot to tens of micrometers, and the power density is more than 5.0×10^6 W/cm^2. A high-energy laser beam completely melts powder under a protective atmosphere, it can avoid shrinkage cavity, porosity, and inclusion to improve the density. Laser absorptivity in AM is essential, showing the parts' quality and transition from keyhole to conduction mode [27]. Rapid cooling inhibits grain growth and segregation of alloy elements, resulting in refined grains and uniform element distribution [28]. At the same time, LPBF also inhibits grain growth to form fine grain strengthening. Its mechanical properties, such as tensile strength, are better than castings and even reach the level of forgings. The Al alloy powder was selected as AlSi$_{10}$Mg, where the density was 2.67 g/cm^3, the modulus of elasticity was 69 GPa, and the Poisson's ratio was 0.33. AlSi$_{10}$Mg is a hardenable Al-based alloy. Cooperating with LPBF, it is applicable for thin-walled components with complex geometries. LPBFed AlSi$_{10}$Mg is highly suitable for processing and characterized by good resistance in corrosive atmospheres and high electrical conductivity [29]. The combination of high achievable strengths while maintaining dynamic load capacity enables it to be used for highly stressed parts. Therefore, the high density, high printing precision, and excellent mechanical properties make LPBF suitable for AM metal mirror manufacturing [30].

The spherical powder with the size of 20–30 μm was selected, and the compactness of the printed parts is better than 99.5% to ensure that its fluidity meets the requirements. SLM Solutions 280 is chosen for the mirror blanks forming. It provides a $280 \times 280 \times 365$ mm^3 build room and patented multi-beam technology. In addition, the patented bidirectional powder coating helps reduce the manufacturing time of individually manufactured metal parts. The thickness of the powder layer is 25 μm, and the scanning speed is 980 mm/s. Argon gas with an oxygen mass fraction of less than 0.1% was used to protect the whole printing process and prevent metal oxide formation. For ensuring good fixation of the mirror to the building platform and preventing collapse and deformation during printing, the complex support structures were designed [31]. The mirror blank is placed at an angle of about 40°, as shown in Figure 8. The self-supporting diamond structures were selected as the support structures with a thickness of 0.2 mm and an angle of 120°, which can save powders and cost. In addition, the removal difficulty of the support structures is reduced via computer numerical control machining (CNC) and wire cutting.

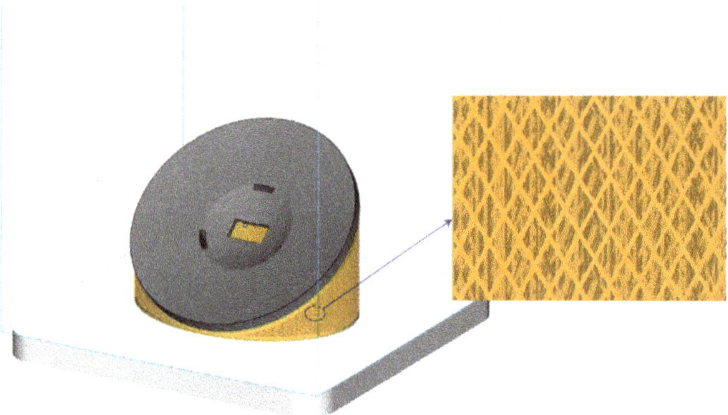

Figure 8. Visualization of support structures.

The final LPBFed mirror blanks are shown in Figure 9a. Technological holes were added to the design model to facilitate the cleaning of residual powders during the AM process. These technological holes are mainly located on the side wall of the stiffeners between the mirror back and the lightweight cells. All the enclosed areas inside the mirror body through these holes relate to the outside. These holes have little impact on the mirror's overall stiffness and mechanical characteristics due to the small sizes and discrete distribution. Figure 9b is an X-ray test image for a 3D-printed mirror blank. The positions of technological holes and lightweight ribs can be seen clearly.

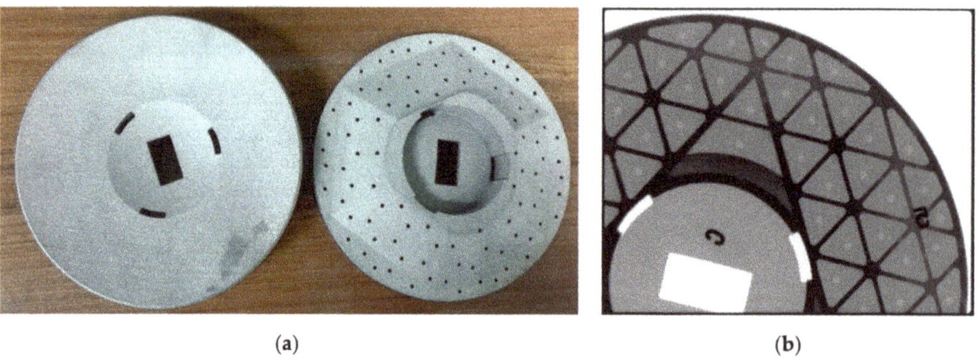

(**a**)   (**b**)

Figure 9. (**a**) AM mirror blanks after cleaning; (**b**) X-ray inspection image for the AM mirror blank (the letter and number are locating datum).

4.1.2. Heat Treatment and Densification

As a result of the unique process of metal AM, the residual stress in the formed parts is enormous. Therefore, heat treatment for stress relieving is the first post-treatment step after forming. In the LPBF process, the powders are not completely melted, especially the Si particles, leading to micropores in the formed mirror. Therefore, densification treatment is also required to eliminate pore defects. Hot isostatic pressing (HIP) is a special heat treatment for reducing the porosity of AM mirrors. In the HIP, the mirror blank was placed in a furnace with a temperature of ~550 °C for 2 h. At the same time, the furnace was filled with argon and pressurized at ~110 MPa [32,33]. Then, the isostatic pressing was applied to the material in all directions until the mirror blank cooled.

4.1.3. Conventional Machining and Thermal Cycling

After densification, the structural elements with specific accuracy requirements need to be finished by traditional processing methods, including the mirror surface, the structural reference plane, the screw hole, and the pinhole. The processing number in this process was controlled by 1 mm. The mounting benchmark after CNC is shown in Figure 10, and the rough (1 mm machining allowance), semi-finishing (0.5 mm machining allowance), and finishing machining (0.1 mm machining allowance) were carried out by CNC until the surface roughness was about 1.6 μm.

Figure 10. The mounting benchmark after CNC.

Thermal cycling and the aging treatment are the processes to improve mechanical properties and reduce the residual stress of mirror blank. Thermal cycling stabilizes components by repeatedly alternating the part temperature between approximately low-temperature and high-temperature conditions at a controlled rate of a few degrees per minute. This obtains a measure of stress relaxation for those small stresses introduced in the rough machining and finishing processes. Heat treatment was carried out once for each machining. The thermal cycling process first placed the AM mirror in liquid N_2 and held it for 30 min. It was then heated to 23 °C for 30 min and heated to 160 °C for 30 min. Finally, the mirror was cooled to 23 °C. The thermal cycling temperature range was −180 °C to +160 °C, with the temperature changing rates less than 2.0 °C/min.

4.2. Optical Manufacturing

4.2.1. SPDT before Plating

The mirror blank was fixed on the main shaft through a particular tooling disk. After the first round of SPDT, the flatness of all mounting surfaces was better than 1.0 μm. Moreover, the mirror surface was poor. These pits and turning scratches can be seen more clearly under white light (Figure 11). The insufficient density of parts causes dimples during AM. When the diamond tip passes through the pit in the high-speed turning process, it produces vibration and leaves fine scratches. These macroscopic surface defects will cause strong scattering of the imaging light and seriously affect the system's signal-to-noise ratio. Thus, surface modification is necessary for optical applications.

Figure 11. Mirror blank surface under white light.

4.2.2. Surface Modification and Second Round SPDT

Electroless nickel (Ni), a coating NiP alloy on the Al-Si substrate by electrochemical deposition, is the most common way for surface modification of Al mirrors. For the homogenous layer composition, the Ni content, pH value, and bath temperature of the electrolyte must be constant during the deposition process. It is necessary to monitor and control them in situ. At constant bath temperature and Ni content, the alloy's final phosphorus (P) contents are different, which is affected by pH value [34]. The P content is the most significant parameter for tailoring the mechanical and thermal properties, such as machining characteristics and coefficient of thermal expansion (CTE). The NiP layer becomes amorphous and isotropic at over 10.5 wt %. Moreover, the tool wear is serious during SPDT for the NiP layer when P content is less than 10.5 wt % [35]. The final range of P content is limited to 10.5 wt % to 14 wt %. The mirror, after surface modification, is shown in Figure 12. Then, the second round of SPDT was carried out on the NiP surface with the machining allowance of 1.0 μm.

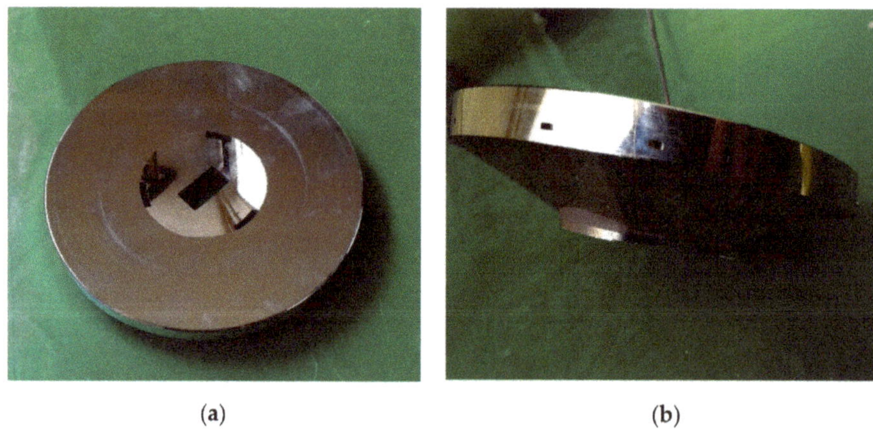

(**a**) (**b**)

Figure 12. The mirror after surface modification: (**a**) front view; (**b**) side view.

4.2.3. Polishing and Figuring

Polishing and figuring were adopted to correct mid- and high-frequency and low-frequency errors of the mirror. The mid- and high-frequency errors are mainly the periodic tool marks during SPDT. Above all, these tool marks have a strong scattering effect on the imaging light and affect imaging quality. Additionally, the low-frequency errors are mainly the surface-shape deviation of the mirror surface. An OptoTech machine was utilized for polishing with a CNC asphalt head (Figure 13a) [36]. After polishing, the turning tool marks of the mirror surface were suppressed dramatically. The final mirror after polishing is shown in Figure 13b. Although ion beam figuring (IBF) positively affects the surface processing of metal mirrors, its time and processing cost are high [37]. Therefore, IBF will be used to verify the influence on the mirror surface shape in the next iteration.

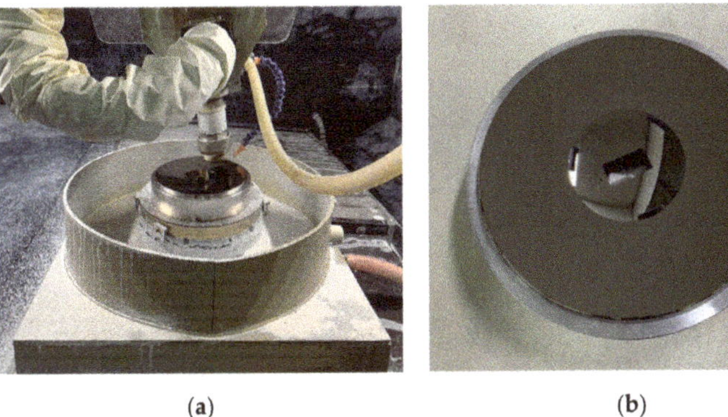

(a)   (b)

Figure 13. (**a**) The mirror on the OptoTech polishing machine; (**b**) the final mirror in the tool disk.

4.3. Final Testing

4.3.1. Surface-Shape Deviation and Roughness Testing

Because of the aspherical optical prescription, a 3D profilometer-Taylorhobson-LUPHOScan (Taylor Hobson Ltd., Leicester, UK) was adopted for the surface-shape testing of the whole process (Figure 14). It is designed to perform ultra-precision 3D-form measurements for rotationally symmetric spherical and aspheric surfaces. LUPHOScan is even capable of free-form surface measurements easily, which dramatically reduces the surface deviation testing difficulty. The Results from non-contact measurements show that it is ideal for applications of high-precision requirements.

Figure 14. The mirror is testing on LUPHOScan.

The form error testing was not carried out after the first round of SPDT due to the poor surface quality. Figures 15 and 16 show the surface-shape testing results after the second round of SPDT and polishing. After surface modification and the second SPDT, the surface shape of the PM and the TM is RMS 173 nm and RMS 87 nm, respectively; after polishing, the surface shape of the PM and the TM is RMS 64 nm and RMS 57 nm, respectively.

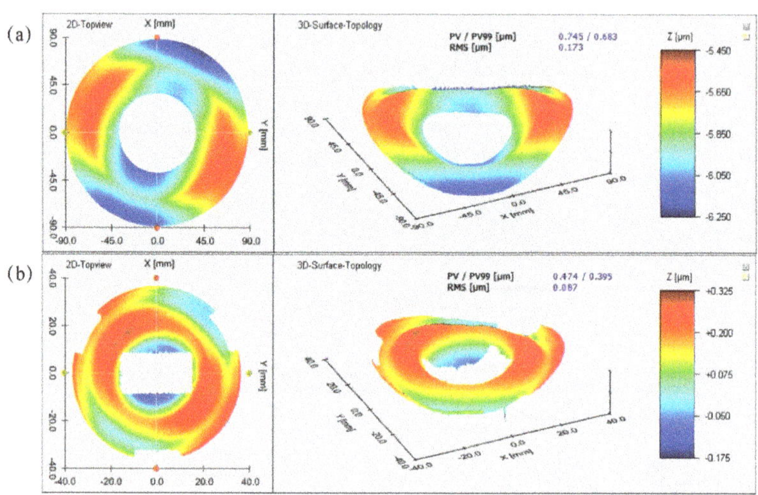

Figure 15. Surface shape deviation after the second Round SPDT: (**a**) PM; (**b**) TM.

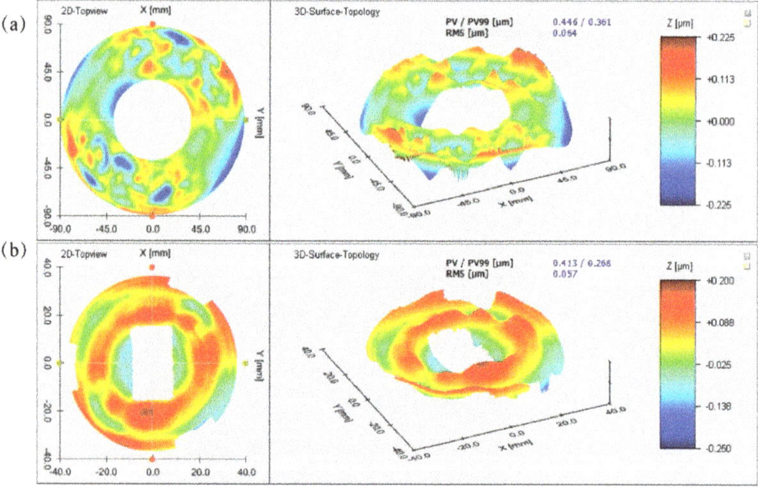

Figure 16. Surface shape deviation after polishing: (**a**) PM; (**b**) TM.

Subsequently, local surface roughness was measured using white light interferometry to assess the homogeneity and distribution of contamination and defects. The result shows that the final roughness was about 1.76 nm, less than 2 nm (Figure 17). Summarily, the surface shape and roughness both meet the parameters and infrared imaging requirements given in Table 2.

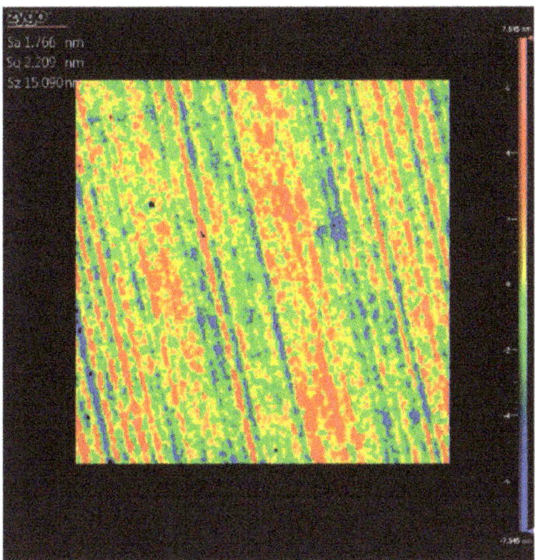

Figure 17. Roughness testing results using white light interferometry.

4.3.2. Thermal Cycle Testing

Generally, the dimensional stability of metal mirrors must be kept within the range of 10^{-6}, amounting to the dimension variation at the meter scale being controlled at the micrometer scale. For AM mirrors, dimensional stability is more critical due to the residual stress caused by the unique forming process. Therefore, it is necessary for the thermal cycle test of AM metal mirrors. Figure 18 shows the AM mirror in the testing furnace. Then, the stability in the extreme environment and whether it meets the practical engineering requirements is verified. The maximum temperature range of the system during operation and storage is −40–60 °C. Thus, this thermal cycle testing set the high-temperature and low-temperature conditions at −50 °C and +70 °C, respectively. The cycle was carried out three times with a temperature change rate of 1.0 °C/min, and the holding time was an hour.

Figure 18. The AM mirror in the test furnace.

After the thermal cycle testing, the mirror surface-shape deviations are shown in Figure 18. The surface shape of the PM (Figure 19a) and the TM (Figure 19b) is RMS 79 nm and RMS 64 nm, respectively. Table 5 summarizes the mirror surface shape under different conditions. Compared with the polished surface, the change of the PM surface shape is RMS 15 nm; the change of the TM surface shape is only RMS 7 nm, which conforms to the basic requirements for optical mirror surface quality, and the slight change in the surface quality could return to its original state. After the thermal cycle testing, the maximum deviation of the surface shape is 15 nm, about RMS $1/42\lambda$ (λ = 632.8 nm). It proved that aging and heat treatment is effective and verified AM's optical grade application.

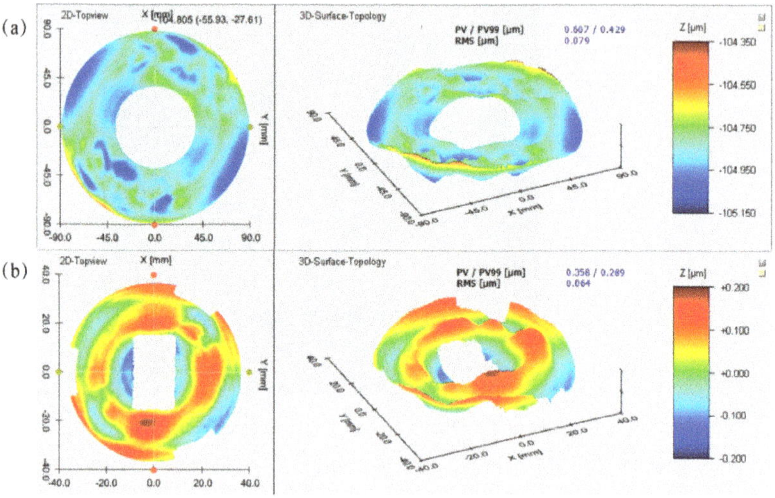

Figure 19. Surface-shape deviation results after the thermal cycle test: (**a**) PM; (**b**) TM.

Table 5. Summary of mirror surface shape under different conditions.

Process Steps	PM-RMS (nm)	TM-RMS (nm)
First Round SPDT	/	/
NiP + Second Round SPDT	173	93
Polishing	64	57
After the thermal cycle test	79	64

5. Conclusions and Future Work

This study looked to develop an AM mirror with compound surfaces. The integrated mirror is designed for use in microsatellites, greatly reducing space occupancy, cycle time, and launch cost. Through FEA, it was found that the performance of the closed-back scheme is better than that of the open-back scheme. The closed-back mirror is printed via LPBF technology. After SPDT, surface modification, and polishing, the surface-shape accuracy of the TM and the PM reached RMS 57 nm and RMS 64 nm, respectively, and the roughness was better than 2 nm. Finally, thermal testing was carried out to further verify the mirror's dimensional stability. Results showed that the surface shape changed within RMS 15 nm, which reached the optical stability requirements.

This work has been successful so far, although the surface shape of the mirror is only about $1/10\lambda$ (λ = 632.8 nm). It can be seen from the thermal cycle testing results that the maximum surface-shape change is about $1/40\lambda$ (λ = 632.8 nm) which explains that the aging treatment is effective. Due to the time limit, IBF was not performed. Subsequently, the effect of IBF on the AM mirror surface will be tested to achieve a high-precision surface shape that meets the visible-light-level application. In addition, the traditional isogrid design was adopted for the mirror design. More advanced structures will be developed to take advantage of the free-form design of AM and trade off the performances well in future.

Author Contributions: Conceptualization, J.Z. and C.W.; methodology, H.Q. and K.Z.; software, C.W. and H.W. (Ha Wang); validation, H.Q., H.W. (Ha Wang) and L.L.; formal analysis, H.G.; investigation, C.W.; resources, H.G.; data curation, X.X.; writing—original draft preparation, J.Z.; writing—review and editing, K.Z.; visualization, H.W. (He Wang); supervision, X.X.; project administration, X.Z.; funding acquisition, X.Z. All authors have read and agreed to the published version of the manuscript.

Funding: This research was supported by the Jilin Provincial Research Foundation for Scientific and Technological Development, China (Grant No. 20200401099GX and Grant No. 20210509056RQ).

Institutional Review Board Statement: Not applicable.

Informed Consent Statement: Not applicable.

Data Availability Statement: Not applicable.

Conflicts of Interest: The authors declare no conflict of interest.

References

1. Sweeting, M.N. Modern small satellites-changing the economics of space. *Proc. IEEE* **2018**, *106*, 343–361. [CrossRef]
2. Erdem, T.; Stefano, S.; Eberhard, G. Autonomous navigation for deep space small satellites: Scientific and technological advances. *Acta Astronaut.* **2022**, *193*, 56–74. [CrossRef]
3. Meng, Q.; Wang, H.; Wang, K.; Wang, Y.; Ji, Z.; Wang, D. Off-axis three-mirror freeform telescope with a large linear field of view based on an integration mirror. *Appl. Opt.* **2016**, *55*, 8962–8970. [CrossRef] [PubMed]
4. Robert, A.W.; William, C.D.; Sara, R.; Heap, T.H.; Stephen, E.K.; Lloyd, R.P.; Michael, S.R.; Eric, M.; Brian, F.; Marty, V.; et al. Optical design for CETUS: A wide-field 1.5-m aperture U.V. payload being studied for a NASA probe class mission study. *J. Astron. Telesc. Instrum. Syst.* **2019**, *5*, 024006. [CrossRef]
5. Atkins, C.; Feldman, C.H.; Brooks, D.; Willingale, R.; Doel, P.; Roulet, M.; Watson, S.; Cochrane, W.; Hugot, E. Additive manufactured X-ray optics for astronomy. In *Optics for EUV, X-ray, and Gamma-ray Astronomy VIII*; International Society for Optics and Photonics: Bellingham, WA, USA, 2017; Volume 10399. [CrossRef]
6. Steinkopf, R.; Gebhardt, A.; Scheiding, S.; Rohde, M.; Stenzel, O.; Gliech, S.; Giggel, V.; Loscher, H.; Ullrich, G.; Rucks, P.; et al. Metal mirrors with excellent figure and roughness. *Int. Soc. Opt. Photonics* **2008**, *7102*, 71020. [CrossRef]
7. Zhao, T.; Hu, H.; Peng, X.; Du, C.; Guan, C.; Yong, J. Study on the surface crystallization mechanism and inhibition method in the CMP process of aluminum alloy mirrors. *Appl. Opt.* **2019**, *58*, 6091–6097. [CrossRef] [PubMed]
8. Zhang, J.; Zhang, X.; Tan, S.; Xie, X. Design and Manufacture of an Off-axis Aluminum Mirror for Visible-light Imaging. *Curr. Opt. Photonics* **2017**, *1*, 364–371.
9. Xie, Y.; Mao, X.; Li, J.; Wang, F.; Wang, P.; Gao, R.; Li, X.; Ren, S.; Xu, Z.; Dong, R. Optical design and fabrication of an all-aluminum unobscured two-mirror freeform imaging telescope. *Appl. Opt.* **2020**, *3*, 833–840. [CrossRef] [PubMed]
10. Liu, F.; Jia, P.; Shen, H.; Xu, Y. Optical performance of aluminum mirror for cryogenic applications. *Optik* **2021**, *231*, 166282. [CrossRef]
11. Vukobratovich, D.; Schaefer, J.P. Large stable aluminum optics for aerospace applications. *Proc. SPIE 8125*. Optomechanics 2011: Innovations and Solutions, 81250T (24 September 2011). Available online: https://ur.art1lib.com/book/40487207/32b77f (accessed on 30 August 2022).
12. Zhang, K.; Xie, X.; Wang, C.; Wang, H.; Xu, F.; Wang, H.; Zhang, X.; Guan, H.; Qu, H.; Zhang, J. Optomechanical Performances of Advanced Lightweight Mirrors Based on Additive Manufacturing. *Micromachines* **2022**, *13*, 1334. [CrossRef]
13. Qiao, G.; Hu, H.; Zhang, X.; Luo, X.; Xue, D.; Zhang, G.; Hu, H.; Yi, L.; Yang, Y.; Deng, W. Stress-induced deformation of the coating on large lightweight freeform optics. *Opt. Express* **2021**, *29*, 4755–4769. [CrossRef] [PubMed]
14. Zhang, K.; Qu, H.; Guan, H.; Zhang, J.; Zhang, X.; Xie, X.; Yan, L.; Wang, C. Design and Fabrication Technology of Metal Mirrors Based on Additive Manufacturing: A Review. *Appl. Sci.* **2021**, *11*, 10630. [CrossRef]
15. Linares, J.-M.; Chaves-Jacob, J.; Lopez, Q.; Sprauel, J.-M. Fatigue life optimization for 17-4Ph steel produced by selective laser melting. *Rapid Prototyp. J.* **2022**, *28*, 1182–1192. [CrossRef]
16. Zhou, Y.; Abbara, E.M.; Jiang, D.; Azizi, A.; Poliks, M.D.; Ning, F. High-cycle fatigue properties of curved-surface AlSi$_{10}$Mg parts fabricated by powder bed fusion additive manufacturing. *Rapid Prototyp. J.* **2022**, *28*, 1346–1360. [CrossRef]
17. Sweeney, M.; Acreman, M.; Vettese, T.; Myatt, R.; Thompson, M. Application and testing of additive manufacturing for mirrors and precision structures. In *Material Technologies and Applications to Optics, Structures, Components, and Sub-Systems II*; International Society for Optics and Photonics: Bellingham, WA, USA, 2015; Volume 9574, p. 957406. [CrossRef]
18. Hartung, J.; Beier, M.; Risse, S. Novel applications based on freeform technologies. *Proc. SPIE 10692*. 2018, Optical Fabrication, Testing, and Metrology VI, 106920K (5 June 2018). Available online: https://neurophotonics.spiedigitallibrary.org/conference-proceedings-of-spie/10692/106920K/Novel-applications-based-on-freeform-technologies/10.1117/12.2313100.short (accessed on 30 August 2022).
19. Snell, R.; Atkins, C.; Schnetler, H.; Todd, I.; Hernández-Nava, E.; Lyle, A.R.; Maddison, G.; Morris, K.; Miller, C.; Roulet, M.; et al. An additive manufactured CubeSat mirror incorporating a novel circular lattice. *Proc. SPIE 11451*. Advances in Optical and Mechanical Technologies for Telescopes and Instrumentation IV, 114510C (13 December 2020). Available online: https://eprints.whiterose.ac.uk/172542/ (accessed on 30 August 2022).
20. Giganto, S.; Martínez-Pellitero, S.; Cuesta, E.; Zapico, P.; Barreiro, J. Proposal of design rules for improving the accuracy of selective laser melting (SLM) manufacturing using benchmarks parts. *Rapid Prototyp. J.* **2022**, *28*, 1129–1143. [CrossRef]
21. Yan, L.; Zhang, X.; Fu, Q.; Wang, L.; Shi, G.; Tan, S.; Zhang, K.; Liu, M. Assembly-level topology optimization and additive manufacturing of aluminum alloy primary mirrors. *Opt. Express* **2022**, *30*, 6258–6273. [CrossRef]

22. Ahmad, A. *Handbook of Optomechanical Engineering*; CRC Press: Boca Raton, FL, USA, 2017.
23. Valente, T.M.; Vukobratovich, D. A Comparison of the merits of open-back, symmetric sandwich, and contoured back mirrors as light-weighted optics. In Proceedings of the Precision Engineering and Optomechanics, San Diego, CA, USA, 10–11 August 1989; pp. 20–36.
24. Zhang, L.; Wang, T.; Zhang, F.; Zhao, H.; Zhao, Y.; Zheng, X. Design and optimization of integrated flexure mounts for unloading lateral gravity of a lightweight mirror for space application. *Appl. Opt.* **2021**, *60*, 417–426. [CrossRef]
25. Shao, M.; Zhang, L.; Jia, X. Optomechanical integrated optimization of a lightweight mirror for space cameras. *Appl. Opt.* **2021**, *60*, 539–546. [CrossRef]
26. Mathieu, O.; Jean-Paul, G.; Guilhem, R.; Camille, F.; Mathieu, S. A solution to the hot cracking problem for aluminium alloys manufactured by laser beam melting. *Acta Mater.* **2020**, *197*, 40–53. [CrossRef]
27. Mahyar, K.; AmirHossein, G.; Martin, L.; Elmira, S.; Laura, C.; Ian, G.; David, D.; Stuart, B.; Milan, B.; Bernard, R. The effect of absorption ratio on meltpool features in laser-based powder bed fusion of IN718. *Opt. Laser Technol.* **2022**, *153*, 108263. [CrossRef]
28. Wang, X.; Xu, S.; Zhou, S.; Xu, W.; Martin, L.; Peter, C.; Qian, M.; Milan, B.; Xie, Y. Topological design and additive manufacturing of porous metals for bone scaffolds and orthopaedic implants: A review. *Biomaterials* **2016**, *83*, 127–141. [CrossRef] [PubMed]
29. Liu, C.; Xu, K.; Zhang, Y.; Hu, H.; Tao, X.; Zhang, Z.; Deng, W.; Zhang, X. Design and Fabrication of Extremely Lightweight Truss-Structured Metal Mirrors. *Materials* **2022**, *15*, 4562. [CrossRef] [PubMed]
30. Bai, Y.; Shi, Z.; Lee, Y.; Wang, H. Optical surface generation on additively manufactured AlSiMg$_{0.75}$ alloys wit ultrasonic vibration-assisted machining. *J. Mater. Process. Technol.* **2020**, *280*, 116597. [CrossRef]
31. Bahaa, S.; Żaneta, G.; Agnieszka, C.; Bartłomiej, W.; Marcin, H.; Maarten, G.; Bart, V.; Koen, B.; Chris, V.; Emilia, C.; et al. Novel design for an additively manufactured nozzle to produce tubular scaffolds via fused filament fabrication. *Addit. Manuf.* **2022**, *49*, 102467. [CrossRef]
32. Tocci, M.; Pola, A.; Gelfi, M. Effect of a New High-Pressure Heat Treatment on Additively Manufactured AlSi$_{10}$Mg Alloy. *Metall. Mater. Trans. A* **2020**, *51*, 4799–4811. [CrossRef]
33. Schneller, W.; Leitner, M.; Springer, S.; Grün, F.; Taschauer, M. Effect of HIP Treatment on Microstructure and Fatigue Strength of Selectively Laser Melted AlSi$_{10}$Mg. *J. Manuf. Mater. Process.* **2019**, *3*, 16. [CrossRef]
34. Bai, Y.; Zhang, Z.; Xue, D.; Zhang, X. Ultra-precision fabrication of a nickel-phosphorus layer on aluminum substrate by SPDT and MRF. *Appl. Opt.* **2018**, *57*, F62–F67. [CrossRef] [PubMed]
35. Kinast, J.; Hilpert, E.; Rohloff, R.-R.; Gebhardt, A.; Tünnermann, A. Thermal expansion coefficient analyses of electroless nickel with varying phosphorous concentrations. *Surf. Coat. Technol.* **2014**, *259*, 500–503. [CrossRef]
36. Xu, C.; Peng, X.; Liu, J.; Hu, H.; Lai, T.; Yang, Q.; Xiong, Y. A High Efficiency and Precision Smoothing Polishing Method for NiP Coating of Metal Mirror. *Micromachines* **2022**, *13*, 1171. [CrossRef] [PubMed]
37. Melanie, U.; Bauer, J.; Frank, F.; Thomas, A. Ion beam planarization of optical aluminum surfaces. *J. Astron. Telesc. Instrum. Syst.* **2020**, *6*, 014001. [CrossRef]

materials

Article

Laser Powder Bed Fusion Process Parameters' Optimization for Fabrication of Dense IN 625

Alexandru Paraschiv [1], Gheorghe Matache [1,2], Mihaela Raluca Condruz [1,*], Tiberius Florian Frigioescu [1] and Laurent Pambaguian [3]

[1] National Research and Development Institute for Gas Turbines COMOTI, 220D Iuliu Maniu Avenue, 061126 Bucharest, Romania
[2] Section IX-Materials Science and Engineering, Technical Sciences Academy of Romania, 030167 Bucharest, Romania
[3] European Space Research and Technology Centre (ESA-ESTEC), Mechanical Department, European Space Agency, 2200 AG Noordwijk, The Netherlands
* Correspondence: raluca.condruz@comoti.ro; Tel.: +40-0768625916

Abstract: This paper presents an experimental study on the influence of the main Laser Powder Bed Fusion (PBF-LB) process parameters on the density and surface quality of the IN 625 superalloy manufactured using the Lasertec 30 SLM machine. Parameters' influence was investigated within a workspace defined by the laser power (150–400 W), scanning speed (500–900 m/s), scanning strategy (90° and 67°), layer thickness (30–70 μm), and hatch distance (0.09–0.12 μm). Experimental results showed that laser power and scanning speed play a determining role in producing a relative density higher than 99.5% of the material's theoretical density. A basic set of process parameters was selected for generating high-density material: laser power 250 W, laser speed 750 mm/s, layer thickness 40 μm, and hatch distance 0.11 mm. The 67° scanning strategy ensures higher roughness surfaces than the 90° scanning strategy, roughness that increases as the laser power increases and the laser speed decreases.

Keywords: IN 625; AM; PBF-LB; density; balling; process parameters

Citation: Paraschiv, A.; Matache, G.; Condruz, M.R.; Frigioescu, T.F.; Pambaguian, L. Laser Powder Bed Fusion Process Parameters' Optimization for Fabrication of Dense IN 625. *Materials* **2022**, *15*, 5777. https://doi.org/10.3390/ma15165777

Academic Editors: Bartłomiej Wysocki, Joseph Buhagiar, Tomasz Durejko and Amir Mostafaei

Received: 7 July 2022
Accepted: 19 August 2022
Published: 21 August 2022

Publisher's Note: MDPI stays neutral with regard to jurisdictional claims in published maps and institutional affiliations.

1. Introduction

Additive Manufacturing (AM) is currently one of the most studied technologies that can be used to produce new high-complexity parts, and its industrialization is currently under discussion, especially due to the increase in maturity.

Since the development of AM, many methods have been improved for different raw materials, and many are available to produce high-end products. The most challenges encountered in the field of AM are related to the production of metallic parts for highly demanding industries such as space and aerospace. One major challenge of all AM methods for developing metallic parts as candidates for space applications is to produce denser materials with reduced porosity and anisotropy levels than conventional manufacturing routes [1,2]. Powder bed methods are used to produce metallic parts with characteristics mainly influenced by the feedstock properties (like shape, surface morphology, size distribution that affect the flowability of powder, and porosity) and process parameter selection [1].

Two types of defects result in metallic parts produced using powder bed methods, feedstock-induced defects and process-induced defects. Powder porosity can result in a high degree of pores in AM-produced materials [3], but the manufacturing route mainly controls the metallic powder properties and particular powders were developed and are commercially available.

However, the most common defects result from laser–material interactions [4,5] and establishing a proper combination of process parameters is a more challenging activity. Typical process parameters in metallic PBF-LB processes are powder layer thickness, beam focus, laser speed, laser beam power, hatch distance, and scanning strategy [1]. Process parameter combinations can result in defects like pores, lack of fusion (LOF), surface roughness, microfractures, delaminated areas, and part dimensional inaccuracies [6–8]. Based on the research available regarding the production of AM parts, the need to tailor the process parameters for each metallic material as well as for each manufacturing equipment is obvious.

Many authors [9–19] have studied the influence of process parameters on metallic material's properties and microstructure, and some of them are referred to as a key-parameter combination such as volumetric energy density (EV or VED) or planar energy density (EP or ED) to define which combination of parameters should be used to obtain high-density materials (low porosity degree) ensuring high manufacturing rates. These key-parameter combinations can be calculated on the bases of Equations (1) and (2) [20,21]:

$$VED = P/vdt \; [J/mm^3] \tag{1}$$

$$ED = P/vd \; [J/mm^2] \tag{2}$$

where: P—laser power; v—scanning/laser speed; d—hatch distance; t—layer thickness.

Based on Equations (1) and (2), it can be stated that the energy density can be increased by increasing the laser powder and by decreasing the scanning speed, hatch distance, or layer thickness. The planar energy density can be used when the layer thickness is kept constant [7,22]. Even if some authors use the energy density for their process parameter optimization, others maintain it can be used in a narrow band of applicability due to its inability to capture the complex physics of the melt pool [20]. Therefore, it can be used to determine how much energy is transferred to the powder bed. Another solution would be to modify the equation to include more process parameters and material characteristics [9].

Many studies have been performed regarding the optimization of process parameters and their influence on the quality of additive manufactured parts from IN 718 or IN 625, but many focused on the influence on the material's mechanical properties [23–27]. Although mechanical performance is crucial, density and surface finish should be carefully inspected, especially in industrial fields such as aerospace, where the highest manufacturing standards regarding part quality are imposed. In the as-built state, additive manufactured metallic parts are characterized by a high surface roughness due to the structure's building angle, which has a stair-stepping effect or balling effect as spherical particles are separated from the unstable melt pool [1,6,7,28–30]. In their extensive paper, DebRoy et al. [1] state that the material's performance can be affected by the roughness degree. The roughness is a result of the physical processes that govern the shape and stability of the melt pool. The process parameters' combination affect these physical processes resulting in balling formation [6] or LOF defects. Balling appearance has also been reported to be influenced by the atmosphere; it could be reduced in atmospheres with low oxygen content (<0.1% O_2) and it could increase in high oxygen content atmospheres (>2% O_2) [30]. In addition to the atmosphere, the shape of the balls is an important aspect, it affects the wettability of the melt pool formed. An increase in layer thickness can result in larger balls, regardless of the hatch distance, which does not impact the balling formation [30]. Process parameter combinations can affect the integrity of deposited layers due to improper penetration and distribution, causing layer delamination and LOF defects; hence, the density and surface quality of the material is affected.

Defect emergence in additive manufactured metallic parts affects the material's densification level. Relative densities between 95.5–99.99% were reported for IN 625 manufactured using different advanced methods [21,31–36]. For example, Marchese et al. [21] studied the densification of IN 625 additive manufactured by selective laser melting (SLM) and laser metal deposition (LMD). In their study, densification was assessed on the bases of porosity measurements on SEM images, which is not the most accurate method that can be applied, but they reported densification degrees up to 99.96% by SLM, and 99.89% by LMD were reported. Gao et al. [31] reported a relative density of 99.7% for electron beam melting (EBM) manufactured IN 625, and Terris et al. [32] obtained relative densities ranging between 95.5–99.8% for SLM manufactured IN 625 using different VED values. High relative densities of laser powder bed fusion (LPBF) manufactured IN 625 were also obtained by Wong et al. [33] and by Poulin et al. [34]. Wong et al. [33] reported a 99.99% relative density, while Poulin et al. [34] obtained the same relative densities for the same specimens, 99.74% relative density determined using the Archimedes method and 99.92% by CT scanning.

Additive manufacturing of metallic materials is considered a time-consuming technology as many trials are made until the part's proper characteristics are achieved. Even if the process parameters are the same for several AM equipment, the specific combination of parameters for each machine and each material must be experimentally determined. A crucial activity before starting the manufacturing of metallic parts is the establishment of a specific set of parameters that ensure different targeted properties. Therefore, when a part with imposed properties is manufactured, the manufacturer will know which combination of parameters will apply, reducing the failure risk and cost associated with it.

The goal of this study was to optimize the PBF-LB process parameters to manufacture high relative density IN 625 superalloy using a Lasertec 30 SLM machine and to assess the influence of different process parameters on the material's quality, focusing on density and surface roughness. The experiments were designed as a progressive analysis consisting of three iterations. First, the influence of laser power and scanning speed on the surface quality and densification were evaluated (process parameters initial optimization). Based on the results obtained during the initial parameters' optimization, other two consecutive iterations were made as a process parameters optimization fine-tuning. For two laser powers, the influence of layer thickness and hatch distance on the densification of IN 625 was assessed. The most commonly used scanning strategies ($90°$ and $67°$) were used throughout the assessment. The bibliographic study showed that some authors used the Archimedes' method to determine the material's density, while others used microscopic methods or micro-CT. In this study, Archimedes and microscopic methods were used for density measurements to assess if a correlation between the two methods can be found. Moreover, the experimental results are discussed as a function of the individual parameters like laser power and laser speed, as well as a function of VED.

2. Materials and Methods

2.1. Specimen Manufacturing

Vacuum gas atomized virgin IN 625 metal powder supplied by LPW Technology Ltd., a subsidiary of Carpenter Technology Corporation, was used to produce specimens for this work. The metal powder exhibits mainly spherical powder particles within the size range of 15–45 μm and particle size distribution of $D_{10} = 20 \pm 2$ μm, $D_{50} = 30 \pm 5$ μm, $D_{90} = 45 \pm 5$ μm (powder size range and size distribution were provided by the manufacturer in the batch test certificate). Figure 1 shows SEM images with the powder particles. In these images, the powder morphology can be observed.

Prismatic specimens were manufactured using a DMG MORI Lasertec 30 SLM machine (first generation) for the experimental procedure. The Lasertec 30 SLM machine operates in a controlled environment (temperature 24 ± 2 °C and humidity $< 60\%$). It has a ytterbium fibre laser and uses the same manufacturing principle as other powder bed machines with the feature that it has a re-plug powder module that includes a sieving unit in a closed loop for metal powder feeding and recirculation. The powder is supplied from the main tank through a sieving system and then to a secondary tank from where a lift supplies the powder into the building chamber. The metallic powder is supplied by the lift and is spread on the building plate by a rubber wiper. The machine has a building volume of $300 \times 300 \times 300$ mm^3 and an Ar gas flow maintains a low oxygen level within the building chamber. Before starting the manufacturing process, the building chamber is flooded with Ar to reduce the oxygen to as low as 0.2%, which is maintained all over the manufacturing process. The flow system prevents material oxidation, reduces the smoke from the building chamber and removes the condensate produced during the powder melting. The Ar is purged through four holes placed at the back of the building chamber and is suctioned through a large slot placed in the equipment's door that is connected to the suction duct. Images of the equipment and the building chamber are presented in Figure 2. The wiper spreads the powder from right to left and afterward, it returns to the initial position. The gas flow was transverse to the direction of the powder spread (Figure 2b).

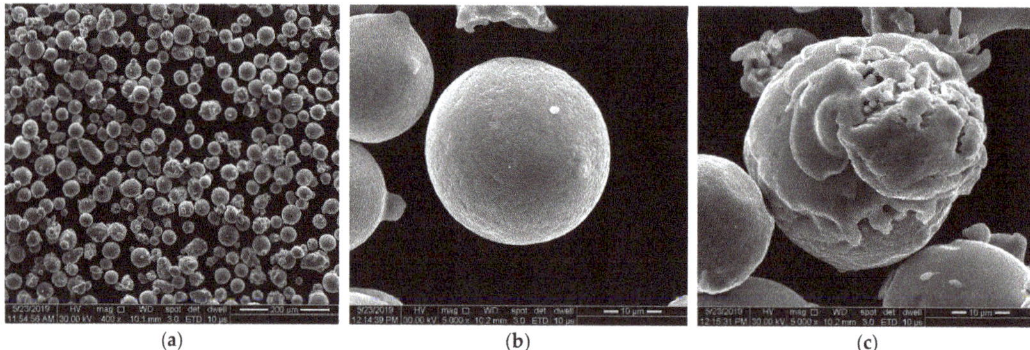

(a) (b) (c)

Figure 1. SEM images with IN 625 powder with a high degree of spheroidization (**a**), characterized by almost spherical particles (**b**) and also by satellite particles (**c**).

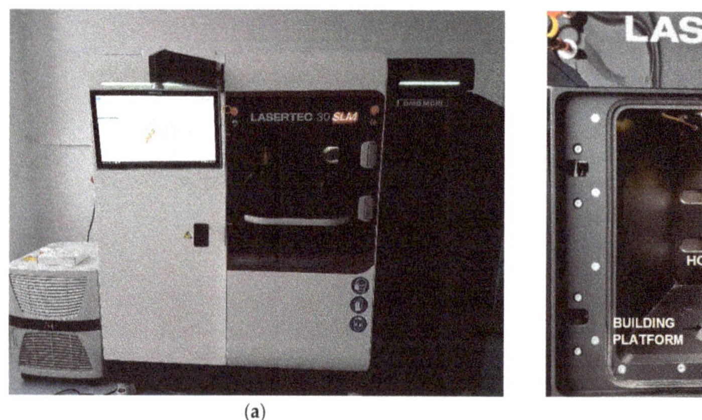

(a)

(b)

Figure 2. Lasertec 30 SLM machine (**a**) and the building chamber (**b**).

Specimens of $10 \times 10 \times 20$ mm^3 in size were built in the vertical position, tilted with $5°$ in the X-Y plane, on a stainless-steel building plate heated at 80 °C to reduce the thermal stresses in the specimens by decreasing thermal gradients during the first layer's deposition. All specimens were built under an argon atmosphere (keeping the oxygen content at a level under 0.2%) using the same support material configuration. All specimens were manufactured without contour.

The process parameters that influence the material density and defects were investigated progressively, starting from a designed workspace. The workspace was defined by the laser power (150–400 W), and laser speed (500–900 mm/s), keeping constant the layer thickness of 50 μm and hatch distance of 0.1 mm, using a 90° scanning strategy with 90° direction change between adjacent layers. Figure 3 presents the workspace definition.

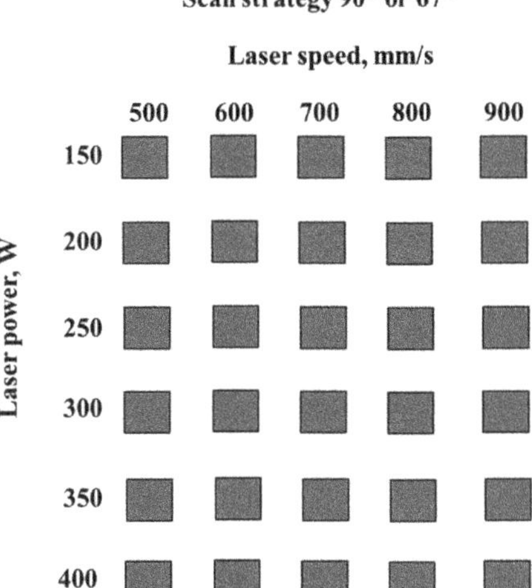

Figure 3. Workspace definition.

The same workspace was used to manufacture specimens with a 67° scanning strategy with a 90° direction change between successive layers. Based on the results obtained during these experiments, a preliminary parameter set was defined, and further parameter fine-tuning was made to select appropriate layer thickness and hatch distance values.

2.2. Density and Porosity Measurement

All specimens were mechanically removed from the building plate, and the support material was removed by grinding and polishing. The bulk density of the specimens was measured by the Archimedes method, according to ISO 3369 standard [37], using an analytical balance Ohaus Pioneer PX224 (accuracy of 0.0001 g) with a density determination kit for non-floating solid specimens. The auxiliary fluid used for the measurements was 99.3% purity ethanol (SC Chimopar Trading SRL, Bucharest, Romania) with temperature-known density variations. The density kit ensures the automatic calculation of the specimen's density considering the temperature variation of the auxiliary liquid (variation with 0.1 °C, between temperatures of 10 °C and 30.9 °C), specimen's mass weighted in air and in liquid, and air buoyancy. Therefore, the balance software uses Equation (3) for the density calculation:

$$\rho = [A/(A - B)] \times (\rho_0 - \rho_L) + \rho_L \ [g/cm^3] \tag{3}$$

where: ρ is the specimen's density, A is the mass of the specimen weighted in air, B is the mass of the specimen weighted in the auxiliary liquid; ρ_0 is the density of the auxiliary liquid and ρ_L is the air density.

All specimens were cleaned and degreased with ethanol before each density measurement. According to the standard [37], for specimens greater than 5 g, the repeatability interval is 0.025 g/cm^3, the reproducibility interval is 0.03 g/cm^3, and the results should round up to 0.01 g/cm^3. In this study, the average bulk density was calculated based on three measurements made on each specimen. Each bulk density value was reported using four decimals, and if there were differences higher than 0.0025 g/cm^3 between values, two more measurements were made. The densification degree of additive manufactured materials is usually reported by the relative density, and therefore, a reference value for a fully dense material is required.

In a very recent paper, Tarasov et al. [38] have shown that so far, no unified and accurate method has been proposed for calculating the density of heat-resistant nickel alloys. This paper reviews some of the available approaches to assess the density of alloys and proposes a new formula to calculate the density of Ni-based alloys with higher accuracy based on the chemical composition. Although their formula has been validated for several Ni-base alloys, all these alloys have a high content of gamma prime forming elements (Al and Ti) and do not contain iron, which is not the case of the investigated IN 625 alloy.

For this study, an analytical method was preferred to calculate the theoretical density of IN 625 based on the chemical composition of the powder batch presented in Table 1 and the densities of the alloying elements.

Table 1. Actual chemical composition of the investigated alloy.

Alloying Element	Specification, wt.% [39]	Metal Powder Chemical Composition, wt.% *
Al	0.40 max	0.06
C	0.10 max	0.02
Co	1.0 max	0.1
Cr	20.0–23.0	20.7
Fe	5.0 max	4.1
Mn	0.50 max	0.01
Mo	8.0–10.0	8.9
Nb	3.15–4.15	3.77
Si	0.50 max	0.01
Ti	0.40 max	0.07
Ni	rem	rem

* Provided by the manufacturer in the powder batch test certificate.

The calculations led to a value of 8.49 g/cm^3 for the theoretical density of the investigated alloy. The theoretical density was used as a reference to calculate the relative density as the degree of densification of the manufactured material.

The material's density is correlated with the material's porosity, thereby a porosity analysis was performed.

Porosity analysis was assessed using light optical microscopy images made at lower magnification (100× g) using a Zeiss Axio Vert.A1 MAT microscope. The images were taken on metallographically prepared, unetched specimens. A binarization technique was achieved by adjusting the brightness and contrast of the optical microscope images to highlight the pores. The technique consisted of converting the images in a 16-bit grayscale format followed by conversion into black and white threshold images. At this point, white represents homogenous material, while black represents the pores. For porosity quantification, Scandium software was used. The porosity was measured in the specimen's cross-section for the entire specimen batch manufactured with the scanning strategy of 67°, as schematically shown in Figure 4a). Measurements were performed for each specimen on 10 micro-areas of 3.5 mm^2 each.

2.3. Roughness Measurement

The as-built specimens' Ra roughness was measured using a MarSurf PS 10 mobile roughness measuring instrument with an evaluation length of 12.5 mm according to the ISO 4288 [40] standard. The measurements were performed on specimens' sides (ZY plane), as shown in Figure 4b. The top surface morphology investigation was performed by scanning electron microscopy using the SEM FEI Inspect F50, on the specimen's face marked with a black arrow in Figure 4c.

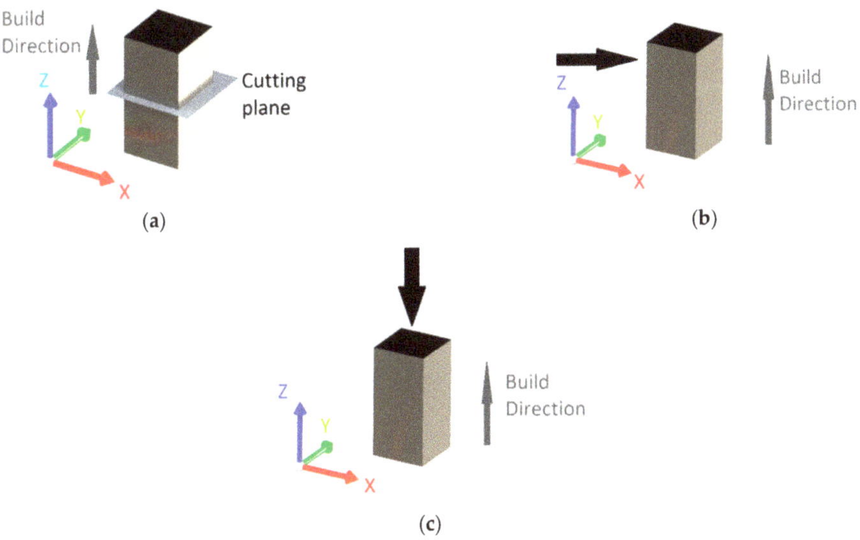

Figure 4. Workspace definition indication of surface for porosity measurements (**a**), roughness measurements (**b**), and top surface morphology investigation (**c**).

3. Results

3.1. Process Parameters Initial Optimization

The initial optimization of process parameters was made considering variable laser powers and laser speeds. Prismatic specimens were manufactured according to a workspace defined by laser power and laser speed, while keeping constant the layer thickness, and hatch distance, and using two scanning strategies. Figure 5 presents the first batch of PBF-LB manufactured specimens, using the 90° scanning strategy.

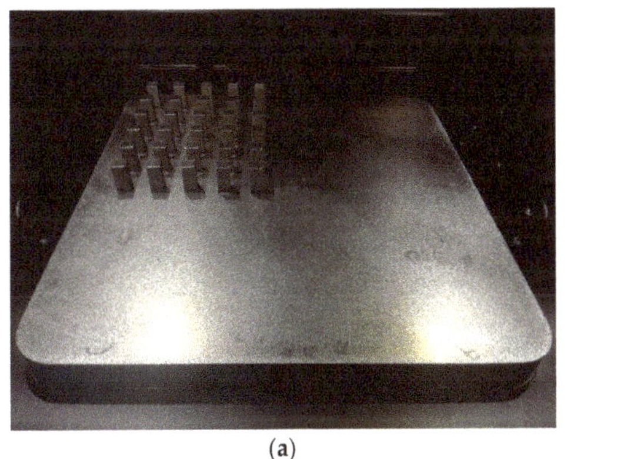

(a)

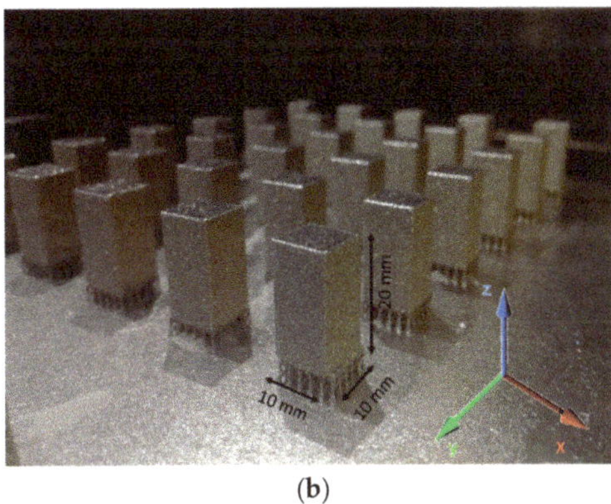

(b)

Figure 5. First batch of PBF-LB manufactured IN 625 specimens: (**a**) specimens on the building plate; (**b**) orientation and dimensions of the specimens.

Figure 6 presents SEM images with the top surface of built specimens, using a 90° scanning strategy. The specimens' top surface morphology showed the influence of laser power and laser speed on the balling effect. The effect is pronounced at low laser powers and laser speeds, and it decreases as the power increases.

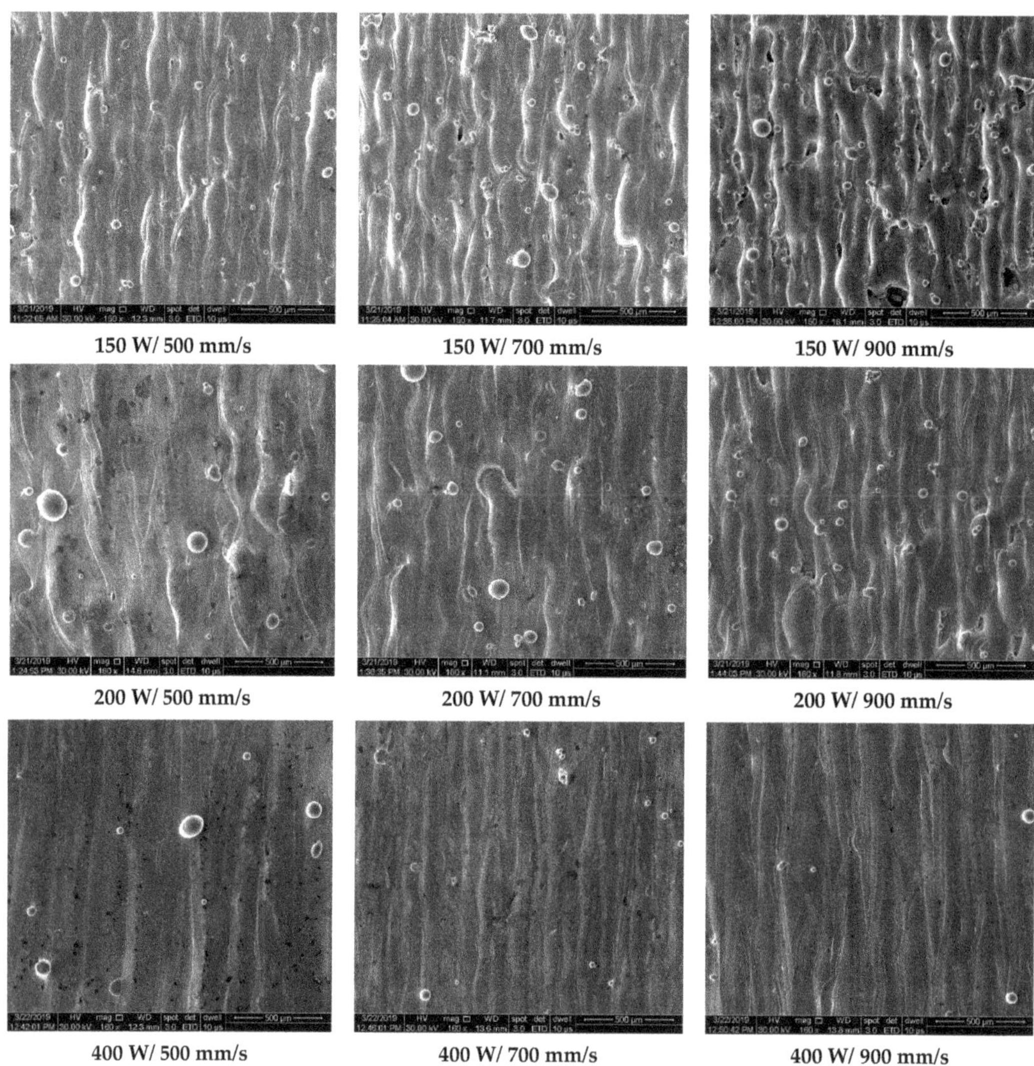

Figure 6. Influence of laser power and laser speed on balling effect (90° scanning strategy).

The morphological investigation has also revealed that the lowest power levels (150 W and 200 W) and increasing the laser speed led to LOF defect formation. In the case of the 900 mm/s laser speed, at lower laser powers, LOF defects were observed, the effect being more pronounced for the lowest laser power. By studying the LOF defects according to the VED values, the SEM analysis highlighted that the best results were obtained within the VED range of 65–90 J/mm^3.

The Ra roughness determined on the ZY surface was within 12–21 µm. The surface roughness measurements on specimen sides have highlighted that the roughness increases as the laser power increases and the laser speed decreases. The scanning strategy also influences the surface roughness. Figure 7 presents the surface roughness variation with laser speed at different laser powers for the two scanning strategies. Moreover, to highlight the evolution of roughness over the workspace, roughness maps were realized for both scanning strategies used.

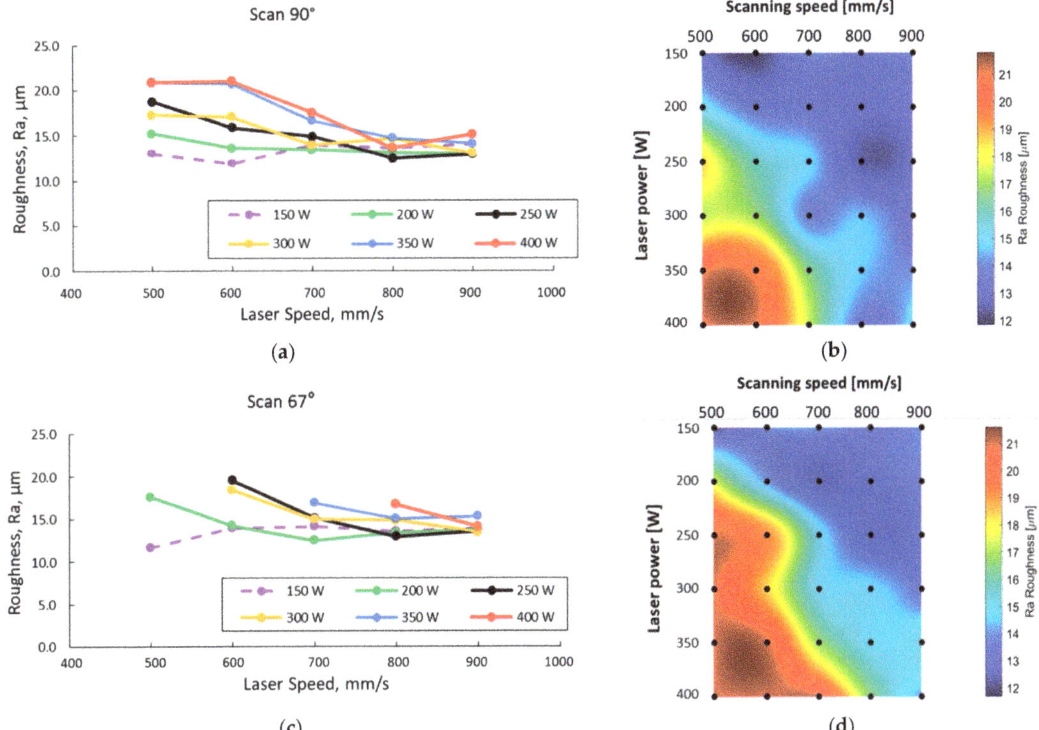

Figure 7. Specimen sides' Ra roughness as a function of laser speed for 90° scanning strategy (**a**) and 67° scanning strategy (**c**), and the corresponding laser power–scanning speed roughness maps for 90° scanning strategy (**b**) and 67° scanning strategy (**d**).

The specimens manufactured with a 67° scanning strategy had a measurable Ra within the range of 12–20 μm, but it was noticed that high energy, high laser power, and low laser speeds generate particle agglomerations of different sizes on the specimens' side surfaces thereby the surface roughness is out of the measuring range. An excessive increase in VED at lower laser speeds causes poor surface roughness.

Such particles may occur either by liquid metal sputtering at high power and low speed or by sticking metal powder particles from the powder bed by the liquid pool, or both. The SEM images presented in Figure 8 compare the side surface morphology for low, medium, and higher laser powers at different laser speeds, respectively. SEM images were taken on the same specimens' sides for roughness measurements.

The specimens' density measurements showed that in both cases (90° and 67° scanning strategy), the density generated at lower laser powers (using 150 W and to a good extent 200 W) sharply drops as the laser speed increases. Images from Figure 9a,c present the relative densities of all specimens as a function of laser speed for different laser power levels, while the images from Figure 9b,d present the relative densities for the specimens manufactured with laser powers over 200 W.

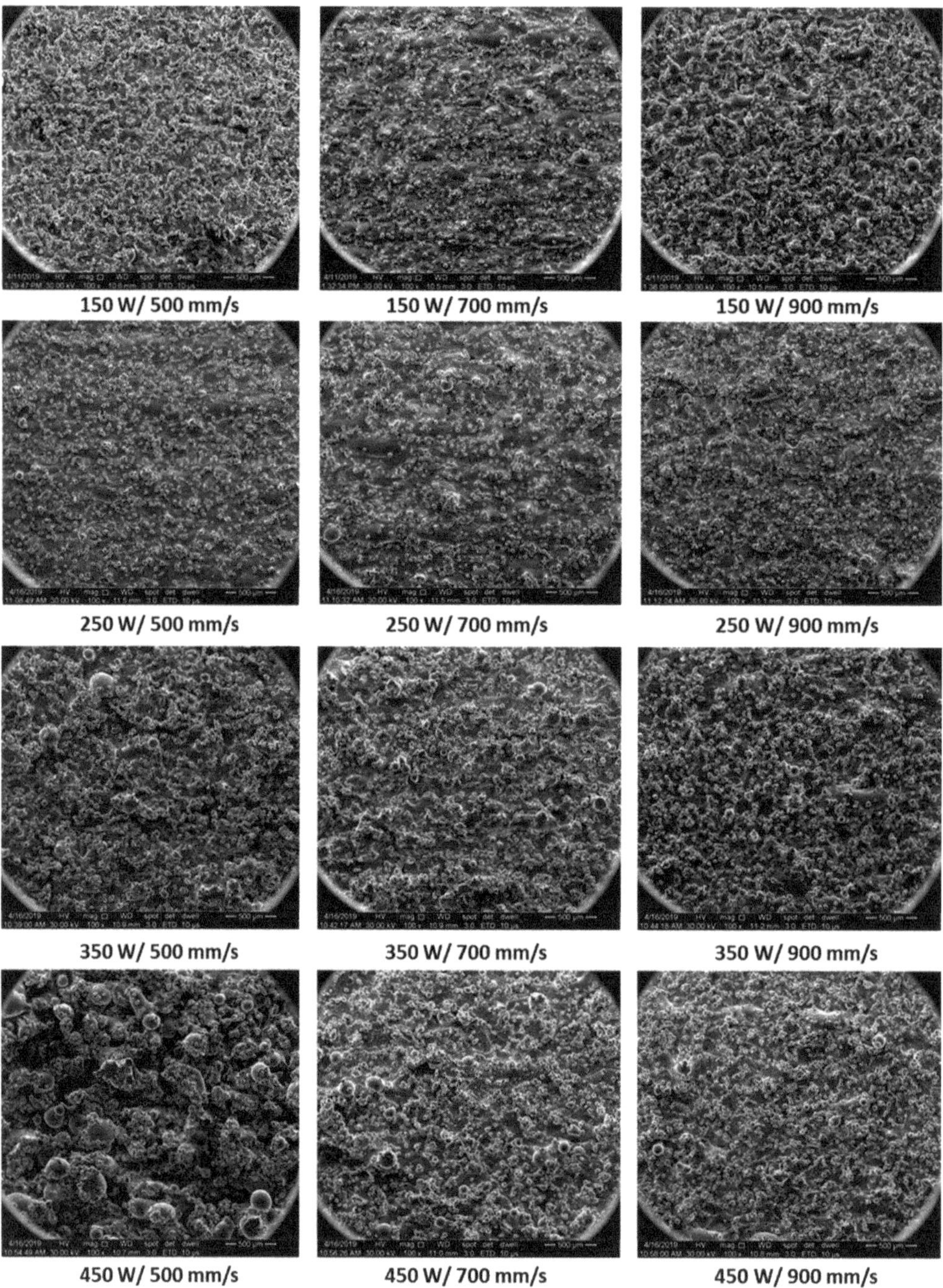

Figure 8. Surface morphology of specimens' sides for low, medium, and high laser power and 500 mm/s, 700 mm/s, and 900 mm/s laser speeds (67° scanning strategy).

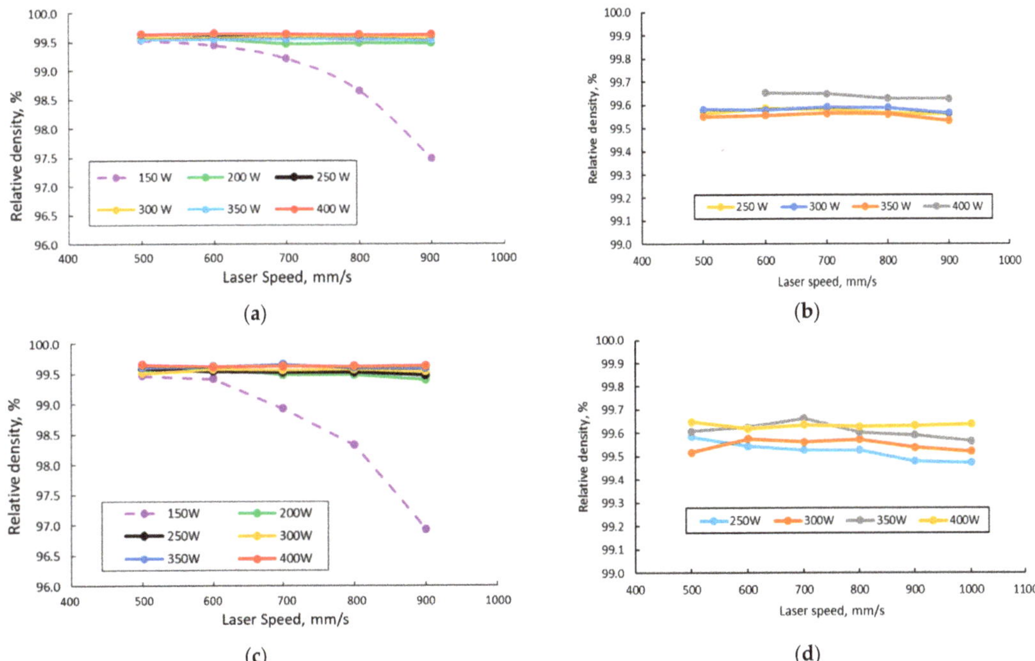

Figure 9. Relative density as a function of laser speed for different laser powers for 90° scanning strategy (**a**,**b**), and 67° scanning strategy (**c**,**d**).

As a general observation, for laser powers above 200 W within the experimental range, the IN 625 relative density increases as the laser power increases and tends to decrease with an increase in laser speed. Experimental results showed that for the specific alloy and process parameters used, laser power and laser speed play a determining role in producing a relative density higher than 99.5% of the theoretical density.

The analysis of the relative density of the as-built alloy as a function of VED (Figure 10) gave evidence that within the laser power–laser speed experimental workspace, the relative density increases rapidly from the lowest VED value until a value of 45 J/mm^3. Furthermore, it increases steadily at a lower rate. It can be observed that the VED values between 60–100 J/mm^3 generate consistent relative densities of approximately 99.60% for the 90° scanning strategy and 99.58% for the 67° scanning strategy. For higher VED values, the results are more spread.

This behaviour may be caused by high energy (high laser power and low laser speed) generated material inhomogeneities mainly by gas entrapment and "keyhole" type porosity formation. At high energies, material inhomogeneities can form due to the changes in the melt pool's dimensions, flow, and stability. When high laser power and low laser speed are used, the melt pool shape is changed from semi-circular to a longer "comet-like" form [41], and it becomes irregular or discontinuous and unstable [42], resulting in defects like porosities, spattering, and balling. These defects can be explained by the Kelvin–Helmholtz hydrodynamic instability or the Plateau Raleigh capillary instability [1].

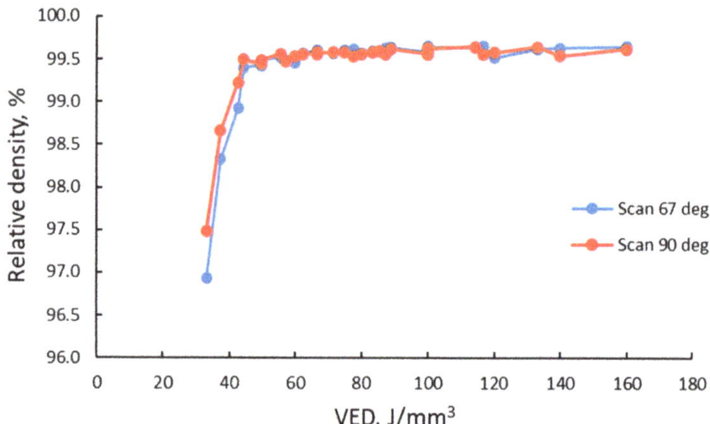

Figure 10. Relative density vs. VED for 90° and 67° scanning strategy.

The careful observation of the manufacturing process highlighted several aspects that affect the production of a high-quality alloy. First, at high laser powers (350 W and 400 W), a large amount of smoke is generated at almost all laser speeds. The smoke production is more pronounced at lower laser speeds. Secondly, at lower laser speeds (500 mm/s and 600 mm/s) and high laser powers, the liquid pools are pulverized, and spatters are projected randomly on the top surface of the built solid parts and on the powder bed.

For a more detailed analysis, porosity measurements were done on the entire batch of specimens manufactured with a 67° scanning strategy. Figure 11 shows the measured porosity as a function of laser power and laser speed.

The results from Figure 11 show that the laser power strongly influences reducing the porosity level as the power increases. At low laser power, the increase in laser speed leads to a sharp increase in porosity, mainly due to the LOF. The increase in laser speeds for the other laser power levels still generated an increase in measured porosity, but to a lesser extent. This can be explained by the mechanism of pore formation. For all low laser powers and laser speeds, porosity is generally generated by incorporating gas bubbles (keyhole porosities) as the AM process implies the use of an inert atmosphere (argon), which is insoluble in the liquid alloy, while at high speeds, the porosities are generated because of lack of fusion, as the melt pool is characterized by an insufficient penetration depth. A comparison of the porosity level in the samples microscopically measured and calculated from density results was made as a function of VED (Figure 12).

From Figure 12, it can be noticed that between the metallographically measured porosity and that based on the relative density calculated from densities measured using Archimedes' method, there are significant differences. No direct correlation can be done between the porosity levels measured/calculated using the two methods.

The metallographic method provides information on the density of specimens in the cutting plane and selected micro-areas. The result may differ for other measurements a few hundred microns below or above this plane. Furthermore, small scratches and indentations of the abrasive used for specimen preparation can be quantified as porosity.

The porosity calculation based on the density measurement by Archimedes' method also considers the possible entrapped un-melted metal powder in the pores resulting from the LOF. Figure 13 presents the SEM image of a pore filled with un-melted powder on a sample manufactured with 150 W and 900 mm/s laser speed. Meanwhile, the porosity calculated using the Archimedes method might be overestimated due to the simplistic theoretical density calculation. Simple density measurement will not give any information on the morphology of the porosity mode of formation (keyhole porosities, LOF, etc.). However, Archimedes' density can provide a rapid assessment of a specimen concerning the porosity level. However, different porosity measurement methods must be conducted properly, and different methods must be used for proper equivalence [43].

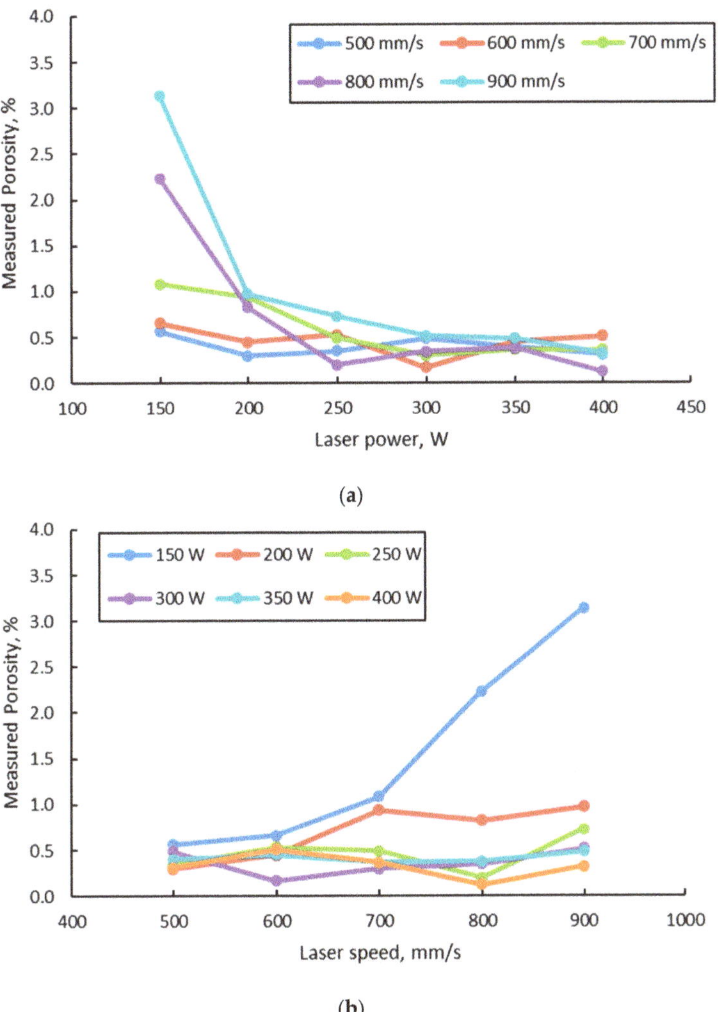

Figure 11. Measured porosity vs. laser power at different laser speeds (**a**) and vs. laser speed for different laser power (**b**).

To manufacture as densely as possible high-quality IN 625, free of contamination, the laser power/laser speed workspace was optimized considering the above density measurement results and the need to eliminate or to reduce as much as possible process sources of contamination. As is schematically presented in Figure 14, this workspace must fulfil several conditions simultaneously:

- Avoid or reduce as much as possible the smoke production and spatters.
- Avoid or reduce the alloy vaporization.
- Avoidance of those parameter sets that generate a sharp decrease in density and lack of fusion.

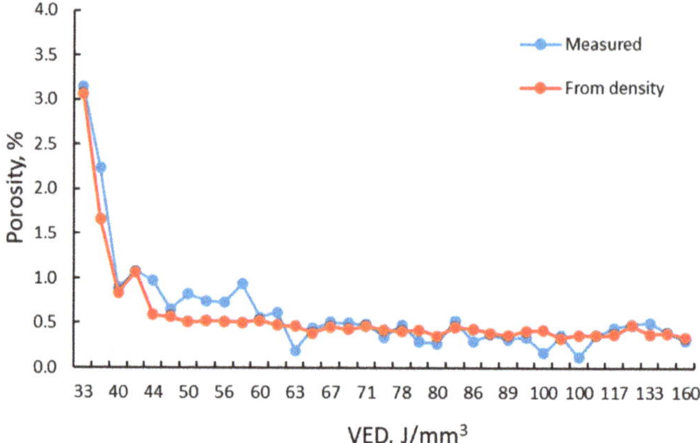

Figure 12. Measured and calculated porosity vs. VED.

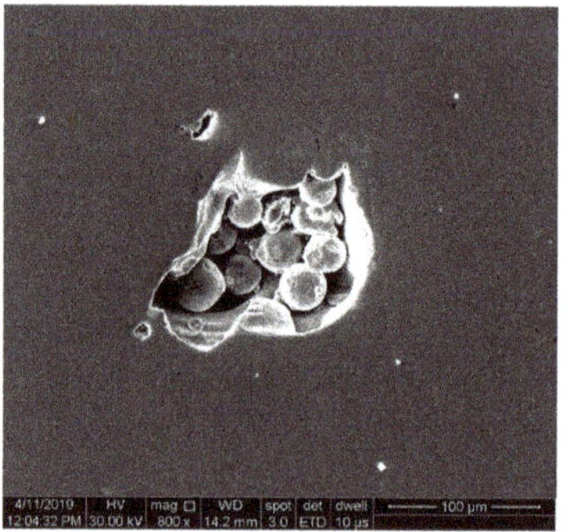

Figure 13. SEM image of entrapped un-melted powder in a specimen manufactured with 150 W laser power and 900 mm/s laser speed.

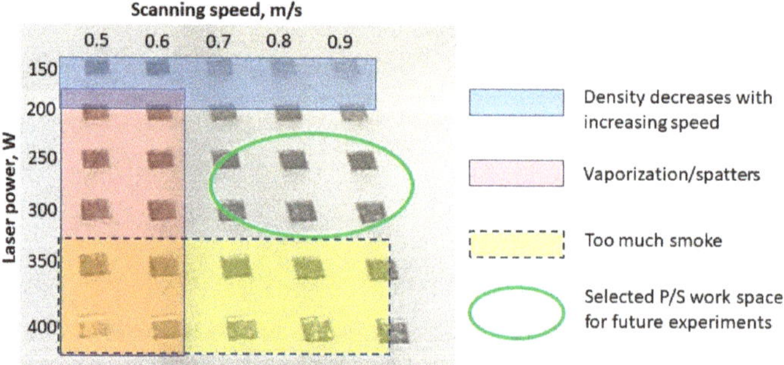

Figure 14. Laser power/laser speed workspace optimization.

Based on the previous density measurements and the process constraints regarding cross-contamination of parts and/or the powder bed, the appropriate workspace for IN 625 was considered for 250 W/300 W laser power and laser speeds greater than 700 mm/s.

3.2. Process Parameters Fine-Tuning

3.2.1. Layer Thickness

For the assessment of layer thickness influence on material density, the following process parameters were selected: laser power 250 and 300 W, laser speeds 700 mm/s,

800 mm/s, and 900 mm/s, scanning strategy 90° and 67°, layer thickness 30 μm, 50 μm, and 70 μm, hatch distance constant 0.1 mm.

The results were analysed along with those previously obtained for a layer thickness of 50 μm for both scanning strategies. The relative densities obtained as a function of laser speed for the two levels of laser power and the three-layer thicknesses (30 μm, 50 μm, and 70 μm) are presented in Figure 15 for both scanning strategies.

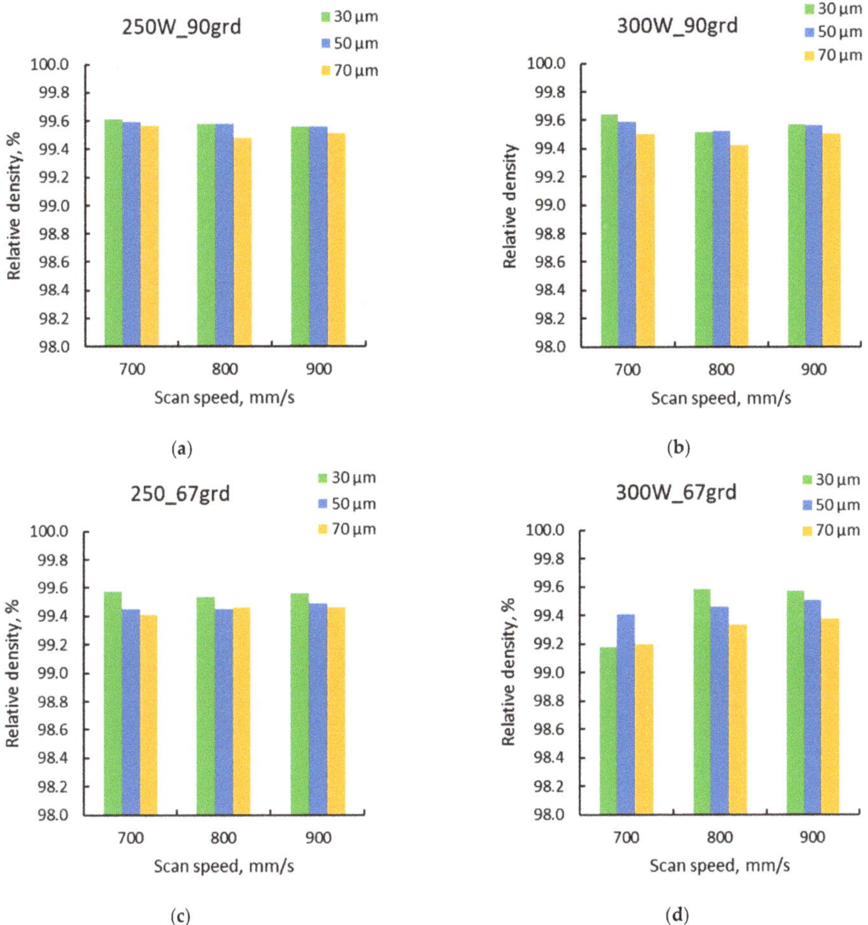

Figure 15. Relative density vs. laser speed for three levels of layer thickness (30 μm, 50 μm, and 70 μm) and laser power powers of 250 W and 300 W for the 90° scanning strategy (**a**,**b**) and 67° scanning strategy (**c**,**d**).

During the manufacturing process, higher contamination of the specimen manufactured with 300 W laser power, 700 mm/s laser speed, and 67° scanning strategy was encountered, so the density measurements were doubtful. For this reason, to have an appropriate image of the influence of layer thickness, this specimen was excluded from the analysis. All other results clearly indicate that the density decreases with increasing the layer thickness and is more accentuated at the largest layer thickness for both scanning strategy and individual laser power.

The results show that most of the results recorded for the 90° scanning strategy are grouped over a relative density of 99.5% for VED values higher than 60 J/mm^3. For the 67°

scanning strategy, for VED values between 60–100 J/mm^3, the relative densities are more spread below and over the dotted line. This observation is consistent with the previous findings within the laser power–laser speed workspace analysis.

3.2.2. Hatch Distance

For assessing the hatch distance influence on material relative density, the following process parameters were selected: laser power 250 and 300 W, laser speeds 700 mm/s, 800 mm/s, and 900 mm/s, scanning strategy 90° and 67°, layer thickness constant 50 μm, hatch distance variable 0.09 mm, 0.10 mm, 0.11 mm, and 0.12 mm.

Hatch distance is another parameter that requires optimization since high-density material can be obtained only by limiting the porosity level. Although each new layer partly melts the previously solidified one, it will inherently remain at points where the fusion of the consecutive layers is lacking, thus generating porosity.

Three additional jobs were performed using hatch spacing of 0.09 mm, 0.11 mm, and 0.12 mm. The density measurements were completed with the results for the 0.1 mm hatch distance used in the first experiment. Figure 16 presents the relative density calculated for the four levels of hatch distance as a function of laser speed for the two laser powers and scanning strategies.

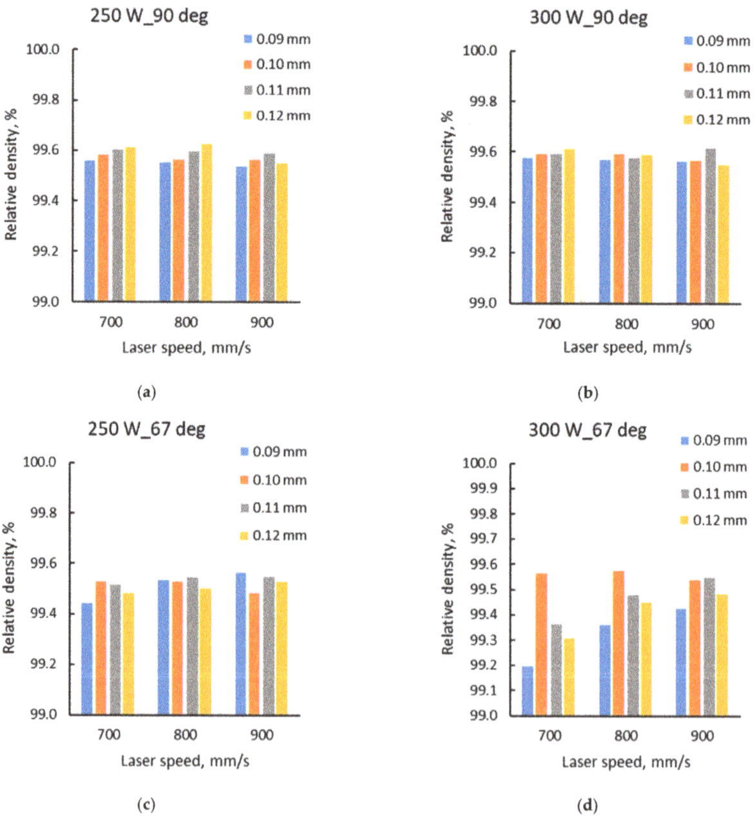

Figure 16. Relative density vs. laser speed for different hatch distances for as-built specimens manufactured with 90° scanning strategy (**a**,**b**), and 67° scanning strategy (**c**,**d**) respectively.

The general trend indicated more clearly for the specimens built with a 67° scanning strategy is that the relative density increases with the increase of hatch distance until a certain level, and then it tends to decrease with the further increase of the hatch distance.

4. Discussions

The PBF-LB process parameters' influence on the densification and surface quality of IN 625 Ni-based superalloy produced using the Lasertec 30SLM machine was assessed using an iterative method, starting by defining a workspace characterized by 6 laser power values (between 150–400 W) and 5 laser scanning speeds (500–900 mm/s), by keeping constant at layer thickness of 50 μm and hatch distance of 0.1 mm, and using two scanning strategies (the 90° and 67° scanning strategy with 90° direction change between adjacent layers). Based on the first results obtained, the parameter tailoring was continued by studying the influence of layer thickness and hatch spacing.

A similar approach was used by Yang et al. [44] for the IN 718 superalloy. In their work, various process windows were studied to assess the parameters that ensure the best characteristics of IN 718. By tailoring the process parameters, they obtained a relative density above 99.9%.

To evaluate the material's density, the Archimedes and micrographic methods were used. The most common approaches used to determine the density of additive manufactured materials are the Archimedes method, micrographic approach and X-ray scanning [45].

The study made by Spierings et al. [45] showed that the Archimedes method shows very high accuracy and repeatability of measurement, while the micrographic approach is not so accurate. Good results were also obtained in the case of X-ray scanning, including the distribution size and form of the pores. The micrographic approach is limited, as it allows the density assessment only on specific planes, while the Archimedes method allows the density assessment of the whole specimen's volume. Moreover, Sufiiarov et al. [46] showed no correlation between the density obtained using the Archimedes method and the micrographic approach in the case of additive manufactured IN 718.

4.1. Specimens' Surface Quality

The first specimen batch was manufactured using the 90° scanning strategy and a surface analysis was made, it highlighted that the combination of low laser powers and laser speeds ensured a pronounced balling effect on the top surface of the specimens. This effect decreases as the laser power increases. In the case of 150 W laser power, the best results were obtained for the lowest scanning speed, this combination ensuring a sufficient penetration depth of the laser, good wettability of the substrate and a continuous scanning track. While increasing the scanning speed, the wettability of the substrate decreases progressively, and spherical balls are formed, until they start to change into ellipsoidal formations, discontinuous tracks that led to the LOF appearance. Similar results were also observed for the 200 W laser power, but to a lower extent. These observations were supported by SEM images. Such defects were also registered by Darvish et al. [47]; they observed that low laser powers (180 W) causes insufficient overlap of the melting track while increasing the laser power causes better overlapping track coverage. Moreover, 300 W powers cause the formation of large size spatters, which was observed in this study as well. Studying the LOF defects according to VED highlighted that the lowest LOF degree was obtained within the VED range of 65–90 J/mm^3. Amirjan et al. [48] also observed LOF defects in the cases where the process parameters did not ensure a melt pool had sufficient depth.

The specimen's surface quality was also investigated on the side faces. The SEM images obtained in these areas showed that in the cases of low and medium laser powers, the balling effect is present, meaning small spherical balls, but to a lower extent than in the case where high laser powers are used. High energies consisting of high laser power and low laser speeds ensure the generation of the highest material inhomogeneities on the side surfaces. This is caused by changes in the melt pool's stability, the material tends to boil and produce high sputter that solidifies at the lateral surfaces of the specimens. These defects can be explained by the Kelvin–Helmholtz hydrodynamic instability or the Plateau Raleigh capillary instability [1]. In the case of high laser powers, by increasing the scanning speed the balling is reduced, both in dimension and degree.

In the case of these prismatic specimens where the stair-stepping effect is not observed, the balling formation and LOF defects are surface defects that cause surface roughness—this defect was measured using the stylus-type method. It was observed that not only the laser power and scanning speed influence the roughness and the scanning speed. In the case of the specimens' batch manufactured using a 90° scanning strategy, Ra values were within the range of 12–21 μm, while the 67° scanning strategy generated more agglomerated particles and the recorded Ra was between 12–20 μm. In the case of the specimens manufactured with the 67° scanning strategy, for high VED values, the surface roughness was out of the apparatus range and could not be measured.

In the SEM images, graphics, and maps, it was observed that for both scanning strategies, the high VED values (high laser powers and low scanning speeds) ensure an increased roughness. As previously explained, it is caused by the instability of the melt pool that produces a high degree of defects. Analyzing the data as a function of VED, it was observed that in the case of the batch manufactured with the 90° scanning strategy, the highest Ra values were recorded for VED values above 80 J/mm^3, while at lower VED values, no association could be made with the Ra value recorded. For the 67° scanning strategy, the lowest Ra values were recorded for VED values under 70 J/mm^3. Koutiri et al. [49] studied the roughness of additive manufactured IN 625 and observed that the laser power and building angle have a great influence on the material's roughness, but they sustain that the association between the roughness and VED is unsatisfactory when considering different laser spot diameters. An increase in roughness caused by balling and defect formation was also registered by Ni et al. [50].

As was observed in the current study, the great influence of process parameters on surface roughness was also observed by Charles et al. [51] and Mumtaz și Hopkinson [28], the highest roughness being a result of different energy levels absorbed by the powder bed and different roughness values could be obtained by process parameter optimization. Similar Ra values (1–20 μm) were registered by Safdar et al. [52] due to the process parameters tailoring. In their case, the Ra values increase along with the layer thickness and beam current, and it decreases with an increase in offset focus and scan speed.

Surface roughness could be improved by applying a contour function that allows the remelting of the surface layers of the specimen; for example, Wang et al. [53] applied different contour strategies to study the influence on the Ra. They also observed the appearance of spattering for the specimen manufactured with the highest energy density (laser power 400 W and 500 mm/s scanning speed), as was registered in the current study. Another solution to improve the surface finish would be applying a laser treatment. Genna et al. [54] studied the laser finishing of EBM manufactured Ti-6Al-4V. In the as-built state, they obtained irregular surfaces with high peaks and valleys with a Ra value of 27 μm. After laser treatment, an 80% reduction in Ra was recorded. The laser treatment ensured a melting and/or vaporization of the highest peaks observed. Another solution to improve the surface roughness is to use a contour function.

4.2. Material Densification

The sputter formed because of the different parameter combinations affect not only the external quality of the material but also the internal quality. The LOF between layers ensures the gas entrapment and formation of porosities. It was observed that at low laser power and high scanning speed, un-melted powder particles were present. Here, the scanning speed is too high to provide sufficient time to melt the powder as the laser power is low and cannot penetrate all the powder layers. Another harmful case is when the laser power is too high, and the scanning speed is too low, the melt pool elongates and becomes unstable, the high energy of the laser melts the material and it can even boil causing evaporation of alloying elements, spattering and keyhole porosities.

Density measurements of the specimens showed that the relative density generated at lower laser powers sharply drops as the laser speed increases. For the laser powers above 200 W, within the experimental range, the alloy relative density increased as the power increased and tend to decrease with the increase in laser speed. This result is consistent with what Amirjan et al. [48] concluded in the case of SLM manufactured IN 718.

Experimental results showed that a densification degree of more than 99.5% was determined from the theoretical density in the case of IN 625. In this case, the maximum values for the as-built material were considered satisfactory. Analyzing the relative density as a function of VED showed that in the designed workspace, the relative density increases rapidly from the lowest VED value until a value of 45 J/mm^3. Furthermore, it increases steadily at a lower rate. It can be observed that the VED values between 60–100 J/mm^3 generate consistent relative densities of approximately 99.60% for the 90° scanning strategy and 99.58% for the 67° scanning strategy. For higher VED values, the results are more spread due to material inhomogeneities, all caused by changes in the stability of the melt pool. When high energy values are used, the melt pool size increases, while as the hatch distance increases, the width and depth of the melt pool decrease [55]. The material inhomogeneities are also influenced by the heat transfer, surface tension, and melt pool flow. In some cases, the Marangoni effect along with the recoil force cause the melt pool to become unstable and to sputter [1].

The results obtained in the case of density measurements were sustained by experimental porosity measurements.

Further, it was observed that not only the laser power, laser speed and the scanning strategy influence the material densification level, but also the layer thickness. The density decreases with the layer thickness increase and this behavior is more accentuated at the largest layer thickness for both scanning strategies. The reason for this decrease is obvious that as the layer thickness increases, in some cases, the laser power is not high enough to completely penetrate the layer, or the laser speed is too high to allow the laser to completely penetrate or melt the powder layer.

The hatch distance is another important parameter in AM processes. It should be high enough to ensure a limited overlapping of adjacent tracks, without the formation of cavities between layers or too much overlapping of the tracks that could result in other defects caused by the uneven distribution of the following powder layer or blocking of the powder coater. The results obtained for the hatch distance tailoring are somewhat spread compared to the other parameters. More pronounced spreading occurred in case of the 67° scanning strategy and at a higher laser power (300 W). The general trend of the samples built with a 67° scanning strategy is that the relative density increases with the increase of hatch distance until a certain level, and then it tends to decrease with the further increase of the hatch distance. In the case of the 90° scanning strategy, the results show in most cases that a higher hatch distance ensures a better densification for the laser speeds of 700 mm/s and 800 mm/s. It was observed that low hatch distance mainly shows a lower densification (regardless of the scanning speed), which is believed to be caused by gas entrapment between tracks and porosity formation.

With a general conclusion regarding the influence of scanning strategy, it can be said that the 90° scanning strategy produces a lower defect percentage that is translated in a higher densification, than the 67° scanning strategy. The 67° scanning strategy is used by the researchers usually to ensure a fine microstructure compared with the microstructure formed in case of a 90° scanning strategy. All these observations are supported by the fact that in the case of the 67° scanning strategy, a higher cooling rate is registered and a higher thermal gradient, this being presented in another study by the authors [56].

A specific set of process parameters was defined at the end of this study to ensure a high-density material with acceptable surface quality. This combination of parameters was established, not only based on the experimental results obtained, but also by observing other aspects of the process that affects the production of a high-quality material. For example, the production of smoke at high laser powers was more pronounced at low laser speeds, as well as the pulverization of liquid alloy that projects spatter on the powder bed and on specimens.

This study advances the knowledge of additive manufacturing of high density IN 625 alloy, manufactured by SLM with Lasertec 30 SLM equipment. The study also provides useful information for selecting the appropriate workspace of process parameters to obtain high densification or low surface roughness or a suitable combination of the two. This basic set of process parameters was used in other studies made by the authors to assess the edge and corner effects during selective laser melting [57], the tensile strength anisotropy of the additive manufactured IN 625 using a Lasertec 30 SLM machine [56], and regarding complex-shaped parts manufacturing [58,59]. Future research will be conducted regarding the appropriate process parameter definition for Lasertec 30 SLM manufactured Ti-based alloys.

5. Conclusions

The goal of this study was to optimize the PBF-LB workspace defined as variables by laser power, laser speed, layer thickness, and hatch distance using two different scanning strategies to improve the densification and surface quality of additive manufactured IN 625. For this research, a workspace was designed and progressively improved until an appropriate set of process parameters was selected for producing a high density IN 625 Ni-based superalloy manufactured using the Lasertec 30 SLM machine. The results obtained revealed that the densification and surface quality of the material are influenced by defects such as LOF, porosity and roughness to different degrees as a function of process parameters combinations. The combination of low laser powers and laser speeds ensure a pronounced balling effect on the top surface of the specimens, which is reduced as the laser power increases. It was found that the scanning strategy influences the material's roughness—values within the range of 12–21 μm were determined for the specimens manufactured with a 90° scanning strategy, while the 67° scanning strategy ensures surfaces characterized by higher roughness (out of the measuring range) in the cases where high laser power and low scan speeds were used.

From all laser power values used, the laser power of 250 W generates more consistent relative densities, regardless of the scanning pattern, layer thickness, or hatch distance values applied. The best relative densities recorded were approximately 99.60% for the 90° scanning strategy and 99.58% for the 67° scanning strategy.

Based on the experimental results, a basic set of parameters that ensure a higher relative density than 99.5% from the theoretical density was defined. The appropriate parameters consisted of a laser power of 250 W, laser speed of 750 mm/s, a layer thickness of 40 μm, and a hatch distance of 0.11 mm.

Based on the analyses performed and results obtained, it was concluded that the process parameters have a significant influence on the stability of the melt pool that causes the internal and external defects.

Author Contributions: Conceptualization, G.M., A.P., M.R.C. and L.P.; methodology, G.M., A.P. and M.R.C.; software, T.F.F.; validation, G.M. and L.P.; formal analysis, G.M.; investigation, A.P., M.R.C. and T.F.F.; resources, G.M. and L.P.; data curation, G.M.; writing—original draft preparation, G.M., A.P., M.R.C., L.P. and T.F.F.; writing—review and editing, G.M., A.P., M.R.C. and T.F.F.; supervision, G.M. and L.P.; project administration, G.M. and L.P.; funding acquisition, G.M. and L.P. All authors have read and agreed to the published version of the manuscript.

Funding: The activity was carried out under a programme of, and was funded by the European Space Agency. The view expressed in this publication can in no way be taken to reflect the official opinion of the European Space Agency. The APC was funded by Ministry of Research, Innovation and Digitization, Program 1—Development of the National Research and Development System, Subprogram 1.2, Institutional Performance—Projects for Financing the Excellence in R&D, Grant no. 30 PFE/2021.

Conflicts of Interest: The authors declare no conflict of interest.

References

1. DebRoy, T.; Wei, H.L.; Zuback, J.S.; Mukherjee, T.; Elmer, J.W.; Milewsk, J.O.; Beese, A.M.; Wilson-Heid, A.; De, A.; Zhang, W. Additive manufacturing of metallic components–Process, structure and properties. *Prog. Mat. Sci.* **2018**, *92*, 112–224. [CrossRef]
2. Kok, Y.; Tan, X.P.; Wang, P.; Nai, M.L.S.; Loh, N.H.; Liu, E.; Tor, S.B. Anisotropy and heterogeneity of microstructure and mechanical properties in metal additive manufacturing: A critical review. *Mater. Des.* **2018**, *139*, 565–586. [CrossRef]
3. Antonysamy, A.A. Microstructure, Texture and Mechanical Property Evolution during Additive Manufacturing of Ti6Al4V Alloy for Aerospace Applications. Ph.D. Thesis, Faculty of Engineering and Physical Sciences, University of Manchester, Manchester, UK, 2012.
4. Yap, C.Y.; Chung, C.K.; Dong, Z.L.; Liu, Z.H.; Zhang, D.Q.; Loh, L.E.; Sing, S.L. Review of selective laser melting: Materials and applications. *Appl. Phys. Rev.* **2015**, *2*, 041101. [CrossRef]
5. Shuai, L.; Wei, Q.; Shi, Y.; Zhang, J.; Wei, L. Micro-crack formation and controlling of Inconel 625 parts fabricated by selective melting, Solid Freeform Fabrication. In Proceedings of the 27th Annual International Solid Freeform Fabrication Symposium 2016–An Additive Manufacturing Conference, Austin, TX, USA, 8–10 August 2016; Bourell, D.L., Crawford, R.H., Seepersad, C.C., Beaman, J.J., Fish, S., Marcus, H., Eds.; University of Texas: Austin, TX, USA, 2016; pp. 520–529.
6. Khairallah, S.A.; Anderson, A.T.; Rubenchik, A.; King, W.E. Laser powder-bed fusion additive manufacturing: Physics of complex melt flow and formation mechanisms of pores, spatter, and denudation zones. *Acta Mater.* **2016**, *108*, 36–45. [CrossRef]
7. Siddiqui, S.F.; Fasoro, A.A.; Gordon, A.P. Selective laser melting (SLM) of Ni-based superalloys. In *Additive Manufacturing Handbook: Product Development for the Defense Industry*, 1st ed.; Badiru, A.B., Valencia, V.V., Liu, D., Eds.; CRC Press: Boca Raton, FL, USA, 2017; pp. 225–250.
8. Sames, W.J.; List, F.A.; Pannala, S.; Dehoff, R.R.; Babu, S.S. The metallurgy and processing science of metal additive manufacturing. *Int. Mater. Rev.* **2016**, *61*, 315–360. [CrossRef]
9. Prashanth, K.G.; Scudino, S.; Maity, T.; Das, J.; Eckert, J. Is the energy density a reliable parameter for materials synthesis by selective laser melting? *Mat. Res. Lett.* **2017**, *5*, 386–390. [CrossRef]
10. Arısoy, Y.M.; Criales, L.E.; Özel, T.; Lane, B.; Moylan, S.; Donmez, A. Influence of scan strategy and process parameters on microstructure and its optimization in additively manufactured nickel alloy 625 via laser powder bed fusion. *Int. J. Adv. Manuf. Technol.* **2017**, *90*, 1393–1417. [CrossRef]
11. Kumar, P.; Farah, J.; Akram, J.; Teng, C.; Ginn, J.; Misra, M. Influence of laser processing parameters on porosity in Inconel 718 during additive manufacturing. *Int. J. Adv. Manuf. Technol.* **2019**, *108*, 1497–1507. [CrossRef]
12. Marchese, G.; Bassini, E.; Calandri, M.; Ambrosio, E.P.; Calignano, F.; Lorusso, M.; Mandredi, D.; Paverse, M.; Biamino, S.; Fino, P. Microstructural investigation of as fabricated and heat-treated Inconel 625 and Inconel 718 fabricated by direct metal laser sintering: Contribution of Politecnico di Torino and Istituto Italiano di Tecnologia (IIT) di Torino. *Met. Powder Rep.* **2016**, *71*, 273–278. [CrossRef]
13. Wang, J.; Wu, W.J.; Jing, W.; Tan, X.; Bi, G.J.; Tor, S.B.; Leong, K.F.; Chua, C.K.; Liu, E. Improvement of densification and microstructure of ASTM A131 EH36 steel samples additively manufactured via selective laser melting with varying laser scanning speed and hatch spacing. *Mat. Sci. Eng. A* **2019**, *746*, 300–313. [CrossRef]
14. Larimian, T.; Kannan, M.; Grzesiak, D.; AlMangour, B.; Borkas, T. Effect of energy density and scanning strategy on densification microstructure and mechanical properties of 316L stainless steel processed via selective laser melting. *Mat. Sci. Eng. A* **2020**, *770*, 138455. [CrossRef]
15. Jia, Q.; Gu, D. Selective laser melting additive manufacturing of Inconel 718 superalloy parts: Densification, microstructure and properties. *J. Alloys Comp.* **2014**, *585*, 713–721. [CrossRef]
16. Sadowski, M.; Ladani, L.; Brindley, W.; Romano, J. Optimizing quality of additively manufactured Inconel 718 using powder bed laser melting process. *Add. Manuf.* **2016**, *11*, 60–70. [CrossRef]
17. Letenneur, M.; Kreitcberg, A.; Brailovski, V. Optimization of Laser Powder Bed Fusion Processing Using a Combination of Melt Pool Modeling and Design of Experiment Approaches: Density Control. *J. Manuf. Mater. Process.* **2019**, *3*, 21. [CrossRef]
18. Peng, T.; Chen, C. Influence of Energy Density on Energy Demand and Porosity of 316L Stainless Steel Fabrication by Selective Laser Melting. *Int. J. Precis. Eng. Manuf.-Green Technol.* **2018**, *5*, 55–62. [CrossRef]
19. Jaskari, M.; Ghosh, S.; Miettunen, I.; Karjalainen, P.; Jarvenpaa, A. Tensile Properties and Deformation of AISI 316L Additively Manufactured with Various Energy Densities. *Materials* **2021**, *14*, 5809. [CrossRef]
20. Bertoli, U.S.; Wolfer, A.J.; Matthews, M.J.; Delplanque, J.P.R.; Schoenung, J.M. On the limitations of Volumetric Energy Density as a design parameter for Selective Laser Melting. *J. Mat. Des.* **2017**, *113*, 331–340.
21. Marchese, G.; Garmendia Colera, X.; Calignano, F.; Lorusso, M.; Biamino, S.; Minetola, P.; Manfredi, D. Characterization and Comparison of Inconel 625 Processed by Selective Laser Melting and Laser Metal Deposition. *Adv. Eng. Mat.* **2016**, *19*, 160063. [CrossRef]
22. Kruth, J.P.; Dadbakhsh, S.; Vracken, B.; Kempen, K.; Vleugels, J.; Van Humbeeck, J. Additive Manufacturign of Metals via Selective Laser Melting: Process Aspects and Material Developments. In *Additive Manufacturing: Innovations, Advances, and Applications*; Srivatsan, T.S., Sudarshan, T.S., Eds.; CRC Press: Boca Raton, FL, USA, 2016; pp. 69–100.

23. Gonzalez, J.A.; Mireles, J.; Stafford, S.W.; Perez, M.A.; Terrazas, C.A.; Wicker, R.B. Characterization of Inconel 625 fabrication using powder-bed-based additive manufacturing technologies. *J. Mater. Process. Technol.* **2019**, *264*, 200–210. [CrossRef]
24. Kreitcberg, A.; Brailovski, V.; Turenne, S. Effect of heat treatment and hot isostatic pressing on the microstructure and mechanical properties of Inconel 625 alloy processed by laser powder bed fusion. *Mater. Sci. Eng. A* **2017**, *689*, 1–10. [CrossRef]
25. Wan, H.Y.; Zhou, Z.J.; Li, C.P.; Chen, G.F.; Zhang, G.P. Effect of scanning strategy on mechanical properties of selective laser melted Inconel 718. *Mater. Sci. Eng. A* **2019**, *753*, 42–48. [CrossRef]
26. Xu, Z.; Hyde, C.J.; Tuck, C.; Clare, A.T. Creep behavior of Inconel 718 processed by laser powder bed fusion. *J. Mater. Process. Technol.* **2018**, *256*, 13–24. [CrossRef]
27. Du, K.; Yang, L.; Xu, C.; Wang, B.; Gao, Y. High Strain Rate Yielding of Additive Manufacturing Inconel 625 by Selective Laser Melting. *Materials* **2021**, *14*, 5408. [CrossRef]
28. Mumtaz, K.; Hopkinson, N. Top surface and side roughness of Inconel 625 parts processed using selective laser melting. *Rapid Prototyp. J.* **2009**, *15*, 96–103. [CrossRef]
29. Fox, J.C.; Moylan, S.P.; Lane, B.M. Effect of Process Parameters on the Surface Roughness of Overhanging Structures in Laser Powder bed Fusion Additive Manufacturing. *Proc. CIRP* **2016**, *45*, 131–134. [CrossRef]
30. Li, R.; Liu, J.; Shi, Y.; Wang, L.; Jiang, W. Balling behavior of stainless steel and nickel powder during selective laser melting process. *Int. J. Adv. Manuf. Technol.* **2012**, *59*, 1025. [CrossRef]
31. Gao, Y.; Zhou, M. Superior Mechanical Behavior and Fretting Wear Resistance of 3D-Printed Inconel 625 Superalloy. *Appl. Sci.* **2018**, *8*, 2439. [CrossRef]
32. Terris, T.; Adamski, F.; Peyre, P.; Dupuy, C. Influence of SLM process parameters on Inconel 625 superalloy samples. In Proceedings of the Lasers in Manufacturing Conference 2017, München, Germany, 26–29 June 2017; p. 12.
33. Wong, H.; Dawson, K.; Ravi, G.A.; Howlett, L.; Jones, R.O.; Sutcliffe, C.J. Multi-Laser Powder Bed Fusion Benchmarking–Initial Trials with Inconel 625. *Int. J. Adv. Manuf. Technol.* **2019**, *105*, 2891–2906. [CrossRef]
34. Poulin, J.-R.; Kreitcberg, A.; Terriault, P.; Brailovski, V. Long fatigue crack propagation behavior of laser powder bed-fused Inconel 625 with intentionally seeded porosity. *Int. J. Fatigue* **2019**, *127*, 144–156. [CrossRef]
35. Mostafaei, A.; Stevens, E.L.; Hughes, E.T.; Biery, S.D.; Hilla, C.; Chmielus, M. Powder bed binder jet printed alloy 625: Densification, microstructure and mechanical properties. *Mater. Des.* **2016**, *108*, 126–135. [CrossRef]
36. Mostafaei, A.; Neelapu, S.H.V.R.; Kisailus, C.; Nath, L.M.; Jacobs, T.D.B.; Chmielus, M. Characterizing surface finish and fatigue behavior in binder-jet 3D-printed nickel-based superalloy 625. *Addit. Manuf.* **2018**, *24*, 565–586. [CrossRef]
37. *ISO 3369*; Impermeable Sintered Metal Materials and Hardmetals–Determination of Density. International Organization for Standardization: Geneva, Switzerland, 2010.
38. Tarasov, D.A.; Milder, O.B.; Tyagunov, A.G. An accurate empirical formula for determining the density of heat-resistant nickel alloys. *Lett. Mater.* **2021**, *11*, 192–197. [CrossRef]
39. *ASTM F3056*; Standard Specification for Additive Manufacturing Nickel Alloy (UNS N06625) with Powder Bed Fusion. ASTM International: West Conshohocken, PA, USA, 2014. [CrossRef]
40. *ISO 4288*; Geometrical Product Specifications (GPS)–Surface Texture: Profile method–Rules and Procedures for the Assessment of Surface Texture. International Organization for Standardization: Geneva, Switzerland, 1996.
41. Cheng, B.; Chou, K. Melt Pool Evolution Study in Selective Laser Melting. In Proceedings of the 26th Annual International Solid Freeform Fabrication Symposium–An Additive Manufacturing Conference, Austin, TX, USA, 10–12 August 2015.
42. Sun, S.; Brandt, M.; Easton, M. *Powder Bed Fusion Processes: An Overview, Laser Additive Manufactur-ing: Materials, Design, Technologies, and Applications*; Woodhead Publishing: Sawston, UK, 2017; pp. 55–77.
43. Slotwinski, J.A.; Garboczi, E.J.; Hebenstreit, K.M. Porosity Measurements and Analysis for Metal Additive Manufacturing Process Control. *J. Res. Natl. Inst. Stand Technol.* **2014**, *119*, 494–528. [CrossRef] [PubMed]
44. Yang, H.; Meng, L.; Luo, S.; Wang, Z. Microstructural evolution and mechanical performances of selective laser melt-ing Inconel 718 for low to high laser power. *J. Alloys Comp.* **2020**, *828*, 154473. [CrossRef]
45. Spierings, A.B.; Schneider, M.; Eggenberer, R. Comparison of density measurement techniques for additive manufac-tured metallic parts. *Rapid Prototyp. J.* **2011**, *17*, 380–386. [CrossRef]
46. Sufiiarov, V.S.; Popovich, A.A.; Borisov, E.V.; Polozov, I.A.; Masaylo, D.V.; Orlov, A.V. The effect of layer thickness at selective laser melting. *Proceedia Eng.* **2017**, *174*, 126–134. [CrossRef]
47. Darvish, K.; Chen, Z.W.; Pasang, T. Reducing lack of fusion during selective laser melting of CoCrMo alloy: Effect of laser power on geometrical features of tracks. *Mater. Des.* **2016**, *112*, 357–366. [CrossRef]
48. Amirjan, M.; Sakiani, H. Effect of scanning strategy and speed on the microstructure and mechanical properties of se-lective laser melted IN718 nickel-based superalloy. *Int. J. Adv. Manufact. Technol.* **2019**, *103*, 1769–1780. [CrossRef]
49. Koutiri, I.; Pessard, E.; Peyre, P.; Amlou, O.; de Terris, T. Influence of SLM process parameters on the surface finish, porosity rate and fatigue behavior of as-built Inconel 625 parts. *J. Mater. Process. Technol.* **2017**, *255*, 536–546. [CrossRef]
50. Ni, C.; Shi, Y.; Liu, J. Effects of inclination angle on surface roughness and corrosion properties of selective laser melted 316L stainless steel. *Mater. Res. Express* **2018**, *6*, 036505. [CrossRef]
51. Charles, A.; Elkaseer, A.; Thijs, L.; Hagenmeyer, V.; Scholz, S. Effect of process Parameters on the Generated Surface Roughness of Down-Facing Surfaces in Selective Laser Melting. *Appl. Sci.* **2019**, *9*, 1256. [CrossRef]

52. Safdar, A.; He, H.Z.; Wei, L.Y.; Snis, A.; Chavez de Paz, L.E. Effect of process parameters settings and thickness on surface roughness of EBM produced Ti-6Al-4V. *Rapid Prototyp. J.* **2012**, *18*, 401–408. [CrossRef]
53. Wang, P.; Sin, W.J.; Nai, M.L.S.; Wei, J. Effects of Processing Parameters on Surface Roughness of Additive Manufactured Ti-6Al-4V via Electron Beam Melting. *Materials* **2017**, *10*, 1121. [CrossRef] [PubMed]
54. Genna, S.; Rubino, G. Laser Finishing of Ti6Al4V Additive Manufacturing Parts by Electron Beam Melting. *Appl. Sci.* **2020**, *10*, 183. [CrossRef]
55. Criales, L.E.; Arisoy, Y.M.; Lane, B.; Moylan, S.; Donmez, A.; Ozel, T. Laser powder bed fusion of nickel alloy 625: Ex-perimental investigations of effects of process parameters on melt pool size and shape with spatter analysis. *Int. J. Mach. Tools Manuf.* **2017**, *121*, 22–36. [CrossRef]
56. Condruz, M.R.; Matache, G.; Paraschiv, A.; Frigioescu, T.F.; Badea, T. Microstructure and Tensile Properties Anisotropy of Selective Laser Melting Manufactured IN 625. *Materials* **2020**, *13*, 4829. [CrossRef]
57. Matache, G.; Vladut, M.; Paraschiv, A.; Condruz, M.R. Edge and corner effects in selective laser melting of IN 625 alloy. *Manuf. Rev.* **2020**, *7*, 8. [CrossRef]
58. Adiaconitei, A.; Vintila, I.S.; Mihalache, R.; Paraschiv, A.; Frigioescu, T.; Vladut, M.; Pambaguian, L. A Study on using the Additive Manufacturing Process for the Development of a Closed Pump Impeller for Mechanically Pumped Fluid Loop Systems. *Materials* **2021**, *14*, 967. [CrossRef]
59. Adiaconitei, A.; Vintila, I.S.; Mihalache, R.; Paraschiv, A.; Frigioescu, T.F.; Popa, I.F.; Pambaguian, L. Manufacturing of Closed Impeller for Mechanically Pump Fluid Loop Systems Using Selective Laser Melting Additive Manufacturing Technology. *Materials* **2021**, *14*, 5908. [CrossRef]

MDPI
St. Alban-Anlage 66
4052 Basel
Switzerland
www.mdpi.com

Materials Editorial Office
E-mail: materials@mdpi.com
www.mdpi.com/journal/materials

Disclaimer/Publisher's Note: The statements, opinions and data contained in all publications are solely those of the individual author(s) and contributor(s) and not of MDPI and/or the editor(s). MDPI and/or the editor(s) disclaim responsibility for any injury to people or property resulting from any ideas, methods, instructions or products referred to in the content.

9 783036 598864